Protection of Human Research Subjects

A Practical Guide to Federal Laws and Regulations

Protection of Human Research Subjects

A Practical Guide to Federal Laws and Regulations

Dennis M. Maloney
Boys Town USA Program
Boys Town, Nebraska

With a chapter by James H. Sweetland

Foreword by Kenneth J. Ryan

Plenum Press · New York and London

Library of Congress Cataloging in Publication Data

Maloney, Dennis M.
Protection of human research subjects.

Bibliography; p.
Includes index.
1. Human experimentation in medicine—Law and legislation—United States. I. Title. [DNLM: 1. Human experimentation—United States—Legislation. 2. Research—United States—Legislation. W 20.5 M257p]
KF3827.M38M34 1984 344.73′0419 84-4873
ISBN 0-306-41522-4 347.304419

A Division of Plenum Publishing Corporation
233 Spring Street, New York, N.Y. 10013

Printed in the United States of America

For my parents,
Bernard and Margaret Maloney,
who taught me very early
to respect the rights of others

Foreword

Regulations on human subjects research have evolved over a 20-year period and now provide a formal set of requirements for the conduct of federally sponsored studies. Over time, government regulations, like taboos in primitive societies, develop a life of their own, seemingly dissociated from their origins and justifications. When the investigator suffers the burdens of trying to comply with complex rules, it is easy for him or her to lapse into a frustration that can be eased by being informed or reminded of why the rules were created in the first place and what they were designed to accomplish. Dennis Maloney's work provides a handy historical record of the processes by which the regulations were created and modified. He also recounts his own experience with research at Boys Town and provides instructions on how to cope with the system. It is difficult to find in one place the current status and appropriate citations for regulations as well as the contacts and know-how to obtain more information on the subject. In this respect, by providing a history and guide to interpretation and compliance, "Protection of Human Subjects" is a reference of importance and utility to the investigator entering into or working in the field of biomedical or behavioral research involving human subjects. Of no less importance, is the fact that the creation and modification of regulations is a continuous process, and investigators as well as an informed public should be prepared to participate in the development of sensible rules that serve the needs of subjects, science, and society.

KENNETH J. RYAN, M.D.
Chairman, 1974–1978
National Commission for the Protection of Human Subjects of Biomedical and Behavioral Research

Acknowledgments

From 1974 to 1982 there have been numerous individuals who have contributed to my interest in this field and supported my work in a variety of ways. Initially, I must thank Larry Larson, Ph.D., who provided assistance through his role as chairman of the Institutional Review Board at Western Carolina Center, Morganton, North Carolina. Certainly, Saleem Shah, Ph.D., Jim Breiling, Ph.D., and Tom Lalley, M.A. (Center for Studies of Crime and Delinquency at the National Institute of Mental Health, Washington, D.C.) guided my early efforts in working with human research subjects and suggested additional source materials in this area.

I gratefully acknowledge the cooperation and untiring efforts of members of the Institutional Review Board at Father Flanagan's Boys' Home (Boys Town): John Barksdale, M.A., Cary Bell, Ph.D., Vaughn Call, Ph.D., David Coughlin, Ph.D., Dennis Culver, M.A., David Cyr, Ph.D., James Garbarino, Ph.D., Kay Graham, Shelton Hendricks, Ph.D., Betty Johnson, the Rev. James Kelly, Richard Lippmann, Ph.D., Luther Otto, Ph.D., James Peck, Ph.D., and Bill Thornton, M.A. As a member of this committee for two years, and then as its chairman, I became more aware of the complexities of the field and the often thankless task of professionals asked to review the work of others with a scarcity of clear guidelines available as support. I offer my congratulations to the committee members, their successors, and our colleagues around the country who struggle to keep abreast of a rapidly changing aspect of research and human services.

I would also like to thank the members of the Legal Issues Committee and the Legislative Committee of the National Teaching-Family Association: Hewitt ("Rusty") Clark, Ph.D., Pam Daly, Ph.D., Jack Freeman, M.A., Paul Gilford, Ph.D., Robert Jones, Ph.D., Kathryn Kirigin Ramp, Ph.D., Michael McManmon, M.A., Michael Pyykola, and Ron Sexton, Ph.D. As chairman of these committees from 1975 to 1979, I benefitted from the counsel and advice of these representatives of research and service organizations from across the country.

My thanks to Joel Kline of the Mental Health Project in Washington, D.C. His consultation to Boys Town at the inception of new programs provided further insight into the legal aspects of human rights in a child care organization.

Of course, I thank Jim Sweetland, M.L.S., Ph.D., for his reference work and citation assistance and for agreeing to write the concluding chapter of this book. If you really want to try to influence those who influence you on human rights in research, reading Dr. Sweetland's chapter is a must. Jim's interest, and the support of my colleagues at Boys Town (especially Drs. Karen Blase, Dean Fixsen, and Elery Phillips), helped maintain my work on this volume over a period of several years.

Finally, it should be pointed out that, unless noted otherwise, the views expressed in this book are those of the author and do not necessarily represent the views of any other person or organization with which the author may be affiliated.

DENNIS M. MALONEY, PH.D.

Contents

Introduction

If you conduct any kind of research with humans or provide treatment to other persons you should already be familiar with much of the material of this book. But it is much more likely that the information will be new. Why is this so, in this age of mass media and rapid communications? There are several reasons why you may lack such vital information, despite the possibility of lawsuits against unwary professionals who may unwittingly abridge the rights of subjects. Lawsuits have already occurred in the area of treatment, where professionals have violated client rights. Perhaps the most salient of the reasons is the sheer bulk of the communications. As a primary example of information overkill, the federal government in recent years has spawned an enormous complex of bureaucratic regulatory offices. This has not only left the human services professional bewildered by a massive array of rules but it has also had an observable effect on our national economy.

An Associated Press news release on January 12, 1980,[1] revealed that this regulatory function directly contributes to inflation in the United States. In fact, the news release noted that the federal government has 90 regulatory offices issuing around 7,000 rules a year! Further, economists estimated that the financial burden of these regulations borne by Americans was a trifling $50 billion to $150 billion a year. Obviously, only a small fraction of these regulations pertain to protection of subjects and clients, but it is little wonder that most health care and human service professionals cannot keep up with new laws and regulations.

This concern must be tempered with attention to the numerous advantages provided by regulations designed to protect American citizens. Indeed, a backlash against regulatory agencies is not only unwarranted but also can hurt consumers in the long run.[2] In the area of human services, however, the regulations

[1]*Omaha World–Herald*, January 12, 1980, p. 3.
[2]Quinn, J. B. "Regulating the regulators." *Newsweek*, January 21, 1980, p. 73.

often deal with philosophical or moral issues rather than safety standards or advertising guidelines. The increase in regulations in human services has paralleled the increasing political visibility of human rights as a separate issue, particularly in the United States.

Although the concept of human rights is not new to twentieth-century America, the role of the federal government in this field definitely is new. By the late 1970s, there had grown a befuddling maze of regulations and legislation originally designed to govern only research with human subjects. As one example, most professionals have heard of "informed consent" in this regard, but the construction of such an agreement can be a tedious and frustrating task for the most conscientious researcher.

Mistakenly, some researchers still consider informed consent as the primary aspect of the researcher–subject relationship. It is not. It is merely the most visible aspect related to protection of human rights. Other issues such as confidentiality of research information are also involved. Furthermore, there are additional considerations which researchers must keep in mind when working with particular types of subjects (e.g., with children).

But what about those of you who are not engaged in research but do provide treatment for clients? What has all this to do with you? The answer is simple: it has a great deal to do with you. Historically, much of the material included in this book grew out of regulations or legislation regarding human research subjects. As we shall discover, however, many guidelines have since moved into the area of *treatment* and are dealt with in several chapters of this book. For example, educational *researchers* may be familiar with the widely publicized Buckley Amendments (although they may be using the wrong version as their guideline), but how many educators are familiar with provisions of that body of regulations that affect access to student records, whether research is intended or not? Does the psychiatrist or psychologist have a right to view education records as part of treatment planning for a student receiving counseling? And what about *your* records—who has the right to see them?

This reference book is constructed in such a fashion that you should be able to answer most such questions. For example, a psychiatrist working with mentally retarded clients might examine Chapter 10 closely ("Persons Institutionalized as Mentally Disabled"), whereas the school counselor may wish to concentrate on Chapter 7 ("Students"). In general, however, it is recommended that you peruse all of Part 1, Chapters 1–5, to understand the general framework of legislation and regulations affecting both research and practice. Part 2, Chapters 6–10, can then best be used as a source of additional material for working with specific populations of subjects or clients.

Even if you only use this book as an occasional reference source, it is strongly urged that you read Chapter 11 ("How to Be Informed and Involved") by James Sweetland, M.L.S., Ph.D. And read *all* of it? Why? Because of the very problem mentioned initially, that of professionals who are unaware of the tremendous amount of legal precedent being set that can so significantly affect their careers. Further, many of these rules come into existence with unbelievably trivial amounts of input from the very professions most likely to be affected by them. This phenomenon can be altered only by professionals who know what to do and when to do it.

For those of you intending to use this entire book as a textbook, an effort has been made to relate material in one chapter to issues presented in another. Where possible, a chronological order of developments has been discussed within the separate topic areas. In particular, Part 1 is presented largely in this fashion, with Chapter 1 containing a brief synopsis of the recent developments in the field, starting with the achievements of the highly influential National Commission for the Protection of Human Subjects of Biomedical and Behavioral Research.

A word of caution: Do not rely on this text as your legal defense in matters of possible litigation. Always seek legal counsel in such situations. This reference work does not contain everything there is to know about the rights of human research subjects and treatment clients. Indeed, it seldom delves into the ethical or moral issues that underlie the guidelines and practical recommendations. Readers interested in the more philosophical aspects of these issues will find examples of such texts listed in various footnotes. In addition, it should be noted that the focus of this work is on federal rules. Readers interested in local applications are urged to contact their state's professional societies or state bar association.

As a reference work, this book should serve your purposes for most topics most of the time. It was written for the use of the author as well as for others, since the author has found it almost impossible to keep track of federal legislation and regulations from the many sources that publish different updates. The author was dismayed also by the increasing bulges in crowded files and decided that one book would be more useful—and would take up less room in his office. Readers who can suggest modifications of this work will find their comments and feedback much appreciated.

Finally, readers should take note of the fact that rules can change rapidly in this area; therefore quoted rules or laws may have changed by the time this volume is published. Nevertheless, the regulatory and legislative history of protection of human research subjects clearly shows that new rules are based on old ones. Thus the quoted texts in this work should serve as reliable guides for the foreseeable future. In addition to the numerous methods listed throughout this book for keeping up to date with relevant rules, readers can telephone the National Institute of Health's Office for Protection from Research Risks (301–496–7005) or the Associate Commission for Health Affairs at the Food and Drug Administration (301–443–6143) for immediate help.

PART I

Guidelines for Research

Part I contains Chapters 1–5, each chapter dealing with a major topic within the field of protection of human subjects. Some of the quoted regulatory text was formulated by government bodies with the explicit intent of affecting research with humans (e.g., regulations covering operation of Institutional Review Boards). Other legislation or regulations are more general in scope but are still related to protection issues. For example, the Privacy Act is less concerned with research or practice than it is with restricting access to information contained in government files. However, a moment's reflection will lead the reader to deduce the natural relationship of the Privacy Act to our topics. Often such information is needed by researchers or practitioners to further their research or directly to benefit their clients in treatment settings.

The chapters of Part I are arranged roughly in order of importance and chronology. Although it is true that Institutional Review Boards (see Chapter 2) existed before national commissions became so active in protection issues (see Chapter 1), it is also true that the boards have never been the same since new regulations were instigated by the commissions.

The reader could read Part I and then skip to Part III ("Keeping Up with Changing Rules"), leaving Part II as a reference source for working with special populations only. But these special populations (e.g., children, students, prisoners) constitute a sizable proportion of all possible subjects and clients. And, as is discussed further in Part II, it is likely that federal agencies will increase, rather than decrease, regulations for as yet unspecified special populations.

Federal agencies and commissions play primary roles in this field, and Part I begins with a discussion of one of the most significant commissions of all in this area, the National Commission for the Protection of Human Subjects of Biomedical and Behavioral Research.

CHAPTER 1

Federal Commissions and Advisory Boards

To date, the single most influential body in the United States involved with the protection of human subjects and clients has been the National Commission for the Protection of Human Subjects of Biomedical and Behavioral Research. For brevity's sake, we will refer to this body simply as the Commission throughout this book. The Commission's activities were public, advance notices of Commission meetings were printed in the *Federal Register*, minutes of the meetings are still available, special reports by persons outside the Commission were authorized and paid for by the Commission, and the Commission intentionally sought opinions of all American citizens before publishing recommendations. The Commission was involved indirectly in establishing rules for the entire Department of Health, Education, and Welfare (HEW)—including regulations for other persons or agencies receiving *any* funds from HEW. The Commission was instrumental in proposing rules regulating research with children, prisoners, mentally infirm persons, and with other special groups, although not all these recommendations have led to final rules yet. As will be seen later, the Commission's work eventually affected practitioners as well as researchers.

Although formally dissolved on October 1, 1978, the Commission's activities were taken over by other federal commissions and agencies. In particular, the President's Commission for the Study of Ethical Problems in Medicine and Biomedical and Behavioral Research actively assumed such a role in 1980. The Ethics Advisory Board of HEW (later retitled HHS as the Department of Health and Human Services) was also active in this area. These two successor bodies are discussed in further detail later in this chapter, but our primary emphasis is on the original Commission.

Knowledge of the Commission's activities, especially during the years

1975–1978, is vital to understanding the current regulations and how they apply to therapy, care, and various types of research today.

The National Commission for the Protection of Human Subjects of Biomedical and Behavioral Research

Despite the effects of the Commission, it is clear that many in the professional human services field (psychologists, physicians, psychiatrists, educators, etc.) have not yet caught up with activities begun years ago. An event as recent as in 1977 suggests that many of us are not yet aware of the tremendous influence that has been exerted on human research and practice in this country by the Commission.

In the instance in 1977, the Commission followed its usual procedure of publishing recommendations in the *Federal Register* and comments from the public were solicited. In this case, the regulations involved research with prisoners. Persons familiar with this regulatory process are aware that publication of the Commissions's recommendations were first considered by HEW along with the citizen comments and then HEW regulations were produced. The details of who is affected by HEW regulations are explained in subsequent chapters of this book, but it constitutes a surprising percentage of the entire population. In this particular instance, however, the Commission's regulations on research with prisoners affected a large number of prisoners in America (for those states that still allowed such research), the attendant physicians, psychologists, social workers, therapists, pharmaceutical representatives and their employees, and prison personnel. Also affected was our present and future general fund of knowledge (or lack thereof) of criminals, criminal behavior, penal institutions, rehabilitation, and the effects of prison incarceration on inmates. Although it is true that these regulations were initially aimed at drug research—research which can pose serious problems from both an ethical and a legal viewpoint—it is also true that the eventual regulations somehow became far more broadly constructed. Therefore, it is important to keep in mind that such regulations have tremendous effects, regardless of whether one happens to agree or disagree with the specific guidelines.

Yet, despite the potential impact of the recommendations of the Commission (and remember that this was only *one* area of the Commission's activities at that time), the citizen response to the recommendations was paltry indeed. The following excerpt is taken from a later edition of the *Federal Register* in which HEW had received a few citizen comments, incorporated them into the Commission's recommendations, and proposed resultant official regulations:

> Pursuant to Section 205 of the National Research Act (Pub. L. 93–348) the recommendations of the National Commission for the Protection of Human Subjects

> of Biomedical and Behavior Research on research involving prisoners were published in the FEDERAL REGISTER (42 FR 3076) on January 14, 1977. *Comments were received from 49 individuals.* (italics added) After reviewing the recommendations and the comments, the Secretary (Joseph A. Califano, Jr., Secretary of HEW) has prepared the notice of proposed rulemaking set forth below, which in essence adopts the (Commission's) recommendations.[1]

Forty-nine individuals in America commented on these regulations. Forty-nine! Perhaps the proposed regulations were routine and there was little response necessary? It is unlikely that the comments of 49 individuals could be expected to represent adequately the opinions and judgments of the thousands affected directly or indirectly by these regulations. Further, HEW itself felt the recommendations were rather detailed:

> The Department (HEW) has concluded that these requirements are so stringent that it is doubtful that any existing prison and few research projects could satisfy them.[2]

More details regarding the prisoner research area can be found in Part II, Chapter 8 of this book. Suggestions on how to exert your influence and that of your colleagues in this process are found in Part III, Chapter 11.

Whether one agrees or disagrees with the Commission's various recommendations in this area, it is clear that the Commission wielded considerable power in guiding the formulation of regulations for protection of human subjects in research. This same power continues to be held by successor commissions.

From 1975 to 1978, the Commission recommended guidelines governing research in a wide variety of areas. Commission proposals continued to be acted upon in 1979 (e.g., the Belmont Report) and beyond. Other such groups have existed in the past and the present (e.g., the President's Biomedical Research Panel), but, to date, the Commission appears to have been tbe most long-lived and active of such bodies.

To understand fully the nature and scope of the Commission, it is necessary to become familiar with its enabling legislation. The Commission came into being in an era of mistrust of public agencies and government officials—a time of worry over privacy, over freedom of the individual from unwarranted government intrusion, and over control of individual lives by authorities. Hence, privacy laws, freedom in inquiry laws, and computer information and banking restrictions were enacted at the federal level. In the abortion case of *Roe* v. *Wade* (1973), the Supreme Court at that time reemphasized the *Constitutional* guarantees of privacy for individuals.[3]

In addition to the effort expended by federal agencies to regulate research

[1]*Federal Register*, Vol. 43, No. 3, Thursday, January 5, 1978, p. 1050.
[2]Ibid.
[3]Friedman, P. R. Legal regulation of applied behavior analysis in mental institutions and prisons. *Arizona Law Review*, 1975, *17* (1), 39–94.

with human subjects, one of the national laws that was passed gave birth to the Commission for the Protection of Human Subjects of Biomedical and Behavioral Research. This law is contained in the legislation known as the National Research Act. Among the Commission's duties was the general responsibility for producing recommendations such that human subjects, ostensibly under some control of researchers and/or federal agencies, would be protected and their individual rights maintained.

The National Research Act

History

The National Research Act (Public Law 93–348) became law on July 12, 1974. The Act was written to amend the Public Health Service Act to: (a) establish a program of National Research Service awards and (b) to provide for the protection of human subjects involved in biomedical and behavioral research and other purposes. Thus, as is so often the case with legislation, several different aims were realized in the same bill.

Unfortunately, this can lead to some confusion when citing a piece of legislation, for the same rules may be referred to by different names. For example, the National Research Act (PL 93–348) actually consists of two separate titles. Title I is labeled "Title I—Biomedical and Behavioral Research Training" in the act itself. Title II is labelled "Title II—Protection of Human Subjects of Biomedical and Behavioral Research" in the act. It is Title II that contained the enabling legislation for the establishment of the Commission, not Title I. To complicate the picture further, Title I of the National Research Act is also known as the National Research Service Award Act of 1974 and is so described in the legislation itself. Thus, for our purposes, Title II of the National Research Act is the most relevant. Readers interested in a more detailed legislative history for this Act should refer to the sources suggested in Chapter 11 of this book.

These distinctions are of interest to more than historians or lawyers. Incorrect citation of legislation or regulations can produce despair in the researcher or practitioner who needs some particular information but is unable to locate rapidly the correct source in the nearest library. This confusion can also complicate communications within professional associations that attempt to influence the legislative process. If this book achieves one of its primary purposes, the reader should be able to locate desired information fairly easily and extract those passages most relevant for his or her professional activities. Additional discussion of this and related problems (and their solutions) can be found in Chapter 11.

Purpose

The purposes of the Commission, as intended in the legislation, were quite broad in scope. This fact usually comes as a surprise to those who, not unreasonably, infer from the Commission's title that it was concerned solely with research. It was not so limited. Indeed, its examination of treatment and services may be its least popularly recognized activity. Its deliberations on treatment primarily involved the physician–patient relationship in instances where medical services were provided by HEW and, to a lesser extent, the practitioner–client relationship in human services (e.g., in psychology). The foundation of this dual interest in research and in practice can be seen in the Belmont Report, a Commission report discussed later in this chapter.

A more complete familiarity with the purposes of the Commission can be obtained from reading the actual legislation as contained at the end of this chapter. The reader is urged to use the actual texts appearing at the end of each chapter as source tools for increased comprehension. In a field that can change as rapidly as human rights, members of the legal profession may wish to consult current legal texts (see Chapter 11).

Although the complete text of Title II of the National Research Act is included later, a specific excerpt here will illustrate the scope of the Commission's activities as intended in 1974 (and the excerpt will confound those readers who thought they could get away with not reading actual legislative text). The Commission was conscientious in pursuing the broad mandate handed them, as we will see in ensuing chapters. This excerpt is taken from the National Research Act, Title II, Part A, Section 202(a):

> Sec. 202. (a) The Commission shall carry out the following:
>
> (1) (A) The Commission shall (i) conduct a comprehensive investigation and study to identify the basic ethical principles which should underlie the conduct of biomedical and behavioral research involving human subjects, (ii) develop guidelines which should be followed in such research to assure that it is conducted in accordance with such principles, and (iii) make recommendations to the Secretary (I) for such administrative action as may be appropriate to apply such guidelines to biomedical and behavioral research conducted or supported under programs administered by the Secretary, and (II) concerning any other matter pertaining to the protection of human subjects of biomedical and behavioral research.
>
> (B) In carrying out subparagraph (A), the Commission shall consider at least the following:
>
> (i) The *boundaries between* biomedical or behavioral *research* involving human subjects and the accepted *and routine practice* of medicine. (italics added)
>
> (ii) The role of assessment of risk–benefit criteria in the determination of the appropriateness of research involving human subjects.
>
> (iii) Appropriate guidelines for the selection of human subjects for participation in biomedical and behavioral research.

(iv) The nature and definition of informed consent in various research settings.
(v) Mechanisms for evaluating and monitoring the performance of Institutional Review Boards established in accordance with section 474 of the Public Health Service Act and appropriate enforcement mechanisms for carrying out their decisions.

(C) The Commission shall consider the appropriateness of applying the principles and guidelines identified and developed under subparagraph (A) to the *delivery of health services* (italics added) to patients under programs conducted or supported by the Secretary (of HEW).

The legislation then goes on to describe the Commission's responsibility with regards to informed consent, work with special populations, and numerous related items.

As can be seen clearly from the quoted legislation, both research and treatment or services came under the rubric of the Commission. What is also apparent, surprisingly so with the benefit of hindsight, is the emphasis on medical research and medical services. Thus, as we see later, a physician was appointed as chairman of the Commission.

Several Professions

This emphasis changed, however, after the Commission was formed and actually began its work. Medical research and services continued to receive considerable scrutiny by the Commission, but the work of other professions also became subject to Commission deliberations and recommendations. Most notably, the fields of social science (e.g., psychology and sociology) and education quickly became involved. This interest can, in part, be traced to the "*Behavioral*" in the Commission's official name. However, it is more likely that at least three other factors led to this increased role of the Commission in fields other than medicine.

First, as we shall soon see, there were nonmedical professionals appointed to serve on the Commission. These appointments were mandated according to diverse professions in the enabling legislation:

Sec. 201. (b) The Commission shall be composed of eleven members appointed by the Secretary of Health, Education, and Welfare (hereinafter in this title referred to as the "Secretary"). The Secretary shall select members of the Commission from individuals distinguished in the fields of medicine, law, ethics, theology, the biological, physical, behavioral and social sciences, philosophy, humanities, health administration, government, and public affairs (excerpted from the National Research Act, Title II, Section 201).

Second, the involvement of the Commission with the operations of Institutional Review Boards around the country inevitably led to consideration of topics as diverse as the ones considered by such boards at institutions receiving any

Federal grant monies. The majority of these boards (see Chapter 2) appear to exist at universities, settings where medical research and services form only one part of the activities supported and brought to boards for consideration.

Third, the broad scope of the Commission may have been due to one attribute of the United States in the 1970s, a decade of disillusion with government, distrust of politicians, and fear for personal liberty and privacy. Vietnam, the resignation of a president, and the "Pentagon Papers" came readily to mind. Thus a *zeitgeist* of the 1970s was fear. It should then come as little surprise to discover that in the same year the Commission was founded the Privacy Act of 1974 was passed independently as a move to place restrictions on use of government-held information (see Chapter 5). This influence was reflected in the meetings of the Commission and it is likely that this added impetus continued to propel Commission efforts into additional arenas of related interests.

But, whatever the causes, the Commission's interests and responsibilities continued to grow throughout its tenure.

At this point it should be clear that: (1) the Commission has played a central role in establishing rules for research with humans in the U.S.; (2) the Commission was started by enabling legislation contained in Title II of the National Research Act (PL 93–348) enacted on July 12, 1974; and (3) the impact of the Commission continues to strengthen and expand into new areas.

Impact

It grows increasingly difficult to assess the overall impact of the activities of the Commission. This is not due to reluctance on the part of the author to assign a value judgment to the role or activity of the Commission so much as it is due to the apparent energy and dedication of Commission members. Indeed, many of the separate chapters of this book are devoted entirely to the related but independent topics investigated by the Commission.

Individual readers may be aware of the Commission's published guidelines regarding Institutional Review Boards (IRBs) as an example of the Commission's broad interests. These boards are sanctioned by HHS, and IRB approval of research is necessary before any HHS agency will award a grant or contract to an applicant organization or individual. Other federal agencies have adopted many of these same procedures. Although IRBs were in existence long before the Commission was developed, the Commission became involved with regulation of IRBs because of their explicit mandate from Congress.

As a further example of the scope of Commission activities, let us look more closely at its study of IRBs. Readers may be familiar with the intricacies of *informed consent* (see Chapter 3), a complicated process that commands much attention from IRBs. The most visible feature of the overall process is an informed consent form—a contract between a researcher or practitioner and a human subject or client that details the precise responsibilities and expectations of both parties. There is a good deal more involved in informed consent than the written contract, but the contract itself is what is usually referred to when informed consent is discussed. As part of the Commission's recommendations for IRB activities, the Commission addressed informed consent at considerable length. This is a quite logical interest for, as those who have applied for IRB approval are aware, IRBs must judge the adequacy of an applicant's informed consent provisions as part of their deliberations. Almost anyone seeking federal grant or contract funds in human research or service must have IRB approval.

In summary, the impact of the Commission has been notable. Not only has it addressed ethical areas (e.g., individual's rights, relationships between doctor and patient) and operational mechanisms (e.g., the policies and procedures of IRBs at universities, foundations, and hospitals), it has also devoted significant attention to protection of special populations in the United States (e.g., children). We may anticipate that the Commission's interest in special populations will continue to be expanded by successor bodies at the federal level.

The Commission became involved in such a diversity of areas and arrived at its recommendations through a straightforward set of procedures. Similar procedures are followed by the Commission's successor boards and commissions in Washington, D.C.

Commission Procedures

Although certain steps may vary in sequence or in the amount of time required for completion, the Commission followed five major steps in disseminating its recommendations and suggestions.

1. *Meetings* open to the public to discuss protection issues
2. *Reports* written by invited experts at the request of the Commission
3. *Recommendations* published by the Commission and distributed to the public
4. *Proposed regulations* published by HEW in the *Federal Register* and public comments requested
5. *Final regulations* published by HEW in the *Federal Register*

Commission Meetings

Commission meetings typically were held in Washington, D.C. In addition, the Commission occasionally met at selected cities around the country to permit more citizen participation. These meetings were public and advance notices were published in the *Federal Register*. The *Federal Register* has an immense amount of information, notices, regulations, and proclamations published every day, and such notices can be missed by even the most conscientious observer. For a deeper examination of the *Federal Register*, please refer to Chapter 11.

The monthly meetings of the Commission were chaired by Dr. Kenneth John Ryan. As of September, 1978, the members of the Commission were:

Kenneth John Ryan, M.D. (Chairman)
Chief of Staff
Boston Hospital for Women

Joseph V. Brady, Ph.D.
Professor of Behavioral Biology
Johns Hopkins University

Robert E. Cooke, M.D.
President, Medical College of Pennsylvania

Dorothy I. Height
President, National Council of Negro Women, Inc.

Albert R. Jonsen, Ph.D.
Associate Professor of Bioethics
University of California at San Francisco

Patricia King, J.D.
Associate Professor of Law
Georgetown University Law Center

Karen Lebacqz, Ph.D.
Associate Professor of Christian Ethics
Pacific School of Religion

Donald W. Seldin, M.D.
Professor and Chairman,
Department of Internal Medicine
University of Texas at Dallas

Eliot Stellar, Ph.D.
Provost of the University and
Professor of Physiological Psychology
University of Pennsylvania

Members since deceased were David W. Louisell, J.D. (Professor of Law, University of California at Berkeley) and Robert H. Turtle, LL.B. (Attorney, VomBauer, Coburn, Simmons and Turtle, Washington, D.C.).

Minutes of these meetings are available in two formats: a mimeographed typed summary (usually one or two pages), and the full minutes are available on microfiche. These minutes can be obtained by writing the Office for Protection from Research Risks, National Institutes of Health, 9000 Rockville Pike, Bethesda, Maryland 20205.

How useful are these minutes? The author has not found them useful if one needs a quick answer to such questions as "How do I write an informed consent form for my research project?" or "Do I need an informed consent form for my type of private practice?" But then, the minutes are not intended to be so used. We hope that this book could help the professional in this regard (see Chapter 3). Alternatively, the researcher or practitioner could attempt to locate the federal regulation in question on his own. In this instance involving interest in informed consent, one would need to start with the most recent regulations published in the *Federal Register*. Such documents can be obtained from local law libraries or directly from federal agency offices (e.g., Office of Protection from Research Risks) as listed throughout this book.

However, the author has found that the minutes of the Commission's meetings can serve three useful purposes. First, examination of the minutes gives a more detailed history of member discussion and provides the reader with a far more complete appraisal of what ethical, legal, and moral issues were addressed for any topic. The resultant regulations that were eventually adopted by HEW could not possibly contain adequate text to reflect the deliberate and often painstaking research and soul-searching by Commission members that led to the clipped official jargon published in the United States Code of Federal Regulations.

Second, the minutes reflect the differences of opinion expressed by the various Commission members. Whether or not any one or two persons can adequately represent the opinions of hundreds or thousands of colleagues is a moot point, but it is obvious that not all votes within the Commission were unanimous. The lack of total agreement underscores both the variety of interests on the part of Commission members and the complexity of the issues involved. Disagreements appeared to exist in greater or lesser degree regardless of the particular issue involved. This is not to say that the same persons disagreed every time, but rather that different lines can be drawn between disciplines and even personalities, depending on the legal, moral, or ethical aspects of the issue at hand. This speaks well for the diversity of the group. Interestingly, the Commission made a special effort to disseminate even the opinions of those on the Commission who disagreed with the majority view. These minority reports were added at the end of the published reports whenever a commissioner disagreed strongly enough to write a separate report.

Finally, Commission minutes could be used to keep up to date regarding Commission deliberations during the Commission's tenure. This is still true for successor commissions. Obviously, the most current meetings reflect the most recent deliberations of any commission. But persons desiring to provide input can more readily forward timely and relevant comments to a federal body by using the most recent minutes as a type of forecast for future developments. This is especially true for those responsible for organizing others to lobby with a commission to put across a certain point of view, for the minutes usually give an edge through advance warning of potential recommendations to be published in the near future. Nevertheless, the 30–90 days posted in the *Federal Register* when recommendations are published is usually a sufficient amount of time in which to forward comments and react to commission proposals.

Commission Reports

In reality, the term *Commission reports* can refer to two different types of reports. First, there were what can be viewed as internal reports. These were written reports contracted for by the Commission to seek expert opinion on a topic. Examination of Commission minutes for 1975–1978 has shown that this approach was used frequently by the Commission to seek information beyond that provided by Commission members themselves. However, in terms of strict frequency of input, the Commission more often received oral reports or arguments from persons appearing at the monthly meetings in Washington, D.C., to espouse one view or another. Summaries of these reports were published in the appendices to the various Commission *Reports and Recommendations*, and the transcripts of discussions of these reports are still available from HHS.

Once again, however, it is likely that the only persons aware of the existence of such reports were staff associated with the Commission or persons receiving copies of the minutes of the Commission's meetings. These minutes usually contained a listing of materials presented to the Commission, and internal reports would be listed. This is not to be taken as any indication for secrecy on the part of the Commission or similar commissions. In fact, a summary of the minutes can typically be obtained free of charge simply by requesting them from a commission and asking to be placed on their mailing list. Unfortunately, most human services professionals appear to be unaware of this procedure.

The second type of Commission report can be considered as their external reports and are more widely known. These reports were compiled by the Commission, printed by the U.S. Government Printing Office, and distributed free of charge by the Commission to hundreds of persons and agencies on the Commission's mailing list. These reports usually also contained recommendations by the Commission regarding the ethical issue at hand. These recommendations were intended by the Commission to be recommendations for *possible* action by one or more federal agencies. These reports, containing recommendations and always accompanied by extensive appendices, were then delivered to the agencies as well as made available to the public.

A third type of report is perhaps the least common and is most similar to the external type. In essence, the entire report can be considered as one complete recommendation or statement of position of the federal government. It is a type of position paper, usually containing no specific procedural guidelines for researchers or practitioners but emphasizing a general approach to an issue that is the basis for preceding, simultaneous, or ensuing legislation or regulations.

An example of this third type of report is the well-publicized Belmont Report,[4] presented in its entirety at the end of this chapter. The Belmont Report was an attempt to summarize the basic ethical principles that surround the conduct of research with human subjects. It also discussed the boundaries between practice and research. The Belmont Report was a product of a four-day conference held in February of 1976 at the Smithsonian Institute's Belmont Conference Center. In addition, various components of the report were influenced by the Commission's regular monthly meetings.

Of the three types of reports described, the latter two most often advanced to the next stage at which HEW actually proposed regulations based on the reports. This same process is followed by the Commission's successor bodies.

Proposed Regulations

The Department of HEW was required by statute to act upon Commission recommendations by reworking them into regulations. These proposed regulations were then printed in the *Federal Register* and comments were solicited from the public (i.e., those who read the *Federal Register*). Although the amount of time allowed for written feedback to be sent to HEW varied, the average appeared to be about 60 days and remains so today. In general, these comments were always to be sent, by a specified date, to:

Office for Protection from Research Risks
National Institutes of Health
5333 Westbard Avenue, Room 303
Bethesda, Maryland 20205

After the deadline, HEW officials decided which, if any, of the received public comments would influence the proposed wording of the regulations. Also, HEW decided which of the received comments would be published along with the final regulations to help explain any modifications of the proposed regulations or any significant contrary opinions. Thus the final regulations were then published in the *Federal Register*.

[4]*Federal Register*, Vol. 44, No. 76, Wednesday, April 18, 1979, p. 23192. Also available as DHEW Publication No. (OS) 78–0012 from the Superintendent of Documents, U.S. Government Printing Office, Washington, D.C. 20402.

Final Regulations

The final regulations were printed in the *Federal Register* and eventually became part of the U.S. Code of Federal Regulations. These steps in the regulatory process are described in greater detail in Chapter 11. In general, the Commission followed these steps closely, as has its primary successor body, the President's Commission for the Study of Ethical Problems in Medicine and Biomedical and Behavioral Research.

The President's Commission for the Study of Ethical Problems in Medicine and Biomedical and Behavioral Research

Referred to as the President's Commission to distinguish it from the original National Commission, this commission has now taken the lead in recommending new regulations regarding research and treatment with humans. Although it will likely be the mid-1980s before its effects are felt as fully as were those of the earlier National Commission, this body has assumed the primary role of carrying on with tasks left unfinished when the National Commission disbanded in late 1978. One of its early accomplishments was to provide the finishing touches on the proposed regulations on the powers of Institutional Review Boards (IRBs) as largely completed by the National Commission. These proposed regulations were published by HHS, or the Department of Health and Human Services (previously known as HEW, the Department of Health, Education, and Welfare). These important regulations deal with the types of research which must be reviewed by local IRBs and those types that may be exempted by the new regulations of 1981. This topic is covered in more detail in Chapter 2.

As noted before, the original National Commission officially terminated on October 1, 1978. Then, on November 9, 1978, Public Law 95–622 was passed, part of which repealed appropriate sections of the National Research Act (Title II, Sections 201–205, subsection f of Section 211, and Section 213) and PL 95–622 established authorization for the new President's Commission with Title III of PL 95–622. However, the new commission was not yet a reality. On December 17, 1979, President Carter issued an Executive Order to start the authorized commission. Still, it was not until January 14, 1980, that the first 11 members of the new commission were sworn in at the White House. Thus approximately 15 months elapsed between the termination of the original commission and the start up of the second federal commission.

The founding members of this second commission were:

Morris B. Abram (Chairperson)
Paul, Weiss, Rifkind, Wharton, and Garrison
New York, New York

Renee Claire Fox, Ph.D.
Department of Sociology
University of Pennsylvania
Philadelphia, Pennsylvania

Mario Garcia-Palmieri, M.D.
Department of Medicine
University of Puerto Rico
Rio Piedras, Puerto Rico

Albert Rupert Jonsen, Ph.D.
University of California School of Medicine
San Francisco, California

Patricia A. King, J.D.
Georgetown University Law School
Washington, D.C.

Mathilde Krim
Sloan-Kettering Institute for Cancer Research
New York, New York

Donald N. Medearis, M.D.
Pediatrics Department
Harvard University
Cambridge, Massachusetts

Arno G. Motulsky, M.D.
School of Medicine
University of Washington
Seattle, Washington

Frederich Carl Redlich, M.D.
Psychiatry Department
University of California at Los Angeles
Los Angeles, California

Anne Scitovsky
Palo Alto Medical Research Foundation
Palo Alto, Calfornia

Charles J. Walker, M.D.
Board of Trustees
Fisk University
Nashville, Tennessee

Organized with a preset termination date of December 31, 1982, the new commission's offices were located at:

2000 K Street, N.W.
Suite 555
Washington, D.C. 20006

Some of these founding members have since been replaced by newer members, and this process will likely continue. The names of current members can be found in the *Federal Register* whenever new *Reports and Recommendations* are issued by the President's Commission.

Although the exact text of the enabling legislation for the new President's Commission is presented at the end of this chapter, it is worth noting that the commission has two main purposes. First, it was designed to finish and expand upon topics previously addressed by the National Commission, such as biomedical and behavioral research. Second, it was to address additional topics such as health care issues (e.g., equitable allocation of health resources geographically and by income in the United States) and to review and monitor implementation of existing regulations on biomedical, behavioral, and social research in all federal agencies.

Because much of the research and treatment with humans is funded by HHS, it is not surprising that the federal agency's own committees would play a role in shaping national regulations for protection of human subjects and clients. Although not as powerful as either of the two federal commissions, the Ethics Advisory Board within HEW initially played a key role in continuing the work of the original Commission, particularly during the interim before the second federal commission became active. This role may well have resulted from the fact that confusion and concern had been at least one of the outcomes of the original Commission's deliberations in a new field; but—during the 15 months when no federal commission was active—there had to be some agency operating within HEW that could maintain a leadership or advisory role while HEW staff responded to public inquiries about regulatory doctrine. These inquiries were natural enough, given that the original Commission terminated before final regulations were published in a number of key areas.

An ongoing coordinating role has also been conducted by staff in the following office within HHS:

Office for Protection from Research Risks
Public Health Service
Department of Health and Human Services
Bethesda, Maryland 20205

Nevertheless, it was the Ethics Advisory Board which played a crucial role during the hiatus between federal commissions, meeting regularly and distributing written accounts of its deliberations to interested people throughout the country.

The Ethics Advisory Board

Meeting monthly in the Washington, D.C., area, the Ethics Advisory Board of HEW followed procedures similar to those of the original National Commission. Although other federal agencies contain similar boards, the fact that the Ethics Advisory Board was in HEW virtually guaranteed that its activities would dominate the field of federal rules dealing with researcher–subject and practitioner–client relationships in the United States.

As of June, 1979, the following persons served as members of the Ethics Advisory Board:[5,6]

James C. Gaither, J.D. (Chairperson)
Cooley, Godward, Castro, Huddleston and Tatum
San Francisco, California

David A. Hamburg, M.D. (Vice Chairman)
President, Institute of Medicine
Washington, D.C.

Sissela Bok, Ph.D.
Lecturer in Medical Ethics
Harvard University
Cambridge, Massachusetts

Jack T. Conway
Senior Vice President,
Government and Labor Movement Relations
United Way of America
Washington, D.C.

Henry W. Foster, M.D.
Professor and Chairman
Department of Obstetrics and Gynecology
Meharry Medical College
Nashville, Tennessee

Donald A. Henderson, M.D.
Dean, School of Hygiene and Public Health
Johns Hopkins University
Baltimore, Maryland

Maurice Lazarus
Chairman, Finance Committee

[5]*Federal Register*, Vol. 44, No. 118, Monday, June 18, 1979, p. 35033.
[6]Supplemental address information for Ethics Advisory Board members was obtained from a September, 1978, listing made available by the Board.

Federated Department Stores, Inc.
Boston, Massachusetts

Richard A. McCormick, S.T.D.
Professor of Christian Ethics
Kennedy Institute for the Study of
Reproduction and Bioethics
Georgetown University
Washington, D.C.

Robert F. Murray, M.D.
Chief, Division of Medical Genetics
College of Medicine
Howard University
Washington, D.C.

Mitchel W. Spellman, M.D.
Dean for Medical Services and
Professor of Surgery
Harvard Medical School
Boston, Massachusetts

Daniel C. Tosteson, M.D.
Dean, Medical School
Harvard University
Boston, Massachusetts

Agnes N. Williams, LL.B.
Potomac, Maryland

Eugene M. Zweiback, M.D.
Omaha, Nebraska

The Board spent a majority of its efforts in 1979 on deliberation of regulations and discussions of the ethical issues underlying human *in vitro* fertilization. These deliberations involved contact with professionals in other countries, including a personal visit by two Board staff members (Charles R. McCarthy, Ph.D., and Philip Halpern, J.D.) to England. The trip included interviews with a number of persons involved in the review of the work of Drs. Steptoe and Edwards, who used the controversial technique in the internationally publicized gestation and birth of a baby girl.

The Board also expanded its deliberations to include other areas (e.g., questions regarding exemptions from the Freedom of Information Act).[7] This Board was disbanded in 1980.

[7]Transcripts of meetings held by the Ethics Advisory Board may be purchased from the National Technical Information Service, 5285 Port Royal Road, Springfield, Virginia 22161.

Who Does What?

One unfortunate side effect arising out of the multiplicity of boards, committees, and commissions on the federal level (ignoring for the moment the various state-level committees) has been confusion over who is responsible for what. In other words, which agency or group is responsible for setting policy on a particular matter? This can be quite difficult in an area where overlap between ethical issues, federal regulations, agency guidelines, professional association rules, reasonable research approaches, and treatment priorities often produces apparently insoluble conflicts.

However, this is more of a problem for federal officials than for the researcher or practitioner. In general, the regulations most relevant for protection issues in research or treatment are published under the auspices of the agency with which one is most closely associated. Usually this is the Department of Health and Human Services but it may be another department (e.g., the Defense Department) for certain researchers. These regulations are published in the *Federal Register*, and the guidelines presented later in Chapter 11 will aid the reader in keeping up-to-date with relevant rules.

The Belmont Report

The actions of the original National Commission, the subsequent President's Commission, and the Ethics Advisory Board have been significantly influenced by ethical positions outlined in what has come to be called the Belmont Report. Although the report was presented to the Senate in September, 1978, it was based on meetings previously held in 1976. It was made generally available to the public when published in the *Federal Register* on April 19, 1979 (44 FR 23192). Thus federal actions in protection of subjects and clients were guided by the policies inherent in the report about three years before publication. It is highly likely that these policies will continue to dominate the positions taken by such commissions in Washington, D.C. in the future.

The Belmont Report is of interest in three respects. First, and most importantly, it provided a basic ethical framework for discussion of protection of human rights and dignity. The report did not attempt to define the applications of ethical guidelines, but rather discussed the general aspects of ethics in both a historical and legal context. The narrative style of the report was apparently aimed at an audience that need not include attorneys and thus is easy to understand for most readers. While one may or may not agree with statements in the report, it is perhaps the most succinct expression of the majority view of the original National Commission members and their successors.

Second, the Belmont Report differs markedly from most reports of the

National Commission in that it did not make specific recommendations to HEW. Instead, the Commission recommended that the *entire* report be adopted by HEW as a policy statement. It is reasonable to hypothesize that the report will long outlive any specific recommendations that may have been proposed by the Commission and will instead form the basis of many individual guidelines for decades to come.

Finally, the Belmont Report is usually described as an ethical statement about *research* with human subjects. It is so described in the cover letters for the report from Dr. Kenneth Ryan (Commission Chairperson) to Walter Mondale (1978 President of the Senate), Thomas P. O'Neill (1978 Speaker of the House), and Joseph Califano (1978 Secretary of Health, Education, and Welfare). However, an often-ignored section of the report ("Boundaries between practice and research") quietly gives credence to the continuing involvement of *both* treatment clients and research subjects in this area. The distinctions drawn in the report between research and practice are interesting, but nonspecific. Thus disagreement regarding the actual application of each new federal regulation in the protection field continues to the present time. We will see this same dichotomy again in our examination of the role and authority of Institutional Review Boards in the next chapter.

Summary

Interest in the protection of subjects and clients is not a recent phenomenon in the United States. However, what is new is the sudden burgeoning of legislation and regulations at all levels of government regarding such protection. Although the focus of these activities has ostensibly been on the researcher–subject relationship, many similarities extend to the practitioner–client relationship as well, particularly in regard to freedom of information and access to various types of files.

Prior to 1974, most governmental monitoring of these relationships was handled through Institutional Review Boards (IRBs), committees established at institutions throughout the country that received research funds from the Department of Health, Education, and Welfare and other federal agencies. IRBs existed to oversee protection of subjects in federally funded grants and contracts, nothing more. As we shall see in Chapter 2, however, the power of IRBs has fluctuated markedly since 1974. For it was on July 12, 1974, that the historic National Research Act (Public Law 93–348) was enacted.

The National Research Act contained, among other things, legislation that led to the formation of the National Commission for the Protection of Human Subjects of Biomedical and Behavioral Research. For roughly four years, 1974–1978, the Commission examined, discussed, recommended, and published opinions on a tremendous variety of issues, all dealing with the protection of humans. The efforts of the Commission have been continued by numerous successor groups, with the President's Commission and the HHS Ethics Advisory Board the most noteworthy nationally from 1978 on. The National Commission's original focus on research subjects has led to interest in the practitioner–client dyad as well. Slowly, perhaps too slowly, human service professionals have become aware of these developing regulations impinging on their work.

Part of the problem facing the professional is the complexity of the regulatory and legislative processes themselves. Part of the problem lies in the variety of areas that are now regulated by government agencies. But part of the dilemma of the ethical researcher or practitioner is simply not knowing what the present situation is in this legal area.

To help alleviate these problems, and to set a pattern for subsequent chapters, this chapter concludes with the exact text of the relevant legislation or regulations for this chapter. In this first instance we find the portion of the National Research Act that led to the formation of the extremely influential Commission for the Protection of Human Subjects of Biomedical and Behavioral Research. Following this excerpt from the National Research Act is the Belmont Report, the position paper of the Commission that supplies the basic ethical framework for the complex issues surrounding protection of subjects and clients. Finally, the text is presented for the enabling legislation that authorized the establishment of the President's Commission for the Study of Ethical Problems in Medicine and Biomedical and Behavioral Research.

APPENDIX 1.1

The National Research Act

[Note: The following excerpts contain the exact texts of the portions of the National Research Act that are most relevant to the discussions presented in Chapter 1.]

Public Law 93–348 July 12, 1974

AN ACT

To amend the Public Health Service Act to establish a program of National Research Service Awards to assure the continued excellence of biomedical and behavioral research and to provide for the protection of human subjects involved in biomedical and behavioral research and for other purposes.

Be it enacted by the Senate and House of Representatives of the United States of America in Congress assembled,

Section 1. This Act may be cited as the "National Research Act."

TITLE I—BIOMEDICAL AND BEHAVIORAL RESEARCH TRAINING

[Note: This section is irrelevent to our purposes; so we will skip to the next title of this Act.]

TITLE II—PROTECTION OF HUMAN SUBJECTS OF BIOMEDICAL AND BEHAVIORAL RESEARCH

Part A—National Commission for the Protection of Human Subjects of Biomedical and Behavioral Research

ESTABLISHMENT OF COMMISSION

Sec. 201. (a) There is established a Commission to be known as the National Commission for the Protection of Human Subjects of Biomedical and Behavioral Research (hereinafter in this title referred to as the "Commission").

(b) (1) The Commission shall be composed of eleven members appointed by the Secretary of Health, Education, and Welfare (hereinafter in this title referred to as the "Secretary"). The Secretary shall select members of the Commission from individuals distinguished in the fields of medicine, law, ethics, theology, the biological, physical, behavioral and social sciences, philosophy, humanities, health administration, government, and public affairs; but five (and not more than five) of the members of the Commission shall be individuals who are or who have been engaged in biomedical or behavioral research involving human subjects. In appointing members of the Commission, the Secretary shall give consideration to recommendations from the National Academy of Sciences and other appropriate entities. Members of the Commission shall be appointed for the life of the Commission. The Secretary shall appoint the members of the Commission within sixty days of the date of the enactment of this Act.

(2) (A) Except as provided in subparagraph (B), members of the Commission shall each be entitled to receive the daily equivalent of the annual rate of the basic pay in effect for grade GS–18 of the General Schedule for each day (including travel time) during which they are engaged in the actual performance of the duties of the Commission.

(B) Members of the Commission who are full-time officers or employees of the United States shall receive no additional pay on account of their service on the Commission.

(C) While away from their homes or regular places of business in the performance of duties of the Commission, members of the Commission shall be allowed travel expenses, including per diem in lieu of subsistence, in the same manner as persons employed intermittently in the Government service are allowed expenses under section 5703(b) of title 5 of the United States Code.

(c) The Chairman of the Commission shall be selected by the members of the Commission from among their number.

(d) (1) The Commission may appoint and fix the pay of such staff personnel as it deems desirable. Such personnel shall be appointed subject to the provisions of title 5, United States Code, governing appointments in the competitive service, and shall be paid in accordance with the provisions of chapter 51 and subchapter III of chapter 53 of such title relating to classification and General Schedule pay rates.

(2) The Commission may procure temporary and intermittent services to the same extent as is authorized by section 3109(b) of title 5 of the United States Code, but at rates for individuals not to exceed the daily equivalent of the annual rate of basic pay in effect for grade GS–18 of the General Schedule.

Sec. 202. (a) The Commission shall carry out the following:

(1) (A) The Commission shall (i) conduct a comprehensive investigation and study to identify the basic ethical principles which should underlie the conduct of biomedical and behavioral research involving human subjects, (ii) develop guidelines which should be followed in such research to assure that it is conducted in accordance with such principles, and (iii) make

recommendations to the Secretary (I) for such administrative action as may be appropriate to apply such guidelines to biomedical and behavioral research conducted or supported under programs administered by the Secretary, and (II) concerning any other matter pertaining to the protection of human subjects of biomedical and behavioral research.

(B) In carrying out subparagraph (A), the Commission shall consider at least the following:

(i) The boundaries between biomedical or behavioral research involving human subjects and the accepted routine practice of medicine.

(ii) The role of assessment of risk–benefit criteria in the determination of the appropriateness of research involving human subjects.

(iii) Appropriate guidelines for the selection of human subjects for participation in biomedical and behavioral research.

(iv) The nature and definition of informed consent in various research settings.

(v) Mechanisms for evaluating and monitoring the performance of Institutional Review Boards established in accordance with section 474 of the Public Health Service Act and appropriate enforcement mechanisms for carrying out their decisions.

(C) The Commission shall consider the appropriateness of applying the principles and guidelines identified and developed under subparagraph (A) to the delivery of health services to patients under programs conducted or supported by the Secretary.

(2) The Commission shall identify the requirements for informed consent to participation in biomedical and behavioral research by children, prisoners, and the institutionalized mentally infirm. The Commission shall investigate and study biomedical and behavioral research conducted or supported under programs administered by the Secretary and involving children, prisoners, and the institutionalized mentally infirm to determine the nature of the consent obtained from such persons or their legal representatives before such persons were involved in such research; the adequacy of the information given them respecting the nature and purpose of the research, procedures to be used, risks and discomforts, anticipated benefits from the research, and other matters necessary for informed consent; and the competence and the freedom of the persons to make a choice for or against involvement in such research. On the basis of such investigation and study the Commission shall make such recommendations to the Secretary as it determines appropriate to assure that biomedical and behavioral research conducted or supported under programs administered by him meets the requirements respecting informed consent identified by the Commission. For purposes of this paragraph, the term "children" means individuals who have not attained the legal age of consent to participate in research as determined under the applicable law of the jurisdiction in which the research is to be conducted; the term "prisoner" means individuals involuntarily confined in correctional institutions or facilities (as defined in section 601 of the Omnibus Crime Control and Safe Streets Act of 1968 (42 U.S.C. 3781)); and the term "institutionalized mentally infirm"

includes individuals who are mentally ill, mentally retarded, emotionally disturbed, psychotic, or senile, or who have other impairments of a similar nature and who reside as patients in an institution.

(3) The Commission shall conduct an investigation and study to determine the need for a mechanism to assure that human subjects in biomedical and behavioral research not subject to regulation by the Secretary are protected. If the Commission determines that such a mechanism is needed, it shall develop and recommend to the Congress such a mechanism. The Commission may contract for the design of such a mechanism to be included in such recommendations.

(b) The Commission shall conduct an investigation and study of the nature and extent of research involving living fetuses, the purposes for which such research has been undertaken, and alternative means for achieving such purposes. The Commission shall, not later than the expiration of the 4-month period beginning on the first day of the first month that follows the date on which all the members of the Commission have taken office, recommend to the Secretary policies defining the circumstances (if any) under which such research may be conducted or supported.

(c) The Commission shall conduct an investigation and study of the use of psychosurgery in the United States during the five-year period ending December 31, 1972. The Commission shall determine the appropriateness of its use, evaluate the need for it, and recommend to the Secretary policies defining the circumstances (if any) under which its use may be appropriate. For purposes of this paragraph, the term "psychosurgery" means brain surgery on (1) normal brain tissue of an individual, who does not suffer from any physical disease, for the purpose of changing or controlling the behavior or emotions of such individual, or (2) diseased brain tissue of an individual, if the sole object of the performance of such surgery is to control, change, or affect any behavioral or emotional disturbance of such individual. Such term does not include brain surgery designed to cure or ameliorate the effects of epilepsy and electric shock treatments.

(d) The Commission shall make recommendations to the Congress respecting the functions and authority of the National Advisory Council for the Protection of Subjects of Biomedical and Behavioral Research to be established under section 217(f) of the Public Health Service Act.

SPECIAL STUDY

Sec. 203. The Commission shall undertake a comprehensive study of the ethical, social, and legal implications of advances in biomedical and behavioral research and technology. Such study shall include—

(1) an analysis and evaluation of scientific and technological advances in past, present, and projected biomedical and behavioral research and services;

(2) an analysis and evaluation of the implications of such advances, both for individuals and for society;

(3) an analysis and evaluation of laws and moral and ethical principles governing the use of technology in medical practice;

(4) an analysis and evaluation of public understanding of and attitudes toward such implications and laws and principles; and

(5) an analysis and evaluation of implications for public policy of such findings as are made by the Commission with respect to advances in biomedical and behavioral research and technology and public attitudes toward such advances.

ADMINISTRATIVE PROVISIONS

Sec. 204. (a) The Commission may for the purpose of carrying out its duties under sections 202 and 203 hold such hearings, sit and act at such times and places, take such testimony, and receive such evidence as the Commission deems advisable.

(b) The Commission may secure directly from any department or agency of the United States information necessary to enable it to carry out its duties. Upon the request of the chairman of the Commission, the head of such department or agency shall furnish such information to the Commission.

(c) The Commission shall not disclose any information reported to or otherwise obtained by it in carrying out its duties which (1) identifies any individual who has been the subject of an activity studied and investigated by the Commission, or (2) which concerns any information which contains or relates to a trade secret or other matter referred to in section 1905 of title 18 of the United States Code.

(d) Except as provided in subsection (b) of section 202, the Commission shall complete its duties under sections 202 and 203 not later than the expiration of the 24-month period beginning on the first day of the first month that follows the date on which all the members of the Commission have taken office. The Commission shall make periodic reports to the President, the Congress, and the Secretary respecting its activities under sections 202 and 203 and shall, not later than ninety days after the expiration of such 24-month period, make a final report to the President, the Congress, and the Secretary respecting such activities and including its recommendations for administrative action and legislation.

(e) The Commission shall cease to exist thirty days following the submission of its final report pursuant to subsection (d).

DUTIES OF THE SECRETARY

Sec. 205. Within 60 days of the receipt of any recommendation made by the Commission under section 202, the Secretary shall publish it in the Federal Register and provide opportunity for interested persons to submit written data, views, and arguments with respect to such recommendation. The Secretary shall consider the Commission's recommendation and relevant matter submitted with respect to it and, within 180 days of the date of its publication in the Federal Register, the Secretary shall (1) determine whether the administrative action proposed by such recommendation is appropriate to assure the protection of human subjects of biomedical and behavioral research conducted or supported under programs administered by him, and (2) if he determines that such action is not so appropriate, publish in the Federal Register such determination together with an adequate statement of the reasons for his determination. If the Secretary determines that administrative action recommended by the Commission should be undertaken by him, he shall undertake such action as expeditiously as is feasible.

Part B—Miscellaneous

NATIONAL ADVISORY COUNCIL FOR THE PROTECTION OF SUBJECTS OF BIOMEDICAL AND BEHAVIORAL RESEARCH

Sec. 211. (a) Section 217 of the Public Health Service Act is amended by adding at the end of the following new subsection:

(f) (1) There shall be established a National Advisory Council for the Protection of Subjects of Biomedical and Behavioral Research (hereinafter in this subsection referred to as the "Council") which shall consist of the Secretary who shall be Chairman and not less than seven nor more than fifteen other members who shall be appointed by the Secretary without regard to the provisions of title 5, United States Code, governing appointments in the competitive service. The Secretary shall select members of the Council from individuals distinguished in the fields of medicine, law, ethics, theology, the biological, physical, behavioral and social sciences, philosophy, humanities, health administration, government, and public affairs; but three (and not more than three) of the members of the Council shall be individuals who are or who have been engaged in biomedical or behavioral research involving human subjects. No individual wbo was appointed to be a member of the National Commission for the Protection of Human Subjects of Biomedical and Behavioral Research (established under title II of the National Research Act) may be appointed to be a member of the Council. The appointed members of the Council shall have terms of office of four years, except that for the purpose of staggering the expiration of the terms of office of the Council members, the Secretary shall, at the time of appointment, designate a term of office of less than four years for members first appointed to the Council.

(2) The Council shall—

(A) advise, consult with, and make recommendations to, the Secretary concerning all matters pertaining to the protection of human subjects of biomedical and behavioral research;

(B) review policies, regulations, and other requirements of the Secretary governing such research to determine the extent to which such policies, regulations, and requirements require and are effective in requiring observance in such research of the basic ethical principles which should underlie the conduct of such research and, to the extent such policies, regulations, or requirements do not require or are not effective in requiring observance of such principles, make recommendations to the Secretary respecting appropriate revision of such policies, regulations, or requirements; and

(C) review periodically changes in the scope, purpose, and types of biomedical and behavioral research being conducted and the impact such changes have on the policies, regulations, and other requirements of the Secretary for the protection of human subjects of such research.

(3) The Council may disseminate to the public such information, recommendations, and other matters relating to its functions as it deems appropriate.

(4) Section 14 of the Federal Advisory Committee Act shall not apply with respect to the Council.

(b) The amendment made by subsection (a) shall take effect July 1, 1976.

INSTITUTIONAL REVIEW BOARDS; ETHICS GUIDANCE PROGRAM

Sec. 212. (a) Part I of title IV of the Public Health Service Act, as amended by section 103 of this Act, is amended by adding at the end the following new section:

INSTITUTIONAL REVIEW BOARDS; ETHICS GUIDANCE PROGRAM

Sec. 474. (a) The Secretary shall by regulation require that each entity which applies for a grant or contract under this Act for any project or program which involves the conduct of biomedical or behavioral research involving human subjects submit in or with its application for such grant or contract assurances satisfactory to the Secretary that it has established (in accordance with regulations which the Secretary shall prescribe) a board (to be known as an "Institutional Review Board") to review biomedical and behavioral research involving human subjects conducted at or sponsored by such entity in order to protect the rights of the human subjects of such research.

(b) The Secretary shall establish a program within the Department under which requests for clarification and guidance with respect to ethical issues raised in connection with biomedical or behavioral research involving human subjects are responded to promptly and appropriately.

(c) The Secretary of Health, Education, and Welfare shall within 240 days of the date of the enactment of this Act promulgate such regulations as may be required to carry out section 474 (a) of the Public Health Service Act. Such regulations shall apply with respect to applications for grants and contracts under such Act submitted after promulgation of such regulations.

LIMITATION ON RESEARCH

Sec. 213. Until the Commission has made its recommendations to the Secretary pursuant to section 202 (b), the Secretary may not conduct or support research in the United States or abroad on a living human fetus, before or after the induced abortion of such fetus, unless such research is done for the purpose of assuring the survival of such fetus.

INDIVIDUAL RIGHTS

Sec. 214. (a) Subsection (c) of section 401 of the Health Programs Extension Act of 1973 is amended (1) by inserting "(1)" after "(c)", (2) by redesignating paragraphs (1) and (2) as subparagraphs (A) and (B), respectively, and (3) by adding at the end the following new paragraph:

(2) No entity which receives after the date of enactment of this paragraph a grant or contract for biomedical or behavioral research under any program administered by the Secretary of Health, Education, and Welfare may—

(A) discriminate in the employment, promotion, or termination of employment of any physician or other health care personnel, or

(B) discriminate in the extension of staff or other privileges to any physician or other health care personnel,

because he performed or assisted in the performance of any lawful health service or research activity, because he refused to perform or assist in the performance of any such service or activity on the grounds that his performance or assistance in the performance of such service or activity would be contrary to his religious beliefs or moral convictions, or because of his religious beliefs or moral convictions respecting any such service or activity.

(b) Section 401 of such Act is amended by adding at the end the following new subsection:

(d) No individual shall be required to perform or assist in the performance of any part of a health service program or research activity funded in whole or in part under a program administered by the Secretary of Health, Education, and Welfare if his performance or assistance in the performance of such part of such program or activity would be contrary to his religious beliefs or moral convictions.

SPECIAL PROJECT GRANTS AND CONTRACTS

Sec. 215. Section 772 (a) (7) of the Public Health Service Act is amended by inserting immediately before the semicolon at the end thereof the following:

or, (C) providing increased emphasis on the ethical, social, legal, and moral implications of advances in biomedical research and technology with respect to the effects of such advances on individuals and society.

APPENDIX 1.2

The Belmont Report

[Note: The following excerpt consists of the complete text of the Belmont Report as discussed in the preceding chapter.]

ETHICAL PRINCIPLES AND GUIDELINES FOR RESEARCH INVOLVING HUMAN SUBJECTS

Scientific research has produced substantial social benefits. It has also posed some troubling ethical questions. Public attention was drawn to these questions by reported abuses of human subjects in biomedical experiments, especially during the Second World War. During the Nuremberg War Crimes Trials, the Nuremberg Code was drafted as a set of standards for judging physicians and scientists who had conducted biomedical experiments on concentration camp prisoners. This code became the prototype of many later codes[1] intended to assure that research involving human subjects would be carried out in an ethical manner.

The codes consist of rules, some general, others specific, that guide the investigators or the reviewers of research in their work. Such rules often are inadequate to cover complex situations; at times they come into conflict, and they are frequently difficult to interpret or apply. Broader ethical principles will provide a basis on which specific rules may be formulated, criticized and interpreted.

Three principles, or general prescriptive judgments, that are relevant to research involving human subjects are identified in this statement. Other principles may also be relevant. These three are comprehensive, however, and are stated at a level of generalization that should assist scientists, subjects, reviewers and interested citizens to understand the ethical issues inherent in research involving human subjects. These principles cannot

[1]Since 1945, various codes for the proper and responsible conduct of human experimentation in medical research have been adopted by different organizations. The best known of these codes are the Nuremberg Code of 1947, the Helsinki Declaration of 1964 (revised in 1975), and the 1971 guidelines (codified into Federal Regulations in 1974) issued by the U.S. Department of Health, Education, and Welfare. Codes for the conduct of social and behavioral research have also been adopted, the best known being that of the American Psychological Association, published in 1973.

always be applied so as to resolve beyond dispute particular ethical problems. The objective is to provide an analytical framework that will guide the resolution of ethical problems arising from research involving human subjects.

This statement consists of a distinction between research and practice, a discussion of the three basic ethical principles, and remarks about the application of these principles.

A. BOUNDARIES BETWEEN PRACTICE AND RESEARCH

It is important to distinguish between biomedical and behavioral research, on the one hand, and the practice of accepted therapy on the other, in order to know what activities ought to undergo review for the protection of human subjects of research. The distinction between research and practice is blurred partly because both often occur together (as in research designed to evaluate a therapy) and partly because notable departures from standard practice are often called "experimental" when the terms "experimental" and "research" are not carefully defined.

For the most part, the term "practice" refers to interventions that are designed solely to enhance the well-being of an individual patient or client and that have a reasonable expectation of success. The purpose of medical or behavioral practice is to provide diagnosis, preventive treatment or therapy to particular individuals.[2] By contrast, the term "research" designates an activity designed to test a hypothesis, permit conclusions to be drawn, and thereby to develop or contribute to generalizable knowledge (expressed, for example, in theories, principles, and statements of relationships). Research is usually described in a formal protocol that sets forth an objective and a set of procedures designed to reach that objective.

When a clinician departs in a significant way from standard or accepted practice, the innovation does not, in and of itself, constitute research. The fact that a procedure is "experimental," in the sense of new, untested or different, does not automatically place it in the category of research. Radically new procedures of this description should, however, be made the object of formal research at an early stage in order to determine whether they are safe and effective. Thus, it is the responsibility of medical practice committees, for example, to insist that a major innovation be incorporated into a formal research project.[3]

Research and practice may be carried on together when research is designed to evaluate the safety and efficacy of a therapy. This need not cause any confusion regarding whether or not the activity requires review; the general rule is that if there is any element

[2]Although practice usually involves interventions designed solely to enhance the well-being of a particular individual, interventions are sometimes applied to one individual for the enhancement of the well-being of another (e.g., blood donation, skin grafts, organ transplants) or an intervention may have the dual purpose of enhancing the well-being of a particular individual, and, at the same time, providing some benefit to others (e.g., vaccination, which protects both the person who is vaccinated and society generally). The fact that some forms of practice have elements other than immediate benefit to the individual receiving an intervention, however, should not confuse the general distinction between research and practice. Even when a procedure applied in practice may benefit some other person, it remains an intervention designed to enhance the well-being of a particular individual or groups of individuals; thus, it is practice and need not be reviewed as research.

[3]Because the problems related to social experimentation may differ substantially from those of biomedical and behavioral research, the Commission specifically declines to make any policy determination regarding such research at this time. Rather, the Commission believes that the problem ought to be addressed by one of its successor bodies.

of research in an activity, that activity should undergo review for the protection of human subjects.

B. BASIC ETHICAL PRINCIPLES

The expression "basic ethical principles" refers to those general judgments that serve as a basic justification for the many particular ethical prescriptions and evaluations of human actions. Three basic principles, among those generally accepted in our cultural tradition, are particularly relevant to the ethics of research involving human subjects: the principles of respect for persons, beneficence and justice.

1. *Respect for Persons*

Respect for persons incorporates at least two basic ethical convictions: first, that individuals should be treated as autonomous agents, and second, that persons with diminished autonomy are entitled to protection. The principle of respect for persons thus divides into two separate moral requirements: the requirement to acknowledge autonomy and the requirement to protect those with diminished autonomy.

An autonomous person is an individual capable of deliberation about personal goals and of acting under the direction of such deliberation. To respect autonomy is to give weight to autonomous persons' considered opinions and choices while refraining from obstructing their actions unless they are clearly detrimental to others. To show a lack of respect for an autonomous agent is to repudiate that person's considered judgments, to deny an individual the freedom to act on those considered judgments, or to withhold information necessary to make a considered judgment, when there are no compelling reasons to do so.

However, not every human being is capable of self-determination. The capacity for self-determination matures during an individual's life, and some individuals lose this capacity wholly or in part because of illness, mental disability, or circumstances that severely restrict liberty. Respect for the immature and the incapacitated may require protecting them as they mature or while they are incapacitated.

Some persons are in need of extensive protection, even to the point of excluding them from activities which may harm them; other persons require little protection beyond making sure they undertake activities freely and with awareness of possible adverse consequences. The extent of protection afforded should depend upon the risk of harm and the likelihood of benefit. The judgment that any individual lacks autonomy should be periodically reevaluated and will vary in different situations.

In most cases of research involving human subjects, respect for persons demands that subjects enter into the research voluntarily and with adequate information. In some situations, however, application of the principle is not obvious. The involvement of prisoners as subjects of research provides an instructive example. On the one hand, it would seem that the principle of respect for persons requires that prisoners not be deprived of the opportunity to volunteer for research. On the other hand, under prison conditions they may be subtly coerced or unduly influenced to engage in research activities for which they would not otherwise volunteer. Respect for persons would then dictate that prisoners be protected. Whether to allow prisoners to "volunteer" or to "protect" them presents a dilemma. Respecting persons, in most hard cases, is often a matter of balancing competing claims urged by the principle of respect itself.

2. *Beneficence*

Persons are treated in an ethical manner not only by respecting their decisions and protecting them from harm, but also by making efforts to secure their well-being. Such treatment falls under the principle of beneficence. The term "beneficence" is often understood to cover acts of kindness or charity that go beyond strict obligation. In this document, beneficence is understood in a stronger sense, as an obligation. Two general rules have been formulated as complementary expressions of beneficent actions in this sense: (1) do not harm and (2) maximize possible benefits and minimize possible harms.

The Hippocratic maxim "do no harm" has long been a fundamental principle of medical ethics. Claude Bernard extended it to the realm of research, saying that one should not injure one person regardless of the benefits that might come to others. However, even avoiding harm requires learning what is harmful; and, in the process of obtaining this information, persons may be exposed to risk of harm. Further, the Hippocratic Oath requires physicians to benefit their patients "according to their best judgment." Learning what will in fact benefit may require exposing persons to risk. The problem posed by these imperatives is to decide when it is justifiable to seek certain benefits despite the risks involved, and when the benefits should be foregone because of the risks.

The obligations of beneficence affect both individual investigators and society at large, because they extend both to particular research projects and to the entire enterprise of research. In the case of particular projects, investigators and members of their institutions are obliged to give forethought to the maximization of benefits and the reduction of risks that might occur from the research investigation. In the case of scientific research in general, members of the larger society are obliged to recognize the longer term benefits and risks that may result from the improvement of knowledge and from the development of novel medical, psychotherapeutic, and social procedures.

The principle of beneficence often occupies a well-defined justifying role in many areas of research involving human subjects. An example is found in research involving children. Effective ways of treating childhood diseases and fostering healthy development are benefits that serve to justify research involving children—even when individual research subjects are not the direct beneficiaries. Research also makes it possible to avoid the harm that may result from the application of previously accepted routine practices that on closer investigation turn out to be dangerous. But the role of the principle of beneficence is not always so unambiguous. A difficult ethical problem remains, for example, about research that presents more than minimal risk without immediate prospect of direct benefit to the children involved. Some have argued that such research is inadmissable, while others have pointed out that this limit would rule out much research promising great benefit to children in the future. Here again, as with all hard cases, the different claims covered by the principle of beneficence may come into conflict and force difficult choices.

3. *Justice*

Who ought to receive the benefits of research and bear its burdens? This is a question of justice, in the sense of "fairness of distribution" or "what is deserved." An injustice occurs when some benefit to which a person is entitled is denied without good reason or when some burden is imposed unduly. Another way of conceiving the principle of justice is that equals ought to be treated equally. However, this statement requires explication. Who is equal and who unequal? What considerations justify departure from equal distribution? Almost all commentators allow that distinctions based on experience, age, deprivation, competence, merit and position do sometimes constitute criteria justifying differential treatment for certain purposes. It is necessary, then, to explain in what respects people should be treated equally. There are several widely accepted formulations of just ways to distribute burdens and benefits. Each formulation mentions some relevant property on the basis of which burdens and benefits should be distributed. These formulations are (1) to each person an equal share, (2) to each person according to individual need, (3) to each person according to individual effort, (4) to each person according to societal contribution, and (5) to each person according to merit.

Questions of justice have long been associated with social practices such as punishment, taxation and political representation. Until recently these questions have not generally been associated with scientific research. However, they are foreshadowed even in the earliest reflections on the ethics of research involving human subjects. For example, during the 19th and early 20th centuries the burdens of serving as research subjects fell largely upon poor ward patients, while the benefits of improved medical care flowed primarily to private patients. Subsequently, the exploitation of unwilling prisoners as research subjects in Nazi concentration camps was condemned as a particularly flagrant injustice. In this country, in the 1940s, the Tuskegee syphilis study used disadvantaged, rural black men to study the untreated course of a disease that is by no means confined to that population. These subjects were deprived of demonstrably effective treatment in order not to interrupt the project, long after such treatment became generally available.

Against this historical background, it can be seen how conceptions of justice are relevant to research involving human subjects. For example, the selection of research subjects needs to be scrutinized in order to determine whether some classes (e.g., welfare patients, particular racial and ethnic minorities, or persons confined in institutions) are being systematically selected simply because of their easy availability, their compromised position, or their manipulability, rather than for reasons directly related to the problem being studied. Finally, whenever research supported by public funds leads to the development of therapeutic devices and procedures, justice demands both that these not provide advantages only to those who can afford them and that such research should not unduly involve persons from groups unlikely to be among the beneficiaries of subsequent applications of the research.

C. APPLICATIONS

Applications of the general principles to the conduct of research leads to consideration of the following requirements: informed consent, risk/benefit assessment, and the selection of subjects of research.

1. *Informed Consent*

Respect for persons requires that subjects, to the degree that they are capable, be given the opportunity to choose what shall or shall not happen to them. This opportunity is provided when adequate standards for informed consent are satisfied.

While the importance of informed consent is unquestioned, controversy prevails over the nature and possibility of an informed consent. Nonetheless, there is widespread agreement that the consent process can be analyzed as containing three elements: information, comprehension and voluntariness.

Information. Most codes of research establish specific items for disclosure intended to assure that subjects are given sufficient information. These items generally include: the research procedure, their purposes, risks and anticipated benefits, alternative procedures (where therapy is involved), and a statement offering the subject the opportunity to ask questions and to withdraw at any time from the research. Additional items have been proposed, including how subjects are selected, the person responsible for the research, etc.

However, a simple listing of items does not answer the question of what the standard should be for judging how much and what sort of information should be provided. One standard frequently invoked in medical practice, namely, the information commonly provided by practitioners in the field or in the locale, is inadequate since research takes place precisely when a common understanding does not exist. Another standard, currently popular in malpractice law, requires the practitioner to reveal the information that reasonable persons would wish to know in order to make a decision regarding their care. This, too, seems insufficient since the research subject, being in essence a volunteer, may wish to know considerably more about risks gratuitously undertaken than do patients who deliver themselves into the hands of a clinician for needed care. It may be that a standard of "the reasonable volunteer" should be proposed: the extent and nature of information should be such that persons, knowing that the procedure is neither necessary for their care nor perhaps fully understood, can decide whether they wish to participate in the furthering of knowledge. Even when some direct benefit to them is anticipated, the subjects should understand clearly the range of risk and the voluntary nature of participation.

A special problem of consent arises where informing subjects of some pertinent aspect of the research is likely to impair the validity of the research. In many cases, it is sufficient to indicate to subjects that they are being invited to participate in research of which some features will not be revealed until the research is concluded. In all cases of research involving incomplete disclosure, such research is justified only if it is clear that (1) incomplete disclosure is truly necessary to accomplish the goals of the research, (2) there are not undisclosed risks to subjects that are more than minimal, and (3) there is an adequate plan for debriefing subjects, when appropriate, and for dissemination of research results to them. Information about risks should never be withheld for the purpose of eliciting the cooperation of subjects, and truthful answers should always be given to direct questions about the research. Care should be taken to distinguish cases in which disclosure would destroy or invalidate the research from cases in which disclosure would simply inconvenience the investigator.

Comprehension. The manner and context in which information is conveyed is as important as the information itself. For example, presenting information in a disorganized and rapid fashion, allowing too little time for consideration or curtailing opportunities for questioning, all may adversely affect a subject's ability to make an informed choice.

Because the subject's ability to understand is a function of intelligence, rationality, maturity and language, it is necessary to adapt the presentation of the information to the subject's capacities. Investigators are responsible for ascertaining that the subject has comprehended the information. While there is always an obligation to ascertain that the information about risk to subjects is complete and adequately comprehended, when the risks are more serious, that obligation increases. On occasion, it may be suitable to give some oral or written test of comprehension.

Special provision may need to be made when comprehension is severely limited—for example, by conditions of immaturity or mental disability. Each class of subjects that one might consider as incompetent (e.g., infants and young children, mentally disabled patients, the terminally ill and the comatose) should be considered on its own terms. Even for these persons, however, respect requires giving them the opportunity to choose, to the extent they are able, whether or not to participate in research. The objections of these subjects to involvement should be honored, unless the research entails providing them a therapy unavailable elsewhere. Respect for persons also requires seeking the permission of other parties in order to protect the subjects from harm. Such persons are thus respected both by acknowledging their own wishes and by the use of third parties to protect them from harm.

The third parties chosen should be those who are most likely to understand the incompetent subject's situation and to act in that person's best interest. The person authorized to act on behalf of the subject should be given an opportunity to observe the research as it proceeds in order to be able to withdraw the subject from the research, if such action appears in the subject's best interest.

Voluntariness. An agreement to participate in research constitutes a valid consent only if voluntarily given. This element of informed consent requires conditions free of coercion and undue influence. Coercion occurs when an overt threat of harm is intentionally presented by one person to another to obtain compliance. Undue influence, by contrast, occurs through an offer of an excessive, unwarranted, inappropriate or improper reward or other overture in order to obtain compliance. Also, inducements that would ordinarily be acceptable may become undue influences if the subject is especially vulnerable.

Unjustifiable pressures usually occur when persons in positions of authority or commanding influence—especially where possible sanctions are involved—urge a course of action for a subject. A continuum of such influencing factors exists, however, and it is impossible to state precisely where justifiable persuasion ends and undue influence begins. But undue influence would include actions such as manipulating a person's choice through the controlling influence of a close relative and threatening to withdraw health services to which an individual would otherwise be entitled.

2. *Assessment of Risks and Benefits*

The assessment of risks and benefits requires a careful arrayal of relevant data, including, in some cases, alternative ways of obtaining the benefits sought in the research. Thus, the assessment presents both an opportunity and a responsibility to gather systematic and comprehensive information about proposed research. For the investigator, it is a means to examine whether the proposed research is properly designed. For a review committee, it is a method for determining whether the risks that will be presented to subjects are justified. For prospective subjects, the assessment will assist the determination whether or not to participate.

The Nature and Scope of Risks and Benefits. The requirement that research be justified on the basis of a favorable risk/benefit assessment bears a close relation to the principle of beneficence, just as the moral requirement that informed consent be obtained is derived primarily from the principle of respect for persons. The term "risk" refers to a possibility that harm may occur. However, when expressions such as "small risk" or "high risk" are used, they usually refer (often ambiguously) both to the chance (probability) of experiencing a harm and the severity (magnitude) of the envisioned harm.

The term "benefit" is used in the research context to refer to something of positive value related to health or welfare. Unlike "risk," "benefit" is not a term that expresses probabilities. Risk is properly contrasted to probability of benefits, and benefits are properly contrasted with harms rather than risks of harm. Accordingly, so-called risk/benefit assessments are concerned with the probabilities and magnitudes of possible harms and anticipated benefits. Many kinds of possible harms and benefits need be taken into account. There are, for example, risks of psychological harm, physical harm, legal harm, social harm and economic harm and the corresponding benefits. While the most likely types of harms to research subjects are those of psychological or physical pain or injury, other possible kinds should not be overlooked.

Risks and benefits of research may affect the individual subjects, the families of the individual subjects, and society at large (or special groups of subjects in society). Previous codes and federal regulations have required that risks to subjects be outweighed by the sum of both the anticipated benefit to the subject, if any, and the anticipated benefit to society in the form of the knowledge to be gained from the research. In balancing these different elements, the risks and benefits affecting the immediate research subject will normally carry special weight. On the other hand, interests other than those of the subject may on some occasions be sufficient by themselves to justify the risks involved in the research, so long as the subjects' rights have been protected. Beneficence thus requires that we protect against risk of harm to subjects and also that we be concerned about the loss of the substantial benefits that might be gained from research.

The Systematic Assessment of Risks and Benefits. It is commonly said that benefits and risks must be "balanced" and shown to be "in a favorable ratio." The metaphorical character of these terms draws attention to the difficulty of making precise judgments. Only on rare occasions will quantitative techniques be available for the scrutiny of research protocols. However, the idea of systematic, nonarbitrary analysis of risks and benefits should be emulated insofar as possible. This ideal requires those making decisions about the justifiability of research to be thorough in the accumulation and assessment of information about all aspects of the research, and to consider alternatives systematically. This procedure renders the assessment of research more rigorous and precise, while making communication between review board members and investigators less subject to misinterpretation, misinformation and conflicting judgments. Thus, there should first be a determination of the validity of the presuppositions of the research; then the nature, probability and magnitude of risk should be distinguished with as much clarity as possible. The method of ascertaining risks should be explicit, especially where there is no alternative to the use of such vague categories as small or slight risk. It should also be determined whether an investigator's estimates of the probability of harm or benefits are reasonable, as judged by known facts or other available studies.

Finally, assessment of the justifiability of research should reflect at least the following considerations: (i) Brutal or inhumane treatment of human subjects is never morally justified. (ii) Risks should be reduced to those necessary to achieve the research objective. It should be determined whether it is in fact necessary to use human subjects at all. Risk can perhaps never be entirely eliminated, but it can often be reduced by careful attention to alternative procedures. (iii) When research involves significant risk of serious impairment, review committees should be extraordinarily insistent on the justification of the risk (looking usually to the likelihood of benefit to the subject—or, in some rare cases, to the manifest voluntariness of the participation). (iv) When vulnerable populations are involved in research, the appropriateness of involving them should itself be demonstrated. A number of variables go into such judgments, including the nature and degree of risk, the condition of the particular population involved, and the nature and level of the anticipated benefits. (v) Relevant risks and benefits must be thoroughly arrayed in documents and procedures used in the informed consent process.

3. *Selection of Subjects*

Just as the principle of respect for persons finds expression in the requirements for consent, and the principle of beneficence in risk/benefit assessment, the principle of justice gives rise to moral requirements that there be fair procedures and outcomes in the selection of research subjects.

Justice is relevant to the selection of subjects of research at two levels: the social and the individual. Individual justice in the selection of subjects would require that researchers exhibit fairness: thus, they should not offer potentially beneficial research only to some patients who are in their favor or select only "undesirable" persons for risky research. Social justice requires that a distinction be drawn between classes of subjects that ought, and ought not, to participate in any particular kind of research, based on the ability of members of that class to bear burdens and on the appropriateness of placing further burdens on already burdened persons. Thus, it can be considered a matter of social justice that there is an order of preference in the selection of classes of subjects (e.g., adults before children) and that some classes of potential subjects (e.g., the institutionalized mentally infirm or prisoners) may be involved as research subjects, if at all, only on certain conditions.

Injustice may appear in the selection of subjects, even if individual subjects are selected fairly by investigators and treated fairly in the course of the research. This injustice arises from social, racial, sexual and cultural biases institutionalized in society. Thus, even if individual researchers are treating their research subjects fairly, and even if IRBs are taking care to assure that subjects are selected fairly within a particular institution, unjust social patterns may nevertheless appear in the overall distribution of the burdens and benefits of research. Although individual institutions or investigators may not be able to resolve a problem that is pervasive in their social setting, they can consider distributive justice in selecting research subjects.

Some populations, especially institutionalized ones, are already burdened in many ways by their infirmities and environments. When research is proposed that involves risks and does not include a therapeutic component, other less burdened classes of persons should be called upon first to accept these risks of research, except where the research is directly related to the specific conditions of the class involved. Also, even though public funds for research may often flow in the same directions as public funds for health care, it seems unfair that populations dependent on public health care constitute a pool of preferred research subjects if more advantaged populations are likely to be the recipients of the benefits.

One special instance of injustice results from the involvement of vulnerable subjects. Certain groups, such as racial minorities, the economically disadvantaged, the very sick, and the institutionalized may continually be sought as research subjects, owing to their ready availability in settings where research is conducted. Given their dependent status and their frequently compromised capacity for free consent, they should be protected against the danger of being involved in research solely for administrative convenience, or because they are easy to manipulate as a result of their illness or socioeconomic condition.

APPENDIX 1.3

The President's Commission for the Study of Ethical Problems in Medicine and Biomedical and Behavioral Research

Public Law 95–622 November 9, 1978

AN ACT

To amend the Community Mental Health Centers Act to revise and extend the programs under that Act, to amend the Public Health Service Act to revise and extend the programs of assistance for libraries of medicine, the programs of the National Heart, Lung, and Blood Institute, and of the National Cancer Institute, and the program for National Research Service Awards, to establish the President's Commission for the Study of Ethical Problems in Medicine and Biomedical and Behavioral Research, and for other purposes.

Be it enacted by the Senate and House of Representatives of the United States of America in Congress assembled,

TITLE I—COMMUNITY MENTAL HEALTH CENTERS EXTENSION

[Note: This title is irrelevant for the purpose of this book.]

TITLE II—[Note: This title is also irrelevant.]

TITLE III—PRESIDENT'S COMMISSION FOR THE STUDY OF ETHICAL PROBLEMS IN MEDICINE AND BIOMEDICAL AND BEHAVIORAL RESEARCH

Sec. 301. The Public Health Service Act is amended by adding after title XVII the following new title:

TITLE XVIII—PRESIDENT'S COMMISSION FOR THE STUDY OF ETHICAL PROBLEMS IN MEDICINE AND BIOMEDICAL AND BEHAVIORAL RESEARCH

ESTABLISHMENT OF COMMISSION

Sec. 1801. (a) Establishment. (1) There is established the President's Commission for the Study of Ethical Problems in Medicine and Biomedical and Behavioral Research (hereinafter in this title referred to as the "Commission") which shall be composed of eleven members appointed by the President. The members of the Commission shall be appointed as follows:

(A) Three of the members shall be appointed from individuals who are distinguished in biomedical or behavioral research.

(B) Three of the members shall be appointed from individuals who are distinguished in the practice of medicine or otherwise distinguished in the provision of health care.

(C) Five of the members shall be appointed from individuals who are distinguished in one or more of the fields of ethics, theology, law, the natural sciences (other than a biomedical or behavioral science), the social sciences, the humanities, health administration, government, and public affairs.

(2) No individual who is a full-time officer or employee of the United States may be appointed as a member of the Commission. The Secretary of Health, Education, and Welfare, the Secretary of Defense, the Director of Central Intelligence, the Director of the Office of Science and Technology Policy, the Administrator of Veterans' Affairs, and the Director of the National Science Foundation shall each designate an individual to provide liaison with the Commission.

(3) No individual may be appointed to serve as a member of the Commission if the individual has served for two terms of four years each as such a member.

(4) A vacancy in the Commission shall be filled in the manner in which the original appointment was made.

(b) Terms. (1) Except as provided in paragraphs (2) and (3), members shall be appointed for terms of four years.

(2) Of the members first appointed—

(A) four shall be appointed for terms of three years, and

(B) three shall be appointed for terms of two years, as designated by the President at the time of appointment.

(3) Any member appointed to fill a vacancy occurring before the expiration of the term for which his predecessor was appointed shall be appointed only for the remainder of such term. A member may serve after the expiration of his term until his successor has taken office.

(c) Chairman. The Chairman of the Commission shall be appointed by the President, by and with the advice and consent of the Senate, from members of the Commission.

(d) Meetings. (1) Seven members of the Commission shall constitute a quorum for business, but a lessor number may conduct hearings.

(2) The Commission shall meet at the call of the Chairman or at the call of a majority of its members.

(e) Compensation. (1) Members of the Commission shall each be entitled to receive the daily equivalent of the annual rate of basic pay in effect for grade GS–18 of the General Schedule for each day (including travel time) during which they are engaged in the actual performance of duties vested in the Commission.

(2) While away from their homes or regular places of business in the performance of services for the Commission, members of the Commission shall be allowed travel expenses, including per diem in lieu of subsistence, in the same manner as persons employed intermittently in the Government service are allowed expenses under section 5703 of title 5 of the United States Code.

DUTIES OF THE COMMISSION

Sec. 1802. (a) Studies. (1) The Commission shall undertake studies of the ethical and legal implications of—

(A) the requirements for informed consent to participate in research projects and to otherwise undergo medical procedures;

(B) the matter of defining death, including the advisability of developing a uniform definition of death;

(C) voluntary testing, counseling, and information and education programs with respect to genetic diseases and conditions, taking into account the essential equality of all human beings, born and unborn;

(D) the differences in the availability of health services as determined by the income or residence of the persons receiving the services;

(E) current procedures and mechanisms designed (i) to safeguard the privacy of human subjects of behavioral and biomedical research, (ii) to ensure the confidentiality of individually identifiable patient records, and (iii) to ensure appropriate access of patients to information continued in such records, and

(F) such other matters relating to medicine or biomedical or behavioral research as the President may designate for study by the Commission.

The Commission shall determine the priority and order of the studies required under this paragraph.

(2) The Commission may undertake an investigation or study of any other appropriate matter which relates to medicine or biomedical or behavioral research (including the protection of human subjects of biomedical or behavioral research) and which is consistent with the purposes of this title on its own initiative or at the request of the head of a Federal agency.

(3) In order to avoid duplication of effort, the Commission may, in lieu of, or as part of, any study or investigation required or otherwise conducted under this subsection, use a study or investigation conducted by another entity if the Commission sets forth its reasons for such use.

(4) Upon the completion of each investigation or study undertaken by the Commission under this subsection (including a study or investigation which merely uses another study or investigation), it shall report its findings (including any recommendation for

legislation or administrative action) to the President and the Congress and to each Federal agency to which a recommendation in the report applies.

(b) Recommendations to Agencies. (1) Within 60 days of the date a Federal agency receives a recommendation from the Commission that the agency take any action with respect to its rules, policies, guidelines, or regulations, the agency shall publish such recommendation in the Federal Register and shall provide opportunity for interested persons to submit written data, views, and arguments with respect to adoption of the recommendation.

(2) Within the 180-day period beginning on the date of such publication, the agency shall determine whether the action proposed by such recommendation is appropriate, and, to the extent that it determines that—

(A) such action is not appropriate, the agency shall, within such time period, provide the Commission with, and publish in the Federal Register, a notice of such determination (including an adequate statement of the reasons for the determination), or

(B) such action is appropriate, the agency shall undertake such action as expeditiously as feasible and shall notify the Commission of the determination and the action undertaken.

(c) Report on Protection of Human Subjects. The Commission shall biennially report to the President, the Congress, and appropriate Federal agencies on the protection of human subjects of biomedical and behavioral research. Each such report shall include a review of the adequacy and uniformity (1) of the rules, policies, guidelines, and regulations of all Federal agencies regarding the protection of human subjects of biomedical or behavioral research which such agencies conduct or support, and (2) of the implementation of such rules, policies, guidelines, and regulations by such agencies, and may include such recommendations for legislation and administrative action as the Commission deems appropriate.

(d) Annual Report. Not later than December 15 of each year (beginning with 1979) the Commission shall report to the President, the Congress, and appropriate Federal agencies on the activities of the Commission during the fiscal year ending in such year. Each such report shall include a complete list of all recommendations described in subsection (b) (1) made to Federal agencies by the Commission during the fiscal year and the actions taken, pursuant to subsection (b) (2), by the agencies upon such recommendations, and may include such recommendations for legislation and administrative action as the Commission deems appropriate.

(e) Publications. The Commission may at any time publish and disseminate to the public reports respecting its activities.

(f) Definitions. For purposes of this section:

(1) The term "Federal agency" means an authority of the government of the United States, but does not include (A) the Congress, (B) the courts of the United States, and (C) the government of the Commonwealth of Puerto Rico, the government of the District of Columbia, or the government of any territory or possession of the United States.

(2) The term "protection of human subjects" includes the protection of the health, safety, and privacy of individuals.

ADMINISTRATIVE PROVISIONS

Sec. 1803. (a) Hearings. The Commission may for the purpose of carrying out this title hold such hearings, sit and act at such times and places, take such testimony, and receive such evidence, as the Commission may deem advisable.

(b) Staff. (1) The Commission may appoint and fix the pay of such staff personnel as it deems desirable. Such personnel shall be appointed subject to the provisions of title 5, United States Code, governing appointments in the competitive service, and shall be paid in accordance with the provisions of chapter 51 and subchapter III of chapter 53 of such title relating to classification and General Schedule pay rates.

(2) The Commission may procure temporary and intermittent services to the same extent as is authorized by section 3109 (b) of title 5 of the United States Code, but at rates for individuals not to exceed the daily equivalent of the annual rate of basic pay in effect for grade GS–18 of the General Schedule.

(3) Upon request of the Commission, the head of any Federal agency is authorized to detail, on a reimbursable basis, any of the personnel of such agency to the Commission to assist it in carrying out its duties under this title.

(c) Contracts. The Commission, in performing its duties and functions under this title, may enter into contracts with appropriate public or nonprofit private entities. The authority of the Commission to enter into such contracts is effective for any fiscal year only to such extent or in such amounts as are provided in advance in appropriate Acts.

(d) Information. (1) The Commission may secure directly from any Federal agency information necessary to enable it to carry out this title. Upon request of the Chairman of the Commission, the head of such agency shall furnish such information to the Commission.

(2) The Commission shall promptly arrange for such security clearances for its members and appropriate staff as are necessary to obtain access to classified information needed to carry out its duties under this title.

(3) The Commission shall not disclose any information reported to or otherwise obtained by the Commission which is exempt from disclosure under subsection (a) of section 552 of title 5, United States Code, by reason of paragraphs (4) and (6) of subsection (b) of such section.

(e) Support Services. The Administrator of General Services shall provide to the Commission on a reimbursable basis such administrative support services as the Commission may request.

AUTHORIZATION OF APPROPRIATIONS; TERMINATION OF COMMISSION

Sec. 1804. (a) Authorizations. To carry out this title there are authorized to be appropriated $5,000,000 for the fiscal year ending September 30, 1979, $5,000,000 for the fiscal year ending September 30, 1980, $5,000,000 for the fiscal year ending September 30, 1981, and $5,000,000 for the fiscal year ending September 30, 1982.

(b) Federal Adivsory Committee Act; Termination. The Commission shall be subject to the Federal Advisory Committee Act, except that, under section 14 (a) (1) (B) of such Act, the Commission shall terminate on December 31, 1982.

Sec. 302. (a) The President shall initially appoint members to the President's Commission for the Study of Ethical Problems in Medicine and Biomedical and Behavioral Research (established under the amendment made by section 301) not later than 90 days after the date of the enactment of this title.

(b) Effective November 1, 1978, part A of title II of the National Research Act, section 213 of such Act, and subsection (f) of section 217 of the Public Health Service Act are repealed.

Approved November 9, 1978

CHAPTER 2

Institutional Review Boards

Institutional Review Boards play an extremely important role in protecting human research subjects in the United States. As we will see, they are a recent result of increased interest in human rights. The designation, "Institutional Review Board" (IRB), was not in general use until the mid-1970s, although such groups have been required since 1966 for research funded by the Public Health Service. Earlier groups of professionals performing the functions of IRBs were simply referred to as committees under various names (e.g., review committee, ethics committee).

In brief, an IRB is a group of professionals and lay persons at an institution such as a university, hospital, or private research center. These professionals have a variety of backgrounds but even some of the earliest IRB regulations by the former federal Department of Health, Education, and Welfare included provisions on which professional disciplines must be represented on an IRB. The primary purpose of an IRB has changed somewhat since the early federal guidelines in the 1970s. Essentially, an IRB must approve certain types of research and, occasionally, treatment procedures as well, especially if the treatment is considered experimental enough to be research also. Approval by an IRB for a research grant proposal is required by the federal Department of Health and Human Services (HHS) and more than a dozen other federal agencies before they will allocate funds to an institution for a research grant or contract. In reality, HHS usually will not even *review* a request for grant funds, much less approve it, unless IRB approval already has been obtained by the grant's principal investigator.

In view of the fact that many institutions—most notably universities and private research centers or companies—rely heavily on federal grant funds for support, it is easy to see why IRBs have become so important. Without IRB approval, badly needed funds simply would not be available for many institutions, funds that often provide at least partial salary support for faculty as well as indirect costs to help defray institutional expenses such as utilities, maintenance, and office space rent.

History of IRB Development

Institutions in the United States did not develop IRBs as a way of improving their chances of obtaining federal funds. Instead, it was the action of the Surgeon General in 1966 and, more recently, DHEW that *required* the formation of local IRBs. These groups partially reduced the burden on federal officials who, prior to the 1970s, were primarily responsible for overseeing the rights of research subjects involved in projects funded by federal dollars. However, it would be an error to conclude that IRBs arose as the result of a federal effort simply to reduce their responsibilities. Indeed, separate offices still exist within federal agencies to monitor a limited number of IRB activities around the country.

We also need to recall that the 1970s witnessed a society-wide increase in interest over human rights and fears about government investigations and procedures. Thus it was natural that government-sponsored research came under public scrutiny. Some abuses received wide media coverage, particularly the dramatic ones involving drug research with either uninformed or insufficiently informed subjects. These developments led to new informed consent requirements (see Chapter 3) for researchers, which in turn meant far more work for federal officials who had to communicate such new procedures to researchers to make sure projects receiving their funds were protecting the rights of any human subjects.

Dual Function for IRBs

IRB as a Government Proxy. First, a properly constituted and organized IRB can serve almost as a proxy of HHS in that it reviews grant and contract proposals in great detail to assure that any human research subjects are treated with dignity and that no rights are violated. The number of areas studied by an IRB and their content are described later in this chapter, but they are numerous. By serving as a proxy, an IRB also could be expected to educate individual researchers and practitioners on various aspects of protecting human rights, whether such rights are ruled by government agencies, mandated by professional ethics, or required in recognition of basic human rights. As noted in the Belmont Report in the previous chapter, the rationale for protecting human rights goes considerably beyond the legislation or regulation of certain procedures described by government officials.

This educational function could be expected to exist both for the investigators with proposals approved by an IRB and forwarded to HHS or other agency as a request for funds as well as for persons whose proposals are rejected by an IRB. Ideally, any human rights problem, whether substantive or technical, should serve as an opportunity for IRB members and the investigator to engage in dialogue that would inform the investigator and certainly benefit potential research subjects. Further, persons who are engaged more in treatment rather than research may be exposed to knowledge new to them, whether their projects end up needing IRB approval or not. This latter case occurs most often in settings in which both research and service may be intertwined as program personnel seek innovative methods of helping clients. In the author's experience, this educational function of an IRB is very real, particularly in service settings where research is not the predominant activity of staff or faculty.

Advocate for Research. The second part of an IRB's dual function is advocacy for the principal investigator of a grant or contract. The language of federal agency regulations can be confusing, and issues surrounding the protection of human subjects and program clients have become quite complex. Local members of an IRB can often assist faculty and staff with proposals since IRB members are often the persons most thoroughly informed on protection rules at the local level.

These two roles of an IRB can combine to make a rather difficult job for a local IRB member. As a proxy for the federal government, the IRB can bear the brunt of faculty or staff resentment over bureaucratic red tape that investigators may feel is not justified. Fortunately, the ambiguity of rules that made IRB responsibilities murky has been eased considerably with regulations published early in 1981. Prior to that time, it was not always clear just what an IRB should look for in a proposal before a grant or contract-seeker sent his or her application to a federal agency.

Furthermore, it has not always been obvious which types of research or treatment must come under IRB scrutiny. It appears that some of this misunderstanding may be due to the fact that professionals sometimes are unaware of changes in the regulations and may be acting according to rules now obsolete. This is particularly true in the area of protection of research subjects because of the many changes in the rules. For example, regulations published in early 1981 *for the first time* exempted from IRB review whole areas of research previously thought to require IRB review.

To familiarize the reader with these developments, a synopsis of the major revisions of IRB rules is presented. Professionals may discover, as they follow the changes, that rules they are using may no longer apply to certain types of research, IRB proposal procedures, or IRB review techniques.

Major Changes for IRBs

Although numerous changes have been made in the rules for how to operate IRBs, for what types of research must be reviewed by IRBs, and for what kind of information investigators must provide to IRBs, only the highlights will be discussed. Thus these historical reports should not be considered a complete legislative or regulatory history. In the same spirit, no attempt will be made to present the full text of the various past regulations that have been approved, since they have been repeatedly modified. These modifications have come about through the action of various professional groups in the United States seeking clarification or changes in HHS rules, or in the normal course of federal agency activities. The only full texts of legislation by the Congress or of federal agency regulations that are appended at the end of this chapter represent the current rules, not previous rules. At the same time, the reader should note that these rules have been changed repeatedly in past years. It would be wise, therefore, to consult the local library or contact federal agency officials in matters of crucial import.

1974

Even as early as 1974, it can be seen that dissatisfaction was widespread among professionals over existing DHEW regulations regarding work with human subjects. On May 30, 1974, there appeared in the *Federal Register* (39 FR 18914) a notice that a new Part 46 was going to be added to Subtitle A of Title 45 of the Code of Federal Regulations (CFR). Part 46 has since become well known to persons familiar with protection of human subjects and, indeed, is now titled as "Part 46—Protection of Human Subjects."

This notice in May of 1974 followed an earlier notice on October 9, 1973 (38 FR 27882) in which DHEW proposed to amend Subtitle A of Title 45 of the CFR by adding a changed version of existing DHEW policy from their DHEW Grants Administration Manual. Essentially, this policy provided that no DHEW funded grant or contract involving subjects at risk could be undertaken unless a committee at the institution reviewed the proposal. Comments were then received in late 1973 by DHEW from more than 140 representatives of grantee and contractor organizations, 20 public groups, and 40 individuals for a total of 500 separate criticisms of individual sections of the rules proposed in October of 1973.

Themes of Early Rules. In the rules then published in May, 1974, several interesting themes emerged. First, DHEW decided not to exempt particular types of research or types of subjects but to leave such authority with the Secretary of DHEW. Thus the regulations would remain applicable to all activities funded by DHEW until the Secretary published a list of exempt activities. Second, a number of changes were made in the rules about informed consent, and subject consent was defined as having six specific elements (later changed to seven; see Chapter 3). Interest was also expressed by respondents for DHEW to develop specific guidelines for work with special populations (e.g., prisoners), and DHEW proposed rules for some special populations as early as 1973.

Interestingly, several respondents did request that DHEW establish a national commission to start a comprehensive investigation of the basic ethical principles underlying the conduct of biomedical and behavioral research. In response to these suggestions, DHEW apparently dismissed the notion and noted that "it is concluded that these suggestions would require changes not properly within the scope of these regulations."[1] Nevertheless, two months later the National Research Act (PL 93–348) was approved by Congress on July 12, 1974, authorizing the formation of the historic National Commission for the Protection of Human Subjects of Biomedical and Behavioral Research. As we have already seen in the previous chapter, that commision became so active that it radically changed DHEW regulations on many aspects of protection of human subjects.

As one measure of how these 1974 rules have come to be revised, it should be noted that the May, 1974, rules contained 22 sections, whereas the most recent rules (January 26, 1981)[2] contain 38 sections, including new provisions for research with special populations.

1975

On March 13, 1975,[3] some technical amendments were published by DHEW in the *Federal Register*, noting that the National Research Act had designated that the institutional committees DHEW had been working with should now be termed "Institutional Review Boards." Further, any grant or contract proposal sent to DHEW was to be accompanied by an assurance that an IRB at the institution, operating under DHEW guidelines, had approved the proposal.

This notice alerted institutions around the country that DHEW agency policy on protection of subjects was now law as well as policy. Much of the current interest in protection issues, informed consent, and IRBs by persons in human service circles can be traced to these *Federal Register* notices of 1974 and 1975.

The changes of March, 1975, were primarily wording changes, however. For example, "Institutional Review Board" was inserted in place of the previous untitled "committee" to start the terminology still in use today.

1976

Since the inception of federal regulatory pronouncements on protection of human subjects, there has been considerable disagreement on the question of for

[1]*Federal Register*, Vol. 39, No. 105, Thursday, May 30, 1974, p. 18917.
[2]*Federal Register*, Vol. 46, No. 16, Monday, January 26, 1981, p. 8366.
[3]*Federal Register*, Vol. 40, No. 50, Thursday, March 13, 1975, p. 11854.

whom the regulations are intended. This disagreement has been between government agencies and professionals in the field, between different professions (e.g., sociologists and medical researchers), and between the various agencies constructing the guidelines. Such an interagency disagreement or source of confusion occurred when the Food and Drug Administration (FDA) proposed rules for IRBs that were separate from the rules proposed by their parent agency, the HHS. Because the FDA rules constitute an area by themselves, they are presented in more detail later in this chapter.

The history of rules on protection is also such that there has been disagreement on what activities are covered by the rules (e.g., research as opposed to treatment), what types of research are covered (e.g., medical, sociological), how important is the nature of the research subject (e.g., age, mental capacity), and how much risk[4] must be present before a subject's consent must be sought for research participation. These disagreements became obvious in the mid-1970s.

Subject at Risk. In 1981, HHS published the current regulations on protection of human research subjects. One of the many areas clarified was the general concept of *at risk.* Instead of the global concept more common in earlier years, we now have a more restricted concept of *minimal risk,* a level of risk that can lead to a quicker, expedited form of IRB review of a project. If more than minimal risk is involved, then an IRB now must conduct a lengthier, more detailed review of the research project before HHS will consider funding the project.

Essentially, this change means that many types of minimal risk research can be reviewed now more easily by an IRB, and less administrative preparatory work is required by an investigator. The full text of the regulations is appended to this chapter, but the definition of *minimal risk* is presented here to show how IRB responsibilities have become—in my opinion—more reasonable and efficient than they were for the 1974–1981 period.

> "Minimal risk" means that the risks of harm anticipated in the proposed research are not greater, considering probability and magnitude, than those ordinarily encountered in daily life or during the performance of routine physical or psychological examinations or tests.[5] (46.102, g)

This area is addressed in more detail in the subsequent chapter on informed consent.

1977

On December 2, 1977, the National Commission for the Protection of Human Subjects of Biomedical and Behavioral Research published its first major draft of recommendations for changes in IRB functioning. In the 25-page type-

[4]*Federal Register*, Vol. 41, No. 125, Monday, June 28, 1976, p. 26572.
[5]*Federal Register*, Vol. 46, No. 16, Monday, January 26, 1981, p. 8387.

written report, the Commission listed nine recommendations for HEW to consider in its ongoing revisions of protection issues.

Because of the central role played by IRBs, the draft received considerable attention. The author obtained a copy in early 1978 as a way of anticipating possible future DHEW proposed regulations. It was then possible to circulate copies of the National Commission draft to colleagues in both research and service settings and discuss how the recommendations might impact various projects with which we were associated. Thus it was possible to spend considerably more time working on our responses to various issues before DHEW actually published proposed guidelines. This advantage becomes clearly evident when we realize that, typically, a federal agency seldom allows more than 90 days following publication of proposed regulations for the public to respond.

Among other changes, the 1977 draft on IRBs proposed a wide scope for IRB responsibilities and more detailed wording on steps researchers must follow (e.g., to avoid undue influence exerted by researchers to encourage subjects to sign informed consent documents) than is now true.

Scope of Power Changed. Examination of this 1977 Commission draft and final HHS regulations published in 1981 reveals a dual change process in the regulations. First, the overall scope (e.g., types of research requiring IRB review) of IRB activities has been reduced since 1977. Second, the language of some parts of the regulations has become more specific to grant more discretionary authority to IRBs and avoid confusion in the meaning of various terms. For example, the first recommendation made by the National Commission in the 1977 draft proposed that *all* research financially supported by the federal government should be required to submit to IRB review:

> *Recommendation (1) Establishment of IRB Requirement.*
> Any agency or department of the federal government that supports, conducts or regulates research involving human subjects, and any institution that receives support for such research from the federal government, should require that all research involving human subjects conducted by persons associated with such agency, department or institution be reviewed by an institutional review board (IRB) and be conducted in accordance with the determinations of the IRB.[6]

This wording, of course, was not in the language of a regulation. Nevertheless, the increase in *precision* can be seen by the following excerpt from the 1981 HHS regulations on what type of research is covered by IRB review:

> (a) Except as provided in paragraph (b) of this section, this subpart applies to all research involving human subjects conducted by the Department of Health and Human Services or funded in whole or in part by a Department grant, contract, cooperative agreement or fellowship.[7]

---

[6]*IRB Recommendations*, draft by the National Commission for the Protection of Human Subjects of Biomedical and Behavioral Research. December 2, 1977, p. 3, mimeographed.

[7]*Federal Register*, Vol. 46, No. 16, Monday, January 26, 1981, p. 8386.

The reduction in scope is also clear in this case, since the previous initial definition as it appeared in the *Federal Register* in 1981 was immediately followed by five sections describing types of research *not* needing IRB review.

Finally, perhaps because it was a draft report of a commission, the 1977 report discussed possible exclusions or special cases where some of the then-approved procedures (e.g., on informed consent) might not be required. Such exclusions, however, were generally not included in any actual federal regulations until 1979, after considerable public comment and public hearings sponsored by HEW.

1978

Two events in 1978 formed the primary advances in protection regulations for that year. First, in August the FDA published a separate set of proposed regulations[8] for operation of IRBs that reviewed clinical investigations regulated by FDA, even though FDA was within the overall Department of Health, Education, and Welfare. In particular, these separate regulations were considered to apply to "food and color additives, cosmetics, drugs for human use, medical devices for human use, biological products for human use, and electronic products."[9] These FDA rules were to supercede previous FDA rules.

Considerable attention was paid in the proposed regulations to the problem of possible conflicts between the overall HEW regulations and the particular FDA regulations, but several reasons were given why FDA officials felt they needed separate guidelines. Perhaps one of the most salient was concern that the HEW guidelines did not require IRB review of investigations involving FDA-regulated substances if that institution was not receiving HEW funds to support the research.

At any rate, the separate proposed FDA regulations were extensive and were to have significantly altered Chapter I of title 21 of the Code of Federal Regulations (mainly under Subchapter A—General, Part 56—Institutional Review Boards). These regulations included numerous specific requirements for investigations involving drugs and biological materials which were not specifically referred to in the overall HEW guidelines.

However, the process of changing FDA regulations was interrupted by the second main event in this area in 1978. In September, the National Commission released its landmark *Report and Recommendations, Institutional Review Boards*[10] and a massive, 501-page companion *Appendix*[11]. The report itself was

[8]*Federal Register*, Vol. 43, No. 153, Tuesday, August 8, 1978, pp. 35186–35208.
[9]Ibid., p. 35199.
[10]*Reports and Recommendations, Institutional Review Boards*, by the National Commission for the Protection of Human Subjects of Biomedical and Behavioral Research. Washington, D.C.: U.S. Government Printing Office, DHEW Publication No. (OS) 78–0008, 1978.
[11]*Appendix to Report and Recommendations, Institutional Review Boards*, by the National Commis-

later published in the *Federal Register*[12] in its entirety in November of 1978, setting the stage for debates over various issues on IRBs, and suggesting sweeping changes in the way HEW had previously governed IRBs. The *Appendix* contains a wealth of information about how IRBs had been functioning and still serves as an excellent resource for data on what investigators think of IRBs, how easy or difficult it is to get IRB approval for a research project at various IRBs around the country, a summary of protection rules from all federal agencies, and related material.

Not surprisingly, there were other conflicts between protection regulations apparent in 1978. Mentioned specifically in the report on IRBs in the *Federal Register*[13] was the fact that the DHEW regulations did not apply to work conducted under the auspices of the National Institute of Education (NIE) and the U.S. Office of Education (OE). Why? Because "The General Education Provisions Act gives to the Director of NIE and the Commissioner of Education authority to issue their own regulations, subject to the approval of Congress."[14]

There were even additional complications. The proposed FDA regulations of August 8, 1978, as they appeared in the *Federal Register*, also noted that other federal agencies had compiled separate guidelines for experimentation that involved human subjects. These included the Energy Research and Development Administration[15] and the Consumer Product Safety Commission.[16]

Because of the problem of duplication and conflicting guidelines, individual researchers were hard pressed to understand which type of IRB guidelines might apply to which type of research. Fortunately, steps were taken in the following year of 1979 to alleviate this problem and to make it easier—at least for researchers working with drug-related projects—to understand which IRB rules applied to their work. As we will also see, the problem for educational researchers in deciding upon IRB rules was solved later in 1981 when final HHS rules specified the types of research exempted from IRB review, including particular types of educational research.

1979

On August 14, 1979, HHS published its proposed rules regarding operation of IRBs, including definitions and procedures for obtaining informed consent. These proposed rules[17] were the result of public comments received after the

---

sion for the Protection of Human Subjects of Biomedical and Behavioral Research. Washington, D.C.: U.S. Government Printing Office, DHEW Publication No. (OS) 78–0009, 1978.

[12]*Federal Register*, Vol. 43, No. 231, Thursday, November 30, 1978, pp. 56174–56198.

[13]Ibid., p. 56184.

[14]Ibid.

[15]*Federal Register*, Vol. 41, November 30, 1976, p. 52434.

[16]*Federal Register*, Vol. 41, September 2, 1976, p. 37120.

[17]*Federal Register*, Vol. 44, No. 158, Tuesday, August 14, 1979, pp. 47688–47699.

publication of the National Commission's recommendations on IRBs the year before. In recognition of the growing controversy and debate within professional groups and government agencies on what types of research should require IRB review, HHS did something unusual. Within the proposed rule, HHS inserted "Alternative A" and "Alternative B" regarding different approaches to defining various types of research (e.g., surveys, educational research) to be exempted from IRB review.

Essentially, the two alternatives varied in the degree to which one believes that research involving solely observations, surveys, or the study of documents contains only minimal or no risk to subjects and thus need not be reviewed by an IRB. Alternative A generally held that such research did involve only minimal risk, whereas B contained specific provisions for IRB review in the belief that such research can, in certain cases, pose significant risks for subjects. For example, collection of information may seem innocuous enough, but what if the information collected is about child abuse or drug addiction? What if observation of behavior (theoretically a minimal risk activity) involved observation of illegal conduct? Conceivably, subjects could then be exposed to more than minimal risk, depending on such factors as who might have access to research data other than the principle investigator. As we shall see later when we look at the final rules of 1981, fully five times as many commentators responded to these proposed alternatives by choosing Alternative A, which was indeed the approach adopted in the final HHS regulations.

On the same day that HHS published its proposed rules in 1979, FDA withdrew its 1978 proposal and, in its place, proposed revised IRB rules that were more in agreement with the new HHS rules published that same day, August 14, 1979.[18] These new FDA rules contained extensive changes and were preceded by an excellent summary describing just how and why FDA and HHS regulations might differ. For example, the explanatory notes pointed out that

> although this proposal will be essentially compatible and consistent with the regulations to be proposed by HEW [now HHS], the two sets of regulations cannot be identical. The statutory authorities under which FDA regulates clinical research are different from the authorities relied upon by HEW to regulate research that it either funds or conducts. In addition, because HEW's regulations will encompass behavioral research (which FDA does not regulate) the scope of coverage and types of review required will be somewhat different.[19]

During the remainder of 1979 and into 1980, comments were received from the public for both the HEW and FDA regulations. Recording of these responses and determination by HEW and FDA on which comments to most heed was conducted during 1980, and the final regulations for both agencies were finally published in 1981.

[18]*Federal Register,* Vol. 44, No. 158, Tuesday, August 14, 1979, pp. 47699–47729.
[19]Ibid., p. 47699.

1981

In January of 1981, the final regulations on operation of IRBs, informed consent, and related issues were published by HHS and FDA. First, on January 26, 1981, HHS published its new and final "Subpart A—Basic HHS Policy for Protection of Human Research Subjects."[20] Then, on January 27, 1981, FDA published its new and final "Part 56—Institutional Review Boards."[21] Each agency also published extensive descriptions of the types of comments it had received on the regulations proposed in 1979, the agency's reaction to the public comments, and the reasoning behind any changes made in the then existing rules to produce the final 1981 versions.

These 1981 regulations became effective on July 27, 1981, and remain as the current regulations today. They are the authoritative rules governing how IRBs operate, how and when informed consent must be obtained from human research subjects, and which types of research now require IRB review. It is the latter of the areas—types of research—that has been modified the most since the rules used in the early 1970s. Indeed, it was the issue of what types of research require IRB review that drew much of the attention in the comments from over 500 individuals and organizations who responded to the proposed rules of 1979. It is worth noting that one of the organizations that provided comments was the successor to the National Commission: the President's Commission for the Study of Ethical Problems in Medicine and Biomedical and Behavioral Research. Other comments came from public and private research organizations, university personnel, and professionals in a variety of disciplines affected by IRB regulations.

A number of categories of research involving minimal risk are now exempted from these regulations, since Alternative A was the one adopted by HHS and published in 1981. For example, the following types of research are now exempted under most circumstances: most research involving normal educational practices, most research involving educational tests, and survey and interview research involving observation.

Further, the 1981 HHS regulations now apply only to research conducted or funded by HHS, although earlier proposals would have covered *all* research at an institution receiving any HHS funds, regardless of the source of funding for the particular research project.

Finally, the concept of *expedited review* was added in the 1981 HHS regulations. As a method of streamlining the responsibilities of IRBs, expedited review allows for a much quicker and easier form of review of a project by an IRB. The details are contained in the regulations included at the end of this chapter; but, in brief, an expedited review means that the IRB chairperson alone (or his or her

[20]*Federal Register*, Vol. 46, No. 16, Monday, January 26, 1981, pp. 8366–8392.
[21]*Federal Register*, Vol. 46, No. 17, Tuesday, January 27, 1981, pp. 8942–8980.

designates from the IRB) may review the research. Researchers familiar with the difficulties in getting a sufficient number of IRB members together to constitute a quorum and obtaining approval after lengthy discussions with many IRB members will applaud this new procedure.

In an expedited review, the reviewer(s) can only approve the research, not disapprove it. Disapproval requires a review by the full IRB. The intent of the expedited review process was to cut down on the overall number of projects requiring full IRB review, with the important distinction that expedited reviews can only be conducted for projects involving no more than minimal risk.

As a general concept, minimal risk was defined earlier, but that definition still left open the question of just what specific types of research can be considered to involve no more than minimal risk. Fortunately, HHS published in the *Federal Register* a list of types of research which may be processed through IRBs under the expedited review procedures.[22] The list contains ten categories and is reprinted in its entirety at the end of this chapter. As examples, note that the following types of activities are included:

> Recording of data from subjects 18 years of age or older using noninvasive procedures routinely employed in clinical practice. This includes . . . weighing, testing sensory acuity, electrocardiography, . . . collection of blood samples . . . voice recordings made for research purposes such as investigations of speech defects . . . moderate exercise by healthy volunteers . . . research on individual or group behavior or characteristics of individuals, such as studies of perception, cognition, game theory, or test development, where the investigator does not manipulate subjects' behavior and the research will not involve stress to subjects.[23]

Health and Human Services and the Food and Drug Administration

As we noted previously, FDA and HHS each published separate IRB rules within one day of each other. FDA went to considerable effort to make their regulations compatible with those of HHS. But, the nature of the types of activities and products regulated by FDA has led to some definite differences. Let us look for a moment at these differences, with an emphasis not on technical discrepancies but on matters that would affect a researcher who would need to follow FDA guidelines rather than those of HHS.

First of all, who must submit their project to IRB review under FDA rules rather than HHS rules? Essentially, the FDA rules of January 27, 1981 (as included at the end of the chapter) provide that any investigator who conducts a clinical investigation of the following types must follow FDA rules:

[22]*Federal Register*, Vol. 46, No. 16, Monday, January 26, 1981, p. 8392.
[23]Ibid.

> any experiment that involves a test article and one or more human subjects and that either is subject to requirements for prior submission to the Food and Drug Administration under section 505(i), 507(d), or 520(g) of the act (author's note: *Act* means the Federal Food, Drug, and Cosmetic Act), or is not subject to requirements for prior submission to the Food and Drug Administration under these sections of the act, but the results of which are intended to be submitted later to, or held for inspection by, the Food and Drug Administration as part of an application for a research or marketing permit.[24]

The test articles noted in the definition were essentially the same as noted in the proposed 1978 regulations (i.e., drugs, medical devices, food additives, and related products).

The FDA rules on informed consent were essentially the same as those included in the HHS rules. However, the rules governing operation of IRBs varied in a number of ways from the rules adopted by HHS. Included in the next chapter are the current FDA regulations on informed consent, and the FDA rules on IRBs appear at the end of this chapter. Note that, in contrast to HHS rules wherein related issues such as IRBs and informed consent are contained in the same regulatory sections, the rules for FDA appear separately. Informed consent rules and related definitions for FDA appear in a variety of sections, depending on the nature of the test article, starting with Part 50 (Protection of Human Subjects) of Subchapter A (General) of Chapter I of title 21 of the Code of Federal Regulations.[25] Rules on operation of IRBs for FDA are in Part 56 (Institutional Review Boards).[26]

Fortunately, the FDA rules *did* contain new rules on expedited reviews similar to those of HHS. Still, the number of differences between HHS rules and FDA rules on operation of IRBs was considerable. Due to the length of the explanatory notes on such discrepancies, the extensive prologues for these sets of regulations are not included at the end of this chapter. Readers interested in more details can obtain this information from most libraries simply by reading the pages of the *Federal Register* immediately before the first page of the regulations reprinted at the end of this chapter.

However, due at least in part to possible public confusion over the exact differences between FDA and HHS rules, FDA published a very useful side-by-side comparison of the HHS and FDA rules on IRBs and informed consent.[27] The exact wording on specific issues can be compared in this side-by-side comparison. The comparison texts for IRBs are included at the end of this chapter, and the comparison texts for informed consent appear at the end of the next

[24]*Federal Register*, Vol. 46, No. 17, Tuesday, January 27, 1981, p. 8950.

[25]Ibid., pp. 8950–8958.

[26]Ibid., pp. 8975–8980.

[27]Personal communication with John C. Petricciani, M.D., Assistant Director for Clinical Research, Food and Drug Administration, Bethesda, Maryland 20205 (August 10, 1981).

chapter. The following summary by Dr. John Petricciani of FDA highlights a brief summary of the differences between the HHS and FDA regulations for *both* IRBs and informed consent:

The comparison focuses only on the structural and functional requirements of the regulations because those are the potential problem areas in the operation of an IRB as opposed to the administrative matters relating to assurance, grants, contracts, and permits for research or marketing.

The following paragraphs discuss the real or apparent differences between the two sets of regulations.

Sec. 56.108 (FDA) Sec. 46.108 (HHS)	The word *proposed* was inadvertently omitted from FDA's regulation, and has been reinserted. HHS omitted the word *human*, but that does not change the meaning of the phrase. HHS requires prompt reporting of unanticipated problems to the Secretary, but FDA does not specify that a similar report be made to the Commissioner because those reporting requirements are contained in FDA's Sponsor/Monitor and Clinical Investigator regulations.
Sec. 56.109 (FDA) Sec. 46.109 (HHS)	FDA does *not* provide that an IRB may waive the requirement for signed consent when the *principal* risk is a breach of confidentiality because FDA does not regulate studies which would fall into that category of research.
Sec. 56.114 (FDA) Sec. 46.114 (HHS)	FDA does not discuss administrative matters dealing with grants and contracts

because they are irrelevant to the scope of the Agency's regulation.

Sec. 56.115 (FDA) Sec. 46.115 (HHS)	FDA does not require the IRB or institution to report changes in membership whereas HHS does make that requirement. Since FDA has neither an assurance mechanism nor files of IRB membership, there is no reason for FDA to want to know about changes in membership. FDA states that if *may* refuse to consider a study in support of research or marketing permit if the IRB or the institution refuses to allow FDA to inspect IRB records. HHS had no such provision because it does not issue research or marketing permits.
Sec. 50.25(a) (5) (FDA) Sec. 46.116(a)(5) (HHS)	FDA explicitly requires that subjects be informed that FDA may inspect the records of the study because FDA has undertaken a substantial surveillance program which includes inspection of records. While HHS has the right to inspect records of studies it funds, it does not impose that same informed consent requirement because of the infrequency with which the Department actually inspects subject records.

Sec. 46.116(c) & (d) (HHS)	The Department provides for waiving or altering elements of informed consent under certain conditions. FDA has no such provision because the types of studies which would qualify for this section are either (1) not regulated by FDA or (2) covered by the emergency treatment provision (50.23).
Sec. 50.23 (FDA)	FDA, but not HHS, provides explicit guidance for the exemption from the informed consent requirements in emergency situations. The provision is based on a statutory requirement in the Medical Device amendments of 1976.

SUMMARY

Regulations governing the operation of Institutional Review Boards have been modified since the original regulations published by the former Department of Health, Education, and Welfare. These modifications have been due, in large part, to the efforts in the mid-1970s of the former National Commission as discussed previously.

Institutional Review Boards are organized to function according to federal regulations and are primarily responsible for insuring that research investigators that use federal funds take adequate steps to protect any human research subjects. Although it was required at one point that organizations receiving any federal funds must secure IRB approval of all research projects—whether the federal funds supported the research or were used by the organization for other purposes—this is no longer true. IRB approval is now required only for research that actually is directly supported by federal dollars.

Although various federal agencies have their own sets of regulations governing how their research projects must satisfy human subject protection provisions, the activities of the Department of Health and Human Services (HHS) are perhaps the best known. All grants and contracts awarded by HHS must be reviewed by the relevant IRB of the organization where the principal investigator is employed. Members of the IRB review the research proposal to make sure the researcher does all that is necessary to protect any human research subjects. Without IRB approval of a project, HHS will not financially support the project.

Within HHS, the Food and Drug Administration (FDA) developed independent guidelines for subject protection to be followed by researchers who submit data to FDA for approval of new drugs and devices. These separate rules were deemed necessary by FDA for a variety of reasons, including the concern over the possible special dangers of research regulated by FDA (e.g., drug research with humans, investigation of various food additives). However, in January of 1981 these two sets of regulations were made far more similar, and both are contained at the end of this chapter.

Although members of an IRB examine a variety of ethical and procedural aspects of a research proposal, the area causing most confusion and difficulty has been informed consent. The informed consent agreement between the researcher and the subject can be complicated and can pose moral and legal problems in research undertakings. It is therefore no great surprise to see that the HHS regulations on IRB operations also include definitions and guidelines for the types of informed consent required of researchers. However, FDA has *separate* regulations for IRBs and for informed consent within their agency. The material at the end of this chapter contains all this information to enable the reader to compare the two approaches. Further, this inclusion should assist individual researchers to pick those guidelines most relevant for their type of research since the regulations start off with statements on who needs to follow which agency's guidelines. As previously noted, there also appears at the end of this chapter the side-by-side comparison of FDA and HHS rules that was obtained through the kind assistance of Dr. John Petricciani of the Food and Drug Administration. It should be noted that if research involving a drug or device that is regulated by FDA is financially supported by HHS, then *both* sets of regulations apply.

One of the controversies that developed in the 1970s concerned the former broad definitions of research that must be submitted for IRB review. To solve this problem, the current regulations (for both HHS and FDA rules) include provisions for an expedited review of specific types of research. Such a review involves less work for the IRB, and the rules for this type of review are included at the end of this chapter.

The parts of IRB rules that may involve informed consent will be repeated at the end of Chapter 3 ("Informed Consent") so it becomes clear to the reader just which parts of the rules apply primarily to IRB members (IRB rules) in contrast to what the researcher needs to know (i.e., how to construct informed consent agreements). After these topics have been dealt with, we will move on to Chapter 4 to learn just how a researcher should present his research proposal to an IRB.

APPENDIX 2.1

Regulations by the Department of Health and Human Services (HHS) for Institutional Review Boards and Informed Consent

[Note: The following text is from the *Federal Register*—Vol. 46, No. 16, Monday, January 26, 1981, starting at page 8366—and contains the current HHS rules governing IRBs and also includes the rules on informed consent.]

DEPARTMENT OF HEALTH AND HUMAN SERVICES

Office of the Secretary
45 CFR Part 46

Final Regulations Amending Basic HHS Policy for the Protection of Human Research Subjects

AGENCY: Department of Health and Human Services
ACTION: Final Rule
SUMMARY: The Department of Health and Human Services (HHS or Department) is amending the HHS policy for the protection of human research subjects and responding to the recommendations of the National Commission for the Protection of Human Subjects of Biomedical and Behavioral Research (National Commission) and the President's Commission for the Study of Ethical Problems in Medicine and Biomedical and Behavioral Research (President's Commission) concerning Institutional Review Boards (IRBs).

These amendments substantially reduce the scope of the existing HHS regulatory coverage by exempting broad categories of research which normally present little or no risk of harm to subjects. Specifically, the new regulations: (1) Exempt from coverage most social, economic and educational research in which the only involvement of human subjects will be in one or more of the following categories: (a) The use of survey and interview procedures; (b) the observation of public behavior; or (c) the study of data, documents, records and specimens. (2) Require IRB review and approval of research involving human subjects if it is supported by Department funds and does not qualify for exemption from coverage by these regulations. (3) Require only expedited review for certain categories of proposed research involving no more than minimal risk and for minor changes in research already approved by an IRB. (4) Provide specific procedures for full IRB review and for expedited IRB review. (5) Designate basic elements of informed consent which are necessary as a prerequisite for humans to participate as subjects in research, and additional elements of informed consent which may be added when they are appropriate. (6) Indicate circumstances under which an IRB may approve withholding or altering some or all of the elements of informed consent otherwise required to be presented to research subjects. (7) Establish IRB membership requirements. (8) Establish regulations which, to the extent possible, are congruent with FDA final regulations to be published on informed consent and IRB activities.

[Note: These final FDA rules were published in the *Federal Register* on the next day—January 27, 1981—and appear after these HHS rules.]

The notice of proposed rulemaking (NPRM) which preceded this final regulation was controversial in two respects: (1) It proposed prior IRB review and approval of human subject research activities not directly funded by the Department, but carried out in institutions which receive HHS funding for certain research activities; and (2) it left open the question of coverage of behavioral and social science research involving little or no risk to the human subjects. The Department expects these controversies to be resolved because the NPRM is replaced with final regulations which do not extend the requirements as described in item (1) and provide broad exemptions for behavioral and social science research described in item (2).

EFFECTIVE DATE: These regulations shall become effective on July 27, 1981.

[Note: Deleted here are subsequent explanatory sections from the *Federal Register* concerning the regulatory history and public comments prior to these final regulations.]

Accordingly, Part 46 of 45 CFR (Code of Federal Regulations) is amended below by:

Sec. 46.205 (Amended)

1. Amending Sec. 46.205(b) by changing the reference in the eighth line from "Sec. 46.115" to "Sec. 46.120."

Sec. 46.304 (Amended)

2. Amending Sec. 46.304 by changing the reference in the second line from "Sec. 46.106" to "Sec. 46.107."

Subparts A and D (Removed)

3. Removing Subparts A and D and adding the following new Subpart A.

Subpart A—Basic HHS Policy for Protection of Human Research Subjects

Authority: 5 U.S.C. 301; sec 474(a), 88 Stat. 352 (42 U.S.C. 289 1-3(a)).

Sec. 46.101 To what do these regulations apply?

(a) Except as provided in paragraph (b) of this section, this subpart applies to all research involving human subjects conducted by the Department of Health and Human Services or funded in whole or in part by a Department grant, contract, cooperative agreement or fellowship.

(1) This includes research conducted by Department employees, except each Principal Operating Component head may adopt such nonsubstantive, procedural modifications as may be appropriate from an administrative standpoint.

(2) It also includes research conducted or funded by the Department of Health and Human Services outside the United States, but in appropriate circumstances, the Secretary may, under paragraph (e) of this section waive the applicability of some or all of the requirements of these regulations for research of this type.

(b) Research activities in which the only involvement of human subjects will be in one or more of the following categories are exempt from these regulations unless the research is covered by other subparts of this part:

(1) Research conducted in established or commonly accepted educational settings, involving normal educational practices, such as (i) research on regular and special education instructional strategies, or (ii) research on the effectiveness of or the comparison among instructional techniques, curricula, or classroom management methods.

(2) Research involving the use of educational tests (cognitive, diagnostic, aptitude, achievement), if information taken from these sources is recorded in such a manner that subjects cannot be identified, directly or through identifiers linked to the subjects.

(3) Research involving survey or interview procedures, except where all of the following conditions exist: (i) responses are recorded in such a manner that the human subjects can be identified, directly or through identifiers linked to the subjects, (ii) the subject's responses, if they became known outside the research, could reasonably place the subject at risk of criminal or civil liability or be damaging to the subject's financial standing or employability, and (iii) the research deals with sensitive aspects of the subject's own behavior, such as illegal conduct, drug use, sexual behavior, or use of alcohol. All research involving survey or interview procedures is exempt, without exception, when the respondents are elected or appointed public officials or candidates for public office.

(4) Research involving the observation (including observation by participants) of public behavior, except where all of the following conditions exist: (i) observations are recorded in such a manner that the human subjects can be identified, directly or through identifiers linked to the subjects, (ii) the observations recorded about the individual, if they became known outside the research, could reasonably place the subject at risk of criminal or civil liability or be damaging to the subject's financial standing or employability, and (iii) the research deals with sensitive aspects of the subject's own behavior such as illegal conduct, drug use, sexual behavior, or use of alcohol.

(5) Research involving the collection or study of existing data, documents, records, pathological specimens, or diagnostic specimens, if these sources are publicly available or if the information is recorded by the investigator in such a manner that subjects cannot be identified, directly or through identifiers linked to the subjects.

(c) The Secretary has final authority to determine whether a particular activity is covered by these regulations.

(d) The Secretary may require that specific research activities or classes of research activities conducted or funded by the Department, but not otherwise covered by these regulations, comply with some or all of these regulations.

(e) The Secretary may also waive applicability of these regulations to specific research activities or classes of research activities, otherwise covered by these regulations. Notices of these actions will be published in the *Federal Register* as they occur.

(f) No individual may receive Department funding for research covered by these regulations unless the individual is affiliated with or sponsored by an institution which assumes responsibility for the research under an assurance satisfying the requirements of this part, or the individual makes other arrangements with the Department.

(g) Compliance with these regulations will in no way render inapplicable pertinent federal, state, or local laws or regulations.

(h) Each subpart of these regulations contains a separate section describing to what the subpart applies. Research which is covered by more than one subpart shall comply with all applicable subparts.

Sec. 46.102 Definitions.

(a) "Secretary" means the Secretary of Health and Human Services and any other officer or employee of the Department of Health and Human Services to whom authority has been delegated.

(b) "Department" or "HHS" means the Department of Health and Human Services.

(c) "Institution" means any public or private entity or agency (including federal, state, and other agencies).

(d) "Legally authorized representative" means an individual or judicial or other body authorized under applicable law to consent on behalf of a prospective subject to the subject's participation in the procedure(s) involved in the research.

(e) "Research" means a systematic investigation designed to develop or contribute to generalizable knowledge. Activities which meet this definition constitute "research" for purposes of these regulations, whether or not they are supported or funded under a program which is considered research for other purposes. For example, some "demonstration" and "service" programs may include research activities.

(f) "Human subject" means a living individual about whom an investigator (whether professional or student) conducting research obtains (1) data through intervention or interaction with the individual, or (2) identifiable private information. "Intervention" includes both physical procedures by which data are gathered (for example, venipuncture) and manipulations of the subject or the subject's environment that are performed for research purposes. "Interaction" includes communication or interpersonal contact between investigator and subject. "Private information" includes information about behavior that occurs in a context in which an individual can reasonably expect that no observation or recording is taking place, and information which has been provided for specific purposes by an individual and which the individual can reasonably expect will not be made public (for example, a medical record). Private information must be individually identifiable (i.e., the identity of the subject is or may readily be ascertained by the investigator or associated with the information) in order for obtaining the information to constitute research involving human subjects.

(g) "Minimal risk" means that the risks of harm anticipated in the proposed research are not greater, considering probability and magnitude, than those ordinarily encountered in daily life or during the performance of routine physical or psychological examinations or tests.

(h) "Certification" means the official notification by the institution to the Department in accordance with the requirements of this part that a research project or activity involving human subjects has been reviewed and approved by the Institutional Review Board (IRB) in accordance with the approved assurance on file at HHS. (Certification is required when the research is funded by the Department and not otherwise exempt in accordance with Sec. 46.101(b)).

Sec. 46.103 Assurances.

(a) Each institution engaged in research covered by these regulations shall provide written assurance satisfactory to the Secretary that it will comply with the requirements set forth in these regulations.

(b) The Department will conduct or fund research covered by these regulations only if the institution has an assurance approved as provided in this section, and only if the institution has certified to the Secretary that the research has been reviewed and approved by an IRB provided for in the assurance, and will be subject to continuing review by the IRB. This assurance shall at a minimum include:

(1) A statement of principles governing the institution in the discharge of its responsibilities for protecting the rights and welfare of human subjects of research conducted at or sponsored by the institution, regardless of source of funding. This may include an appropriate existing code, declaration, or statement of ethical principles, or a statement formulated by the institution itself. This requirement does not preempt provisions of these regulations applicable to Department-funded research and is not applicable to any research in an exempt category listed in Sec. 46.101.

(2) Designation of one or more IRBs established in accordance with the requirements of this subpart, and for which provisions are made for meeting space and sufficient staff to support the IRB's review and recordkeeping duties.

(3) A list of the IRB members identified by name; earned degrees; representative capacity; indications of experience such as board certifications, licenses, etc., sufficient to describe each member's chief anticipated contributions to IRB deliberations; and any employment or other relationship between each member and the institution; for example: full-time employee, part-time employee, member of governing panel or board, stockholder, paid or unpaid consultant. Changes in IRB membership shall be reported to the Secretary.

(4) Written procedures which the IRB will follow (i) for conducting its initial and continuing review of research and for reporting its findings and actions to the investigator and the institution; (ii) for determining which projects require review more often than annually and which projects need verification from sources other than the investigators that no material changes have occurred since previous IRB review; (iii) for insuring prompt reporting to the IRB of proposed changes in a research activity, and for insuring that changes in approved research, during the period for which IRB approval has already been given, may not be initiated without IRB review and approval except where necessary to eliminate apparent immediate hazards to the subject; and (iv) for insuring prompt reporting to the IRB and to the Secretary of unanticipated problems involving risks to subjects or others.

(c) The assurance shall be executed by an individual authorized to act for the institution and to assume on behalf of the institution the obligations imposed by these regulations, and shall be filed in such form and manner as the Secretary may prescribe.

(d) The Secretary will evaluate all assurances submitted in accordance with these regulations through such officers and employees of the Department and such experts or consultants engaged for this purpose as the Secretary determines to be appropriate. The Secretary's evaluation will take into consideration the adequacy of the proposed IRB in light of the anticipated scope of the institution's research activities and the types of subject populations likely to be involved, the appropriateness of the proposed initial and continuing review procedures in light of the probable risks, and the size and complexity of the institution.

(e) On the basis of this evaluation, the Secretary may approve or disapprove the assurance, or enter into negotiations to develop an approvable one. The Secretary may limit the period during which any particular approved assurance or class of approved assurances shall remain effective or otherwise condition or restrict approval.

(f) Within 60 days after the date of submission to HHS of an application or proposal, an institution with an approved assurance covering the proposed research shall certify that the application or proposal has been reviewed and approved by the IRB. Other institutions shall certify that the application or proposal has been approved by the IRB within 30 days after receipt of a request for such a certification from the Department. If the certification is not submitted within these time limits, the application or proposal may be returned to the institution.

Sec. 46.104 (Reserved)

Sec. 46.105 (Reserved)

Sec. 46.106 (Reserved)

Sec. 47.107 IRB membership.

(a) Each IRB shall have at least five members, with varying backgrounds to promote complete and adequate review of research activities commonly conducted by the institution. The IRB shall be sufficiently qualified through the experience and expertise of its members, and the diversity of the members' backgrounds including consideration of the racial and cultural backgrounds of members and sensitivity to such issues as community attitudes, to promote respect for its advice and counsel in safeguarding the rights and welfare of human subjects. In addition to possessing the professional competence necessary to review specific research activities, the IRB shall be able to ascertain the acceptability of proposed research in terms of institutional commitments and regulations, applicable law, and standards of professional conduct and practice. The IRB shall therefore include persons knowledgeable in these areas. If an IRB regularly reviews research that involves a vulnerable category of subjects, including but not limited to subjects covered by other subparts of this part, the IRB shall include one or more individuals who are primarily concerned with the welfare of these subjects.

(b) No IRB may consist entirely of men or women, or entirely of members of one profession.

(c) Each IRB shall include at least one member whose primary concerns are in nonscientific areas; for example: lawyers, ethicists, members of the clergy.

(d) Each IRB shall include at least one member who is not otherwise affiliated with the institution and who is not part of the immediate family of a person who is affiliated with the institution.

(e) No IRB may have a member participating in the IRB's initial or continuing review of any project in which the member has a conflicting interest, except to provide information requested by the IRB.

(f) An IRB may, in its discretion, invite individuals with competence in special areas to assist in the review of complex issues which require expertise beyond or in addition to that available on the IRB. These individuals may not vote with the IRB.

Sec. 46.108 IRB functions and operations.

In order to fulfill the requirements of these regulations each IRB shall:

(a) Follow written procedures as provided in Sec. 46.103(b)(4).

(b) Except when an expedited review procedure is used (see Sec. 46.110), review proposed research at convened meetings at which a majority of the members of the IRB are present, including at least one member whose primary concerns are in nonscientific areas. In order for the research to be approved, it shall receive the approval of a majority of those members present at the meeting.

(c) Be responsible for reporting to the appropriate institutional officials and the Secretary any serious or continuing noncompliance by investigators with the requirements and determinations of the IRB.

Sec. 46.109 IRB review of research.

(a) An IRB shall review and have authority to approve, require modifications in (to secure approval), or disapprove all research activities covered by these regulations.

(b) An IRB shall require that information given to subjects as part of informed consent is in accordance with Sec. 46.116. The IRB may require that information, in addition to that specifically mentioned in Sec. 46.116, be given to the subjects when in the IRB's judgment the information would meaningfully add to the protection of the rights and welfare of subjects.

(c) An IRB shall require documentation of informed consent or may waive documentation in accordance with Sec. 46.117.

(d) An IRB shall notify investigators and the institution in writing of its decision to approve or disapprove the proposed research activity, or of modifications required to secure IRB approval of the research activity. If the IRB decides to disapprove a research activity, it shall include in its written notification a statement of the reasons for its decision and give the investigator an opportunity to respond in person or in writing.

(e) An IRB shall conduct continuing review of research covered by these regulations at intervals appropriate to the degree of risk, but not less than once per year, and shall have authority to observe or have a third party observe the consent process and the research.

Sec. 46.110 Expedited review procedures for certain kinds of research involving no more than minimal risk, and for minor changes in approved research.

(a) The Secretary has established, and published in the *Federal Register*, a list of categories of research that may be reviewed by the IRB through an expedited review procedure. The list will be amended, as appropriate, through periodic republication in the *Federal Register*.

(b) An IRB may review some or all of the research appearing on the list through an expedited review procedure, if the research involves no more than minimal risk. The IRB may also use the expedited review procedure to review minor changes in previously approved research during the period for which approval is authorized. Under an expedited review procedure, the review may be carried out by the IRB chairperson or by one or more experienced reviewers designated by the chairperson from among members of the IRB. In reviewing the research, the reviewers may exercise all of the authorities of the IRB except that the reviewers may not disapprove the research. A research activity may be disapproved only after review in accordance with the non-expedited procedure set forth in Sec. 46.108(b).

(c) Each IRB which uses an expedited review procedure shall adopt a method for keeping all members advised of research proposals which have been approved under the procedure.

(d) The Secretary may restrict, suspend, or terminate an institution's or IRB's use of the expedited review procedure when necessary to protect the rights or welfare of subjects.

Sec. 46.111 Criteria for IRB approval of research.

(a) In order to approve research covered by these regulations the IRB shall determine that all of the following requirements are satisfied:

(1) Risks to subjects are minimized: (i) By using procedures which are consistent with sound research design and which do not unnecessarily expose subjects to risk, and (ii) whenever appropriate, by using procedures already being performed on the subjects for diagnostic or treatment purposes.

(2) Risks to subjects are reasonable in relation to anticipated benefits, if any, to subjects, and the importance of the knowledge that may reasonably be expected to result. In evaluating risks and benefits, the IRB should consider only those risks and benefits that may result from the research (as distinguished from risks and benefits of therapies subjects would receive even if not participating in the research). The IRB should not consider possible long-range effects of applying knowledge gained in the research (for example, the possible effects of the research on public policy) as among those research risks that fall within the purview of its responsibility.

(3) Selection of subjects is equitable. In making this assessment the IRB should take into account the purposes of the research and the setting in which the research will be conducted.

(4) Informed consent will be sought from each prospective subject or the subject's legally authorized representative, in accordance with, and to the extent required by Sec. 46.116.

(5) Informed consent will be appropriately documented, in accordance with, and to the extent required by Sec. 46.117.

(6) Where appropriate, the research plan makes adequate provision for monitoring the data collected to insure the safety of subjects.

(7) Where appropriate, there are adequate provisions to protect the privacy of subjects and to maintain the confidentiality of data.

(b) Where some or all of the subjects are likely to be vulnerable to coercion or undue influence, such as persons with acute or severe physical or mental illness, or persons who are economically or educationally disadvantaged, appropriate additional safeguards have been included in the study to protect the rights and welfare of these subjects.

Sec. 46.112 Review by institution.

Research covered by these regulations that has been approved by an IRB may be subject to further appropriate review and approval or disapproval by officials of the institution. However, those officials may not approve the research if it has not been approved by an IRB.

Sec. 46.113 Suspension or termination of IRB approval of research.

An IRB shall have authority to suspend or terminate approval of research that is not being conducted in accordance with the IRB's requirements or that has been associated with unexpected serious harm to subjects. Any suspension or termination of approval shall include a statement of the reasons for the IRB's action and shall be reported promptly to the investigator, appropriate institutional officials, and the Secretary.

Sec. 46.114 Cooperative research.

Cooperative research projects are those projects, normally supported through grants, contracts, or similar arrangements, which involve institutions in addition to the grantee or prime contractor (such as a contractor with the grantee, or a subcontractor with the prime contractor). In such instances, the grantee or prime contractor remains responsible to the Department for safeguarding the rights and welfare of human subjects. Also, when cooperating institutions conduct some or all of the research involving some or all of these subjects, each cooperating institution shall comply with these regulations as though it received funds for its participation in the project directly from the Department, except that in complying with these regulations institutions may use joint review, reliance upon the review of another qualified IRB, or similar arrangements aimed at avoidance of duplication of effort.

Sec. 46.115 IRB records.

(a) An institution, or where appropriate an IRB, shall prepare and maintain adequate documentation of IRB activities, including the following:

(1) Copies of all research proposals reviewed, scientific evaluations, if any, that accompany the proposals, approved sample consent documents, progress reports submitted by investigators, and reports of injuries to subjects.

(2) Minutes of IRB meetings which shall be in sufficient detail to show attendance at the meetings; actions taken by the IRB; the vote on these actions including the number of members voting for, against, and abstaining; the basis for requiring changes in or disapproving research; and a written summary of the discussion of controverted issues and their resolution.

(3) Records of continuing review activities.

(4) Copies of all correspondence between the IRB and the investigators.

(5) A list of IRB members as required by Sec. 46.103(b)(3).

(6) Written procedures for the IRB as required by Sec. 46.116(b)(5).

(7) Statements of significant new findings provided to subjects, as required by Sec. 46.116(b)(5).

(b) The records required by this regulation shall be retained for at least 3 years after completion of the research, and the records shall be accessible for inspection and copying by authorized representatives of the Department at reasonable times and in a reasonable manner.

Sec. 46.116 General requirements for informed consent.

Except as provided elsewhere in this or other subparts, no investigator may involve a human being as a subject in research covered by these regulations unless the investigator has obtained the legally effective informed consent of the subject or the subject's legally authorized representative. An investigator shall seek such consent only under circumstances that provide the prospective subject or the representative sufficient opportunity to consider whether or not to participate and that minimize the possibility of coercion or undue influence. The information that is given to the subject or the representative shall be in language understandable to the subject or the representative. No informed consent, whether oral or written, may include any exculpatory language through which the subject or the representative is made to waive or appear to waive any of the subject's legal rights, or releases or appears to release the investigator, the sponsor, the institution or its agents from liability for negligence.

(a) Basic elements of informed consent. Except as provided in paragraph (c) or paragraph (d) of this section, in seeking informed consent the following information shall be provided to each subject:

(1) A statement that the study involves research, an explanation of the purposes of the research and the expected duration of the subject's participation, a description of the procedures to be followed, and identification of any procedures which are experimental;

(2) A description of any reasonably foreseeable risks or discomforts to the subject;

(3) A description of any benefits to the subject or to others which may reasonably be expected from the research;

(4) A disclosure of appropriate alternative procedures or courses of treatment, if any, that might be advantageous to the subject;

(5) A statement describing the extent, if any, to which confidentiality of records identifying the subject will be maintained;

(6) For research involving more than minimal risk, an explanation as to whether any compensation and an explanation as to whether any medical treatments are available if injury occurs and, if so, what they consist of, or where further information may be obtained;

(7) An explanation of whom to contact for answers to pertinent questions about the research and research subjects' rights, and whom to contact in the event of a research-related injury to the subject; and

(8) A statement that participation is voluntary, refusal to participate will involve no penalty or loss of benefits to which the subject is otherwise entitled, and the subject may discontinue participation at any time without penalty or loss of benefits to which the subject is otherwise entitled.

(b) Additional elements of informed consent. When appropriate, one or more of the following elements of information shall also be provided to each subject:

(1) A statement that the particular treatment or procedure may involve risks to the subject (or to the embryo or fetus, if the subject is or may become pregnant) which are currently unforeseeable;

(2) Anticipated circumstances under which the subject's participation may be terminated by the investigator without regard to the subject's consent;

(3) Any additional costs to the subject that may result from participation in the research;

(4) The consequences of a subject's decision to withdraw from the research and procedures for orderly termination of participation by the subject;

(5) A statement that significant new findings developed during the course of the research which may relate to the subject's willingness to continue participation will be provided to the subject; and

(6) The approximate number of subjects involved in the study.

(c) An IRB may approve a consent procedure which does not include, or which alters, some or all of the elements of informed consent set forth above, or waive the requirement to obtain informed consent provided the IRB finds and documents that:

(1) The research is to be conducted for the purpose of demonstrating or evaluating: (i) Federal, state, or local benefit or service programs which are not themselves research programs, (ii) procedures for obtaining benefits or services under these programs, or (iii) possible changes in or alternatives to these programs or procedures; and

(2) The research could not practicably be carried out without the waiver or alteration.

(d) An IRB may approve a consent procedure which does not include, or which alters, some or all of the elements of informed consent set forth above, or waive the requirements to obtain informed consent provided the IRB finds and documents that:

(1) The research involves no more than minimal risk to the subjects;

(2) The waiver or alteration will not adversely affect the rights and welfare of the subjects;

(3) The research could not practicably be carried out without the waiver or alteration; and

(4) Whenever appropriate, the subjects will be provided with additional pertinent information after participation.

(e) The informed consent requirements in these regulations are not intended to preempt any applicable Federal, state, or local laws which require additional information to be disclosed in order for informed consent to be legally effective.

(f) Nothing in these regulations is intended to limit the authority of a physician to provide emergency medical care, to the extent the physician is permitted to do so under applicable Federal, state, or local law.

Sec. 46.117 Documentation of informed consent.

(a) Except as provided in paragraph (c) of this section, informed consent shall be documented by the use of a written consent form approved by the IRB and signed by the subject or the subject's legally authorized representative. A copy shall be given to the person signing the form.

(b) Except as provided in paragraph (c) of this section, the consent form may be either of the following:

(1) A written consent document that embodies the elements of informed consent required by Sec. 46.116. This form may be read to the subject or the subject's legally authorized representative, but in any event, the investigator shall give either the subject or the representative adequate opportunity to read it before it is signed; or

(2) A "short form" written consent document stating that the elements of informed consent required by Sec. 46.116 have been presented orally to the subject or the subject's legally authorized representative. When this method is used, there shall be a witness to the oral presentation. Also, the IRB shall approve a written summary of what is to be said to the subject or the representative. However, the witness shall sign both the short form and a copy of the summary, and the person actually obtaining consent shall sign a copy of the summary. A copy of the summary shall be given to the subject or the representative, in addition to a copy of the "short form."

(c) An IRB may waive the requirement for the investigator to obtain a signed consent form for some or all subjects if it finds either:

(1) That the only record linking the subject and the research would be the consent document and the principal risk would be potential harm resulting from a breach of confidentiality. Each subject will be asked whether the subject wants documentation linking the subject with the research, and the subject's wishes will govern; or

(2) That the research presents no more than minimal risk of harm to subjects and involves no procedures for which written consent is normally required outside of the research context.

In cases where the documentation requirement is waived, the IRB may require the investigator to provide subjects with a written statement regarding the research.

Sec. 46.118 Applications and proposals lacking definite plans for involvement of human subjects.

Certain types of applications for grants, cooperative agreements, or contracts are submitted to the Department with the knowledge that subjects may be involved within the period of funding, but definite plans would not normally be set forth in the application or proposal. These include activities such as institutional type grants (including bloc grants) where selection of specific projects is the institution's responsibility; research training grants where the activities involving subjects remain to be selected; and projects in which human subjects' involvement will depend upon completion of instruments, prior animal studies, or purification of compounds. These applications need not be reviewed by an IRB before an award may be made. However, except for research described in Sec. 46.101(b), no human subjects may be involved in any project supported by these awards until the project has been reviewed and approved by the IRB, as provided in these regulations, and certification submitted to the Department.

Sec. 46.119 Research undertaken without the intention of involving human subjects.

In the event research (conducted or funded by the Department) is undertaken without the intention of involving human subjects, but it is later proposed to use human subjects in the research, the research shall first be reviewed and approved by an IRB, as provided in these regulations, a certification submitted to the Department, and final approval given to the proposed change by the Department.

Sec. 46.120 Evaluation and disposition of applications and proposals.

(a) The Secretary will evaluate all applications and proposals involving human subjects submitted to the Department through such officers and employees of the Department and such experts and consultants as the Secretary determines to be appropriate. This evaluation will take into consideration the risks to the subjects, the adequacy of protection against these risks, the potential benefits of the proposed research to the subjects and others, and the importance of the knowledge to be gained.

(b) On the basis of this evaluation, the Secretary may approve or disapprove the application or proposal, or enter into negotiations to develop an approvable one.

Sec. 46.121 Investigational new drug or device 30-day delay requirement.

When an institution is required to prepare or to submit a certification with an application or proposal under these regulations, and the application or proposal involves an investigational new drug (within the meaning of 21 U.S.C. 355(i) or 357(d)) or a significant risk device (as defined in 21 CFR 812.3(m)), the institution shall identify the drug or device in the certification. The institution shall also state whether the 30-day interval required for investigational new drugs by 21 CRF 312.1(a) and for significant risk devices by 21 CFR 812.30 has elapsed, or whether the Food and Drug Administration has waived that requirement. If the 30-day interval has expired, the institution shall state whether the Food and Drug Administration has requested that the sponsor continue to withhold or restrict the use of the drug or device in human subjects. If the 30-day interval has not expired, and a waiver has not been received, the institution shall send a statement to the Department upon expiration of the interval. The Department will not consider a certification acceptable until the institution has submitted a statement that the 30-day interval has elapsed, and the Food and Drug Administration has not requested it to limit the use of the drug or device, or that the Food and Drug Administration has waived the 30-day interval.

Sec. 46.122 Use of Federal funds.

Federal funds administered by the Department may not be expended for research involving human subjects unless the requirements of these regulations, including all subparts of these regulations, have been satisfied.

Sec. 46.123 Early termination of research funding; evaluation of subsequent applications and proposals.

(a) The Secretary may require that Department funding for any project be terminated or suspended in the manner prescribed in applicable program requirements, when the Secretary finds an institution has materially failed to comply with the terms of these regulations.

(b) In making decisions about funding applications or proposals covered by these regulations the Secretary may take into account, in addition to all other eligibility requirements and program criteria, factors such as whether the applicant has been subject to a termination or suspension under paragraph (a) of this section and whether the applicant or the person who would direct the scientific and technical aspects of an activity has in the judgment of the Secretary materially failed to discharge responsibility for the protection of the rights and welfare of human subjects (whether or not Department funds were involved).

Sec. 46.124 Conditions.

With respect to any research project or any class of research projects the Secretary may impose additional conditions prior to or at the time of funding when in the Secretary's judgment additional conditions are necessary for the protection of human subjects.

APPENDIX 2.2

Regulations by HHS for "Expedited Review"

[Note: The following text from the *Federal Register* (Vol. 46, No. 16, Monday, January 26, 1981, p. 8392) contains the new regulations regarding types of research under HHS regulation that can be reviewed by IRBs with the new *expedited*, more rapid procedures.]

DEPARTMENT OF HEALTH AND HUMAN SERVICES

Public Health Service
Research Activities Which May Be
Reviewed through Expedited Review Procedures
Set Forth in HHS Regulations for Protection of
Human Research Subjects

AGENCY: Department of Health and Human Services
ACTION: Notice
SUMMARY: This notice contains a list of research activities which Institutional Review Boards may review through the expedited review procedures set forth in HHS regulations for the protection of human subjects.

EFFECTIVE DATE: This notice shall become effective on July 27, 1981.

[Note: Deleted here is supplementary information that preceded the text of the actual regulation.]

Section 46.110 of the new final regulations provides that: "The Secretary will publish in the *Federal Register* a list of categories of research activities, involving no more than minimal risk, that may be reviewed by the Institutional Review Board, through an expedited review procedure. . ." This notice is published in accordance with Sec. 46.110.

Research activities involving no more than minimal risk and in which the only involvement of human subjects will be in one or more of the following categories (carried out through standard methods) may be reviewed by the Institutional Review Board through the expedited review procedure authorized in Sec. 46.110 of 45 CFR Part 46.

(1) Collection of: hair and nail clippings, in a nondisfiguring manner; deciduous teeth; and permanent teeth if patient care indicates a need for extraction.

(2) Collection of excreta and external secretions including sweat, uncannulated saliva, placenta removed at delivery, and amniotic fluid at the time of rupture of the membrane prior to or during labor.

(3) Recording of data from subjects 18 years of age or older using noninvasive procedures routinely employed in clinical practice. This includes the use of physical sensors that are applied either to the surface of the body or at a distance and do not involve input of matter or significant amounts of energy into the subject or an invasion of the subject's privacy. It also includes such procedures as weighing, testing sensory acuity, electrocardiography, electroencephalography, thermography, detection of naturally occurring radioactivity, diagnostic echography, and electroretinography. It does not include exposure to electromagnetic radiation outside the visible range (for example, X-rays, microwaves).

(4) Collection of blood samples by venipuncture, in amounts not exceeding 450 milliliters in an eight-week period and no more often than two times per week, from subjects 18 years of age or older and who are in good health and not pregnant.

(5) Collection of both supra- and subgingival dental plague and calculus, provided the procedure is not more invasive than routine prophylactic scaling of the teeth and the process is accomplished in accordance with accepted prophylactic techniques.

(6) Voice recordings made for research purposes such as investigations of speech defects.

(7) Moderate exercise by healthy volunteers.

(8) The study of existing data, documents, records, pathological specimens, or diagnostic specimens.

(9) Research on individual or group behavior or characteristics of individuals, such as studies of perception, cognition, game theory, or test development, where the investigator does not manipulate subjects' behavior and the research will not involve stress to subjects.

(10) Research on drugs or devices for which an investigational new drug exemption or an investigational device exemption is not required.

DATED: January 14, 1981

JULIUS B. RICHMOND
Assistant Secretary for Health
and Surgeon General

APPENDIX 2.3

Regulations by the Food and Drug Administration (FDA) for Operation of Institutional Review Boards

[Note: The following regulations are taken from the *Federal Register*—Vol. 46, No. 17, Tuesday, January 27, 1981, starting at page 8942—and deal with how FDA rules affect IRBs. Remember that these rules differ in some ways from the IRB rules published by HHS and the discrepancies are covered in the side-by-side comparisons appearing in later supplementary materials at the end of this chapter.]

DEPARTMENT OF HEALTH AND HUMAN SERVICES

Food and Drug Administration
21 CFR Parts 50, 71, 171, 180, 310,
312, 314, 320, 330, 361, 430, 431, 601, 630,
812, 813, 1003, 1010
(Docket No. 78N–0400)
Protection of Human Subjects; Informed Consent

AGENCY: Food and Drug Administration
ACTION: Final Rule
SUMMARY: The Food and Drug Administration (FDA) is issuing regulations to provide protection for human subjects of clinical investigations conducted pursuant to requirements for prior submission to FDA or conducted in support of applications for permission to conduct further research or to market regulated products. The regulations clarify existing FDA requirements governing informed consent and provide protection of the rights and welfare of human subjects involved in research activities that fall within FDA's jurisdiction.

EFFECTIVE DATE: July 27, 1981

[Note: Deleted here are several pages of discussion about the development of the FDA regulations. Also, please note that the sections on informed consent have been placed at the end of Chapter 3 of this book for ease of understanding and that only the new Part 56 of these regulations is reprinted here. For the technical changes to the 17 sections listed as Parts 71–1010 as noted previously, the reader is referred to the original *Federal Register* text.]

Therefore, under the Federal Food, Drug, and Cosmetic Act (secs. 406, 408, 409, 501, 502, 503, 505, 506, 507, 510, 513–516, 518–520, 701(a), 706, and 801, 52 Stat. 1049–1054 as amended, 1055, 1058 as amended, 55 Stat. 851 as amended, 59 Stat. 463 as amended, 68 Stat. 511–518 as amended, 72 Stat. 1785–1788 as amended, 74 Stat. 399–407 as amended, 76 Stat. 794–795 as amended, 90 Stat. 540–546, 560, 562–574 (21 U.S.C. 346, 346a, 348, 351, 352, 353, 355, 356, 357, 360, 360c–360f, 360h–360j, 371(a), 376, and 381)) and the Public Health Service Act (secs. 215, 351, 354–360F, 58 Stat. 690, 702 as amended, 82 Stat. 1173–1186 as amended (42 U.S.C. 216, 241, 262, 263b–263n)) and under authority delegated to the Commissioner of Food and Drugs (21 CFR 5.1), Chapter I of Title 21 of the Code of Federal Regulations is amended as follows:

PART 16—REGULATORY HEARING BEFORE THE FOOD AND DRUG ADMINISTRATION

1. In Part 16, Sec. 16.1 is amended by adding a new regulatory provision under paragraph (b)(2) to read as follows:

Sec. 16.1 Scope.

(b)
(c)
Section 56.121(a), Relating to disqualifying an institutional review board or an institution.
2. By adding new Part 56, to read as follows:

PART 56—INSTITUTIONAL REVIEW BOARDS

Subpart A—General
Sec.
56.101 Scope.
56.102 Definitions.
56.103 Circumstances in which IRB review is required.
56.104 Exemptions from IRB requirement.
56.105 Waiver of IRB requirement.

Subpart B—Organization and Personnel
56.107 IRB membership.

AUTHORITY: Secs. 406, 408, 409, 501, 502, 503, 505, 506, 507, 510, 513–516, 518–520, 701(a), 706, and 801, Pub. L. 717, 52 Stat. 1049–1054 as amended, 1055, 1058 as amended, 55 Stat. 851 as amended, 59 Stat. 463 as amended, 68 Stat. 511–518 as amended, 72 Stat. 1785–1788 as amended, 74 Stat. 399–407 as amended, 76 Stat. 794–795 as amended, 90 Stat. 540–546, 560, 562–574 (21 U.S.C. 346, 346a, 348, 351, 352, 353, 355, 356, 357, 360, 360c–360f, 360h–360j, 371(a), 376, and 381), secs. 215, 301, 351, 354–360f, Pub. L. 410, 58 Stat. 690, 702 as amended, 82 Stat. 1173–1186 as amended (42 U.S.C. 216, 241, 262, 263b–263n).

Subpart A—General Provisions

Sec. 56.101 Scope.

(a) This part contains the general standards for the composition, operation, and responsibility of an Institutional Review Board (IRB) that reviews clinical investigations regulated by the Food and Drug Administration under sections 505(i), 507(d), and 520(g) of the act, as well as clinical investigations that support applications for research or marketing permits for products regulated by the Food and Drug Administration, including food and color additives, drugs for human use, medical devices for human use, biological products for human use, and electronic products. Compliance with this part is intended to protect the rights and welfare of human subjects involved in such investigations.

(b) References in this part to regulatory sections of the Code of Federal Regulations are to Chapter I of Title 21, unless otherwise noted.

Sec. 56.102 Definitions.

As used in this part:

(a) "Act" means the Federal Food, Drug, and Cosmetic Act, as amended (secs. 201–902, 52 Stat. 1040 et seq., as amended (21 U.S.C. 321–392)).

(b) "Application for research or marketing permit" includes:

(1) A color additive petition, described in Part 71.

(2) Data and information regarding a substance submitted as part of the procedures for establishing that a substance is generally recognized as safe for a use which results or may reasonably be expected to result, directly or indirectly, in its becoming a component or otherwise affecting the characteristics of any food, described in Sec. 170.35.

(3) A food additive petition, described in Part 171.

(4) Data and information regarding a food additive submitted as part of the procedures regarding food additives permitted to be used on an interim basis pending additional study, described in Sec. 180.1.

(5) Data and information regarding a substance submitted as part of the procedures for establishing a tolerance for unavoidable contaminants in food and food-packaging materials, described in section 406 of the act.

(6) A "Notice of Claimed Investigational Exemption for a New Drug" described in Part 312.

(7) A new drug application, described in Part 314.

(8) Data and information regarding the bioavailability or bioequivalence of drugs for human use submitted as part of the procedures for issuing, amending, or repealing a bioequivalence requirement, described in Part 320.

(9) Data and information regarding an over-the-counter drug for human use submitted as part of the procedures for classifying such drugs as generally recognized as safe and effective and not misbranded, described in Part 330.

(10) Data and information regarding an antibiotic drug submitted as part of the procedures for issuing, amending, or repealing regulations for such drugs, described in Part 430.

(11) An application for a biological product license, described in Part 601.

(12) Data and information regarding a biological product submitted as part of the procedures for determining that licensed biological products are safe and effective and not misbranded, as described in Part 601.

(13) An "Application for an Investigational Device Exemption," described in Parts 812 and 813.

(14) Data and information regarding a medical device for human use submitted as part of the procedures for classifying such devices, described in Part 860.

(15) Data and information regarding a medical device for human use submitted as part of the procedures for establishing, amending, or repealing a standard for such device, described in Part 861.

(16) An application for premarket approval of a medical device for human use, described in section 515 of the act.

(17) A product development protocol for a medical device for human use, described in section 515 of the act.

(18) Data and information regarding an electronic product submitted as part of the procedures for establishing, amending, or repealing a standard for such products, described in section 358 of the Public Health Service Act.

(19) Data and information regarding an electronic product submitted as part of the procedures for obtaining a variance from any electronic product performance standard, as described in Sec. 1010.4.

(20) Data and information regarding an electronic product submitted as part of the procedures for granting, amending, or extending an exemption from a radiation safety performance standard, as described in Sec. 1010.5.

(21) Data and information regarding an electronic product submitted as part of the procedures for obtaining an exemption from notification of a radiation safety defect or failure of compliance with a radiation safety performance standard, described in Subpart D of Part 1003.

(c) "Clinical investigation" means any experiment that involves a test article and one or more human subjects, and that either must meet the requirements for prior submission to the Food and Drug Administration under section 505(i), 507(d), or 520(g) of the act, or need not meet the requirements for prior submission to the Food and Drug Administration under these sections of the act, but the results of which are intended to be later submitted to, or held for inspection by, the Food and Drug Administration as part of an application for a research or marketing permit. The term does not include experiments that must meet the provisions of Part 58, regarding nonclinical laboratory studies. The terms "research," "clinical research," "clinical study," "study," and "clinical investigation" are deemed to be synonymous for purposes of this part.

(d) "Emergency use" means the use of a test article on human subjects in a life-threatening situation in which no standard acceptable treatment is available, and in which there is not sufficient time to obtain IRB approval.

(e) "Human subject" means an individual who is or becomes a participant in research, either as a recipient of the test article or as a control. A subject may be either a healthy individual or a patient.

(f) "Institution" means any public or private entity or agency (including Federal, State, and other agencies). The term "facility" as used in section 520(g) of the act is deemed to be synonymous with the term "institution" for purposes of this part.

(g) "Institutional Review Board (IRB)" means any board, committee, or other group formally designated by an institution to review, to approve the initiation of, and to conduct periodic review of, biomedical research involving human subjects. The primary purpose of such review is to assure the protection of the rights and welfare of the human subjects. The term has the same meaning as the phrase "institutional review committee" as used in section 520(g) of the act.

(h) "Investigator" means an individual who actually conducts a clinical investigation (i.e., under whose immediate direction the test article is administered or dispensed to, or used involving, a subject) or, in the event of an investigation conducted by a team of individuals, is the responsible leader of that team.

(i) "Minimal risk" means that the risks of harm anticipated in the proposed research are not greater, considering probability and magnitude, than those ordinarily encountered in daily life or during the performance of routine physical or psychological examinations or tests.

(j) "Sponsor" means a person or other entity that initiates a clinical investigation, but that does not actually conduct the investigation, i.e., the test article is administered or dispensed to, or used involving, a subject under the immediate direction of another individual. A person other than an individual (e.g., a corporation or agency) that uses one or more of its own employees to conduct an investigation that it has initiated is considered to be a sponsor (not a sponsor–investigator), and the employees are considered to be investigators.

(k) "Sponsor–investigator" means an individual who both initiates and actually conducts, alone or with others, a clinical investigation, i.e., under whose immediate direction the test article is administered or dispensed to, or used involving, a subject. The term does not include any person other than an individual, e.g., it does not include a corporation or agency. The obligations of a sponsor–investigator under this part include both those of a sponsor and those of an investigator.

(l) "Test article" means any drug for human use, biological product for human use, medical device for human use, human food additive, color additive, electronic product, or any other article subject to regulation under the act or under sections 351 or 354–360F of the Public Health Service Act.

Sec. 56.103 Circumstances in which IRB review is required.

(a) Except as provided in Sections 56.104 and 56.105, any clinical investigation which must meet the requirements for prior submission (as required in Parts 312, 812, and 813) to the Food and Drug Administration shall not be initiated unless that investigation has been reviewed and approved by, and remains subject to continuing review by, an IRB meeting the requirements of this part.

(b) Except as provided in Sections 56.104 and 56.105, the Food and Drug Administration may decide not to consider in support of an application for a research or marketing permit any data or information that has been derived from a clinical investigation that has not been approved by, and that was not subject to initial and continuing review by, an IRB meeting the requirements of this part. The determination that a clinical investigation may not be considered in support of an application for a research or marketing permit does not, however, relieve the applicant for any such a permit of any obligation under any other applicable regulations to submit the results of the investigation to the Food and Drug Administration.

(c) Compliance with these regulations will in no way render inapplicable pertinent Federal, State, or local laws or regulations.

Sec. 56.104 Exemptions from IRB requirement.

The following categories of clinical investigations are exempt from the requirements of this part for IRB review:

(a) Any investigation which commenced before July 27, 1981, and was subject to requirements for IRB review under FDA regulations before that date, provided that the investigation remains subject to review of an IRB which meets the FDA requirements in effect before July 27, 1981.

(b) Any investigation commenced before July 27, 1981, and was not otherwise subject to requirements for IRB review under Food and Drug Administration regulations before that date.

(c) Emergency use of a test article, provided that such emergency use is reported to the IRB within 5 working days. Any subsequent use of the test article at the institution is subject to IRB review.

Sec. 56.105 Waiver of IRB requirement.

On the application of a sponsor or sponsor–investigator, the Food and Drug Administration may waive any of the requirements contained in these regulations, including the requirements for IRB review, for specific research activities or for classes of research activities, otherwise covered by these regulations.

Subpart B—Organization and Personnel

Sec. 56.107 IRB membership.

(a) Each IRB shall have at least five members, with varying backgrounds to promote complete and adequate review of research activities commonly conducted by the institution. The IRB shall be sufficiently qualified through the experience and expertise of its members, and the diversity of the members' backgrounds including consideration of the racial and cultural backgrounds of members and sensitivity to such issues as community attitudes, to promote respect for its advice and counsel in safeguarding the rights and welfare of human subjects. In addition to possessing the professional competence necessary to review specific research activities, the IRB shall be able to ascertain the acceptability of proposed research in terms of institutional commitments and regulations, applicable law, and standards of professional conduct and practice. The IRB shall therefore include persons knowledgeable in these areas. If an IRB regularly reviews research that involves a vulnerable category of subjects, including but not limited to subjects covered by other parts of this chapter, the IRB should include one or more individuals who are primarily concerned with the welfare of these subjects.

(b) No IRB may consist entirely of men, or entirely of women, or entirely of members of one profession.

(c) Each IRB shall include at least one member whose primary concerns are in nonscientific areas; for example: lawyers, ethicists, members of the clergy.

(d) Each IRB shall include at least one member who is not otherwise affiliated with the institution and who is not part of the immediate family of a person who is affiliated with the institution.

(e) No IRB may have a member participate in the IRB's initial or continuing review of any project in which the member has a conflicting interest, except to provide information requested by the IRB.

(f) An IRB may, in its discretion, invite individuals with competence in special areas to assist in the review of complex issues which require expertise beyond or in addition to that available on the IRB. These individuals may not vote with the IRB.

Subpart C—IRB Functions and Operations

Sec. 56.108 IRB functions and operations.

In order to fulfill the requirements of these regulations, each IRB shall:

(a) Follow written procedures (1) for conducting its initial and continuing review of research and for reporting its findings and actions to the investigator and the institution, (2) for determining which projects require review more often than annually and which projects need verification from sources other than the investigators that no material changes have occurred since previous IRB review, (3) for insuring prompt reporting to the IRB for changes in a research activity, (4) for insuring that changes in approved research, during the period for which IRB approval has already been given, may not be initiated without IRB review and approval except where necessary to eliminate apparent immediate hazards to the human subjects; and (5) for insuring prompt reporting to the IRB of unanticipated problems involving risks to subjects or others.

(b) Except when an expedited review procedure is used (see Sec. 56.110), review proposed research at convened meetings at which a majority of the members of the IRB are present, including at least one member whose primary concerns are in nonscientific areas. In order for the research to be approved, it shall receive the approval of a majority of those members present at the meeting.

(c) Be responsible for reporting to the appropriate institutional officials and the Food and Drug Administration any serious or continuing noncompliance by investigators with the requirements and determinations of the IRB.

Sec. 56.109 IRB review of research.

(a) An IRB shall review and have authority to approve, require modifications in (to secure approval), or disapprove all research activities covered by these regulations.

(b) An IRB shall require that information given to subjects as part of informed consent is in accordance with Sec. 50.25. The IRB may require that information, in addition to that specifically mentioned in Sec. 50.25, be given to the subjects when in the IRB's judgment the information would meaningfully add to the protection of the rights and welfare of subjects.

(c) An IRB shall require documentation of informed consent in accordance with Sec. 50.27, except that the IRB may, for some or all subjects, waive the requirement that the subject or the subject's legally authorized representative sign a written consent form if it finds that the research presents no more than minimal risk of harm to subjects and involves no procedures for which written consent is normally required outside the research context. In cases where the documentation requirement is waived, the IRB may require the investigator to provide subjects with a written statement regarding the research.

(d) An IRB shall notify investigators and the institution in writing of its decision to approve or disapprove the proposed research activity, or of modifications required to secure IRB approval of the research activity. If the IRB decides to disapprove a research activity, it shall include in its written notification a statement of the reasons for its decision and give the investigator an opportunity to respond in person or in writing.

(e) An IRB shall conduct continuing review of research covered by these regulations at intervals appropriate to the degree of risk, but not less than once per year, and shall have authority to observe or have a third party observe the consent process and the research.

Sec. 56.110 Expedited review procedures for certain kinds of research involving no more than minimal risk, and for minor changes in approved research.

(a) The Food and Drug Administration has established, and published in the *Federal Register*, a list of categories of research that may be reviewed by the IRB through an expedited review procedure. The list will be amended, as appropriate, through periodic republication in the *Federal Register*.

(b) An IRB may review some or all of the research appearing on the list through an expedited review procedure, if the research involves no more than minimal risk. The IRB may also use the expedited review procedure to review minor changes in previously approved research during the period for which approval is authorized. Under an expedited review procedure, the review may be carried out by the IRB chairperson or by one or more experienced reviewers designated by the chairperson from among members of the IRB. In reviewing the research, the reviewers may exercise all of the authorities of the IRB except that the reviewers may not disapprove the research. A research activity may be disapproved only after review in accordance with the non-expedited procedure set forth in Sec. 56.108(b).

(c) Each IRB which uses an expedited review procedure shall adopt a method for keeping all members advised of research proposals which have been approved under the procedure.

(d) The Food and Drug Administration may restrict, suspend, or terminate an institution's or IRB's use of the expedited review procedure when necessary to protect the rights or welfare of subjects.

Sec. 56.111 Criteria for IRB approval of research.

(a) In order to approve research covered by these regulations the IRB shall determine that all of the following requirements are satisfied:

(1) Risks to subjects are minimized: (i) by using procedures which are consistent with sound research design and which do not unnecessarily expose subjects to risk, and (ii) whenever appropriate, by using procedures already being performed on the subjects for diagnostic or treatment purposes.

(2) Risks to subjects are reasonable in relation to anticipated benefits, if any, to subjects, and the importance of the knowledge that may be expected to result. In evaluating risks and benefits, the IRB should consider only those risks and benefits that may result from the research (as distinguished from risks and benefits of therapies that subjects would receive even if not participating in the research). The IRB should not consider possible long-range effects of applying knowledge gained in the research (for example, the possible effects of the research on public policy) as among those research risks that fall within the purview of its responsibility.

(3) Selection of subjects is equitable. In making this assessment, the IRB should take into account the purposes of the research and the setting in which the research will be conducted.

(4) Informed consent will be sought from each prospective subject or the subject's legally authorized representative, in accordance with and to the extent required by Part 50.

(5) Informed consent will be appropriately documented, in accordance with and to the extent required by Sec. 50.27.

(6) Where appropriate, the research plan makes adequate provision for monitoring the data collected to ensure the safety of subjects.

(7) Where appropriate, there are adequate provisions to protect the privacy of subjects and to maintain the confidentiality of data.

(b) Where some or all of the subjects are likely to be vulnerable to coercion or undue influence, such as persons with acute or severe physical or mental illness, or persons who are economically or educationally disadvantaged, appropriate additional safeguards have been included in the study to protect the rights and welfare of these subjects.

Sec. 56.112 Review by institution.

Research covered by these regulations that has been approved by an IRB may be subject to further appropriate review and approval or disapproval by officials of the institution. However, those officials may not approve the research if it has not been approved by an IRB.

Sec. 56.113 Suspension or termination of IRB approval of research.

An IRB shall have authority to suspend or terminate approval of research that is not being conducted in accordance with the IRB's requirements or that has been associated with unexpected serious harm to subjects. Any suspension or termination of approval shall include a statement of the reasons for the IRB's action and shall be reported promptly to the investigator, appropriate institutional officials, and the Food and Drug Administration.

Sec. 56.114 Cooperative research.

In complying with these regulations, institutions involved in multiinstitutional studies may use joint review, reliance upon the review of another qualified IRB, or similar arrangements aimed at avoidance of duplication of effort.

Subpart D—Records and Reports

Sec. 56.115 IRB records.

(a) An institution, or where appropriate an IRB, shall prepare and maintain adequate documentation of IRB activities, including the following:

(1) Copies of all research proposals reviewed, scientific evaluations, if any, that accompany the proposals, approved sample consent documents, progress reports submitted by investigators, and reports of injuries to subjects.

(2) Minutes of IRB meetings which shall be in sufficient detail to show attendance at the meetings; actions taken by the IRB; the vote on these actions including the number of members voting for, against, and abstaining; the basis for requiring changes in or disapproving research; and a written summary of the discussion of controverted issues and their resolution.

(3) Records of continuing review activities.

(4) Copies of all correspondence between the IRB and the investigators.

(5) A list of IRB members identified by name; earned degrees; representative capacity; indications of experience such as board certifications, licenses, etc., sufficient to describe each member's chief anticipated contributions to IRB deliberations; and any employment or other relationship between each member and the institution; for example:

full-time employee, part-time employee, a member of governing panel or board, stockholder, paid or unpaid consultant.

(6) Written procedures for the IRB as required by Sec. 56.108(a).

(7) Statements of significant new findings provided to subjects, as required by Sec. 50.25.

(b) The records required by this regulation shall be retained for at least 3 years after completion of the research, and the records shall be accessible for inspection and copying by authorized representatives of the Food and Drug Administration at reasonable times and in a reasonable manner.

(c) The Food and Drug Administration may refuse to consider a clinical investigation in support of an application for a research or marketing permit if the institution or the IRB that reviewed the investigation refuses to allow an inspection under this section.

Subpart E—Administrative Actions for Noncompliance

Sec. 56.120 Lesser administrative actions.

(a) If apparent noncompliance with these regulations in the operation of an IRB is observed by an FDA investigator during an inspection, the inspector will present an oral or written summary of observations to an appropriate representative of the IRB. The Food and Drug Administration may subsequently send a letter describing the noncompliance to the IRB and to the parent institution. The agency will require that the IRB or the parent institution respond to this letter within a time period specified by FDA and describe the corrective actions that will be taken by the IRB, the institution, or both to achieve compliance with these regulations.

(b) On the basis of the IRB's or the institution's response, FDA may schedule a reinspection to confirm the adequacy of corrective actions. In addition, until the IRB or the parent institution takes appropriate corrective action, the agency may:

(1) Withhold approval of new studies subject to the requirements of this part that are conducted at the institution or reviewed by the IRB;

(2) Direct that no new subjects be added to ongoing studies subject to this part;

(3) Terminate ongoing studies subject to this part when doing so would not endanger the subjects; or

(4) When the apparent noncompliance creates a significant threat to the rights and welfare of human subjects, notify relevant State and Federal regulatory agencies and other parties with a direct interest in the agency's action of the deficiencies in the operation of the IRB.

(c) The parent institution is presumed to be responsible for the operation of an IRB, and the Food and Drug Administration will ordinarily direct any administrative action under this subpart against the institution. However, depending on the evidence of responsibility for deficiencies, determined during the investigation, the Food and Drug Administration may restrict its administrative actions to the IRB or to a component of the parent institution determined to be responsible for formal designation of the IRB.

Sec. 56.121 Disqualification of an IRB or an institution.

(a) Whenever the IRB or the institution has failed to take adequate steps to correct the noncompliance stated in the letter sent by the agency under Sec. 56.120(a), and the

Commissioner of Food and Drugs determines that this noncompliance may justify the disqualification of the IRB or of the parent institution, the Commissioner will institute proceedings in accordance with the requirements for a regulatory hearing set forth in Part 16.

(b) The Commissioner may disqualify an IRB or the parent institution if the Commissioner determines that:

(1) The IRB has refused or repeatedly failed to comply with any of the regulations set forth in this part, and

(2) The noncompliance adversely affects the rights or welfare of the human subjects in a clinical investigation.

(c) If the Commissioner determines that disqualification is appropriate, the Commissioner will issue an order that explains the basis for the determination and that prescribes any actions to be taken with regard to ongoing clinical research conducted under the review of the IRB. The Food and Drug Administration will send notice of the disqualification to the IRB and the parent institution. Other parties with a direct interest, such as sponsors and clinical investigators, may also be sent a notice of the disqualification. In addition, the agency may elect to publish a notice of its action in the *Federal Register*.

(d) The Food and Drug Administration will not approve an application for a research permit for a clinical investigation that is to be under the review of a disqualified IRB or that it is to be conducted at a disqualified institution, and it may refuse to consider in support of a marketing permit the data from a clinical investigation that was reviewed by a disqualified IRB as conducted at a disqualified institution, unless the IRB or the parent institution is reinstated as provided in Sec. 56.123.

Sec. 56.122 Public disclosure of information regarding revocation.

A determination that the Food and Drug Administration has disqualified an institution and the administrative record regarding that determination are disclosable to the public under Part 20.

Sec. 56.123 Reinstatement of an IRB or an institution.

An IRB or an institution may be reinstated if the Commissioner determines, upon an evaluation of a written submission from the IRB or institution that explains the corrective action that the institution or IRB plans to take, that the IRB or institution has provided adequate assurance that it will operate in compliance with the standards set forth in this part. Notification of reinstatement shall be provided to all persons notified under Sec. 56.121(c).

Sec. 56.124 Actions alternative or additional to disqualification.

Disqualification of an IRB or of an institution is independent of, and neither in lieu of nor a precondition to, other proceedings or actions authorized by the act. The Food and Drug Administration may, at any time, through the Department of Justice institute any appropriate judicial proceedings (civil or criminal) and any other appropriate regulatory action, in addition to or in lieu of, and before, at the time of, or after, disqualification. The agency may also refer pertinent matters to another Federal, State, or local government agency for any action that that agency determines to be appropriate.

EFFECTIVE DATE: This regulation shall become effective July 27, 1981.

(Secs. 406, 408, 409, 501, 502, 503, 505, 506, 507, 510, 513–516, 518–520, 701(a), 706, and 801, 52 Stat. 1049–1054 as amended, 1055, 1058 as amended, 55 Stat. 851 as amended, 59 Stat. 463 as amended, 68 Stat. 511–517 as amended, 72 Stat. 1785–1788 as amended, 74 Stat. 399–407 as amended, 76 Stat. 794–795 as amended, 90 Stat. 540–560, 562–574 (21 U.S.C. 346, 346a, 348, 351, 352, 353, 355, 356, 357, 360, 360c–360f, 360h–360j, 371(a), 376, and 381); secs. 215, 301, 351, as amended (42 U.S.C. 216, 241, 262, 263b–263n)).

DATED: January 19, 1981

JERE E. GOYAN
Commissioner of Food and Drugs

APPENDIX 2.4

Regulations by FDA for "Expedited Review"

[Note: The following text is from the *Federal Register*—Vol. 46, No. 17, Tuesday, January 27, 1981, p. 8980—and consists of the FDA rules for rapid review by an IRB of certain types of FDA research.]

DEPARTMENT OF HEALTH AND HUMAN SERVICES

Food and Drug Administration
(Docket No. 77N–0350)
Protection of Human Research Subjects;
Clinical Investigations Which May Be Reviewed
Through Expedited Review Procedure Set Forth
in FDA Regulations

AGENCY: Food and Drug Administration
ACTION: Notice.
SUMMARY: This notice contains a list of research activities which institutional review boards may review through the expedited review procedures set forth in FDA regulations for the protection of human research subjects.

[Note: Deleted here is supplementary information.]

The agency concludes that research activities with human subjects involving no more than minimal risk and involving one or more of the following categories (carried out through standard methods), may be reviewed by an IRB through the expedited review procedure authorized in Sec. 56.110.

(1) Collection of hair and nail clippings in a non-disfiguring manner; of deciduous teeth; and of permanent teeth if patient care indicates a need for extraction.

(2) Collection of excreta and external secretions including sweat and uncannulated saliva; of placenta at delivery; and of amniotic fluid at the time of rupture of the membrane before or during labor.

(3) Recording of data from subjects who are 18 years of age using noninvasive procedures routinely employed in clinical practice. This category includes the use of physical sensors that are applied either to the surface of the body or at a distance and do not involve input of matter or significant amounts of energy into the subject or an invasion of the subject's privacy. It also includes such procedures as weighing, electrocardiography, electroencephalography, thermography, detection of naturally occurring radioactivity, diagnostic echography, and electroretinography. This cateogry does not include exposure to electromagnetic radiation outside the visible range (for example, x-rays or microwaves).

(4) Collection of blood samples by venipuncture, in amounts not exceeding 450 milliliters in an eight-week period and no more often than two times per week, from subjects who are 18 years of age or older and who are in good health and not pregnant.

(5) Collection of both supra- and subgingival dental plaque and calculus, provided the procedure is not more invasive than routine prophylactic scaling of the teeth, and the process is accomplished in accordance with accepted prophylactic techniques.

(6) Voice recordings made for research purposes such as investigations of speech defects.

(7) Moderate exercise by healthy volunteers.

(8) The study of existing data, documents, records, pathological specimens, or diagnostic specimens.

(9) Research on drugs or devices for which an investigational new drug exemption or an investigational device exemption is not required.

This list will be amended as appropriate and a current list will be published periodically to the *Federal Register*.

DATED: January 19, 1981

JERE E. GOYAN
Commissioner of Food and Drugs

APPENDIX 2.5

A Comparison of HHS and FDA Rules Governing Operation of Institutional Review Boards

[Note: The following text is an adaptation of a side-by-side comparison of FDA and HHS rules provided by Dr. John Petricciani of the Food and Drug Administration. Differences between the rules are noted in italics.]

Sec. 56.107 IRB membership.

(a) Each IRB shall have at least five members, with varying backgrounds to promote complete and adequate review of research activities commonly conducted by the institution. The IRB shall be sufficiently qualified through the experience and expertise of its members, and the diversity of the members' backgrounds including consideration of the racial and cultural backgrounds of members and sensitivity of such issues as community attitudes, to promote respect for its advice and counsel in safeguarding the rights and welfare of human subjects. In addition to possessing the professional competence necessary to review specific research activities, the IRB shall be able to ascertain the acceptability of proposed research in terms of institutional commitments and regulations, applicable law, and standards of professional conduct and practice. The IRB shall therefore include persons knowledgeable in these areas. If an IRB regularly reviews research that involves a vulnerable category of subjects, including but not limited to subjects covered by other parts of this chapter, the IRB should include one or more individuals who are primarily concerned with the welfare of these subjects.

(b) No IRB may consist entirely of men, or entirely of women, or entirely of members of one profession.

(c) Each IRB shall include at least one member whose primary concerns are in nonscientific areas; for example: lawyers, ethicists, members of the clergy.

(d) Each IRB shall include at least one member who is not otherwise affiliated with the institution and who is not part of the immediate family of a person who is affiliated with the institution.

(e) No IRB may have a member participate in the IRB's initial or continuing review of any project in which the member has a conflicting interest, except to provide information requested by the IRB.

(f) An IRB may, in its discretion, invite individuals with competence in special areas to assist in the review of complex issues which require expertise beyond or in addition to that available on the IRB. These individuals may not vote with the IRB.

Sec. 56.108 IRB functions and operations.

In order to fulfill the requirements of these regulations, each IRB shall:

(a) Follow written procedures (1)

HHS

Sec. 46.107 IRB membership.

(a) Each IRB shall have at least five members, with varying backgrounds to promote complete and adequate review of research activities commonly conducted by the institution. The IRB shall be sufficiently qualified through the experience and expertise of its members, and the diversity of the members' backgrounds including consideration of the racial and cultural backgrounds of members and sensitivity of such issues as community attitudes, to promote respect for its advice and counsel in safeguarding the rights and welfare of human subjects. In addition to possessing the professional competence necessary to review specific research activities, the IRB shall be able to ascertain the acceptability of proposed research in terms of institutional commitments and regulations, applicable law, and standards of professional conduct and practice. The IRB shall therefore include persons knowledgeable in these areas. If an IRB regularly reviews research that involves a vulnerable category of subjects, including but not limited to subjects covered by other parts of this chapter, the IRB should include one or more individuals who are primarily concerned with the welfare of these subjects.

(b) No IRB may consist entirely of men, or entirely of women, or entirely of members of one profession.

(c) Each IRB shall include at least one member whose primary concerns are in nonscientific areas; for example: lawyers, ethicists, members of the clergy.

(d) Each IRB shall include at least one member who is not otherwise affiliated with the institution and who is not part of the immediate family of a person who is affiliated with the institution.

(e) No IRB may have a member participate in the IRB's initial or continuing review of any project in which the member has a conflicting interest, except to provide information requested by the IRB.

(f) An IRB may, in its discretion, invite individuals with competence in special areas to assist in the review of complex issues which require expertise beyond or in addition to that available on the IRB. These individuals may not vote with the IRB.

Sec. 46.108 IRB functions and operations.

In order to fulfill the requirements of these regulations, each IRB shall:

(a) Follow written procedures *as provided in Sec. 46.103(b)(4). [From 46.103(b)(4)] written procedures which the IRB will will follow* (i) for conducting its initial and continuing review of research and for reporting its findings and actions to the investigator and the institution, (2) for determining which projects require review more often than annually and which projects need verification from sources other than the investigators that no material changes have occurred since previous IRB review, (3) for insuring prompt reporting to the IRB of changes in a research activity, (4) for insuring that changes in approved research, during the period for which IRB approval has already been given, may not be initiated without IRB review and approval except where necessary to eliminate apparent immediate hazards to the *human subjects*; and (5) for insuring prompt reporting to the IRB of unanticipated problems involving risks to subjects or others.

(b) Except when an expedited review procedure is used (see Sec. 56.110), review proposed research at convened meetings at which a majority of the members of the IRB are present, including at least one member whose primary concerns are in nonscientific areas. In order for the research to be approved, it shall receive the approval of a majority of those members present at the meeting.

(c) Be responsible for reporting to the appropriate institutional officials and the Food and Drug Administration any serious or continuing noncompliance by investigators with the requirements and determinations of the IRB.

Sec. 56.109 IRB review of research.

(a) An IRB shall review and have authority to approve, require modifications in (to secure approval), or disapprove all research activities covered by these regulations.

(b) An IRB shall require that information given to subjects as part of informed consent is in accordance with Sec. 50.25. The IRB may require that information, in addition to that specifically mentioned in Sec. 50.25, be given to the subjects when in the IRB's judgment the information would meaningfully add to the protection of the rights and welfare of subjects.

(c) An IRB shall require documentation of informed consent in accordance with Sec. 50.27, except that the IRB may, for some or all subjects, waive the requirement that the subject or the subject's legally authorized representative sign a written consent form if it finds for conducting its initial and continuing review of research and for reporting its findings and actions to the investigator and the institution; (ii) for determining which projects require review more often than annually and which projects need verification from sources other than the investigators that no material changes have occurred since previous IRB review; (iii) for insuring prompt reporting to the IRB of *proposed* changes in a research activity, *and* for insuring that changes in approved research, during the period for which IRB approval has already been given, may not be initiated without IRB review and approval except where necessary to eliminate apparent immediate hazards to the subject; and (iv) for insuring prompt reporting to the IRB *and to the secretary* of unanticipated problems involving risks to subjects or others.

(b) Except when an expedited review procedure is used (see Sec. 46.110), review proposed research at convened meetings at which a majority of the members of the IRB are present, including at least one member whose primary concerns are in nonscientific areas. In order for the research to be approved, it shall receive the approval of a majority of those members present at the meeting.

(c) Be responsible for reporting to the appropriate institutional officials and the *secretary* any serious or continuing noncompliance by investigators with the requirements and determinations of the IRB.

Sec. 46.109 IRB review of research.

(a) An IRB shall review and have authority to approve, require modifications in (to secure approval), or disapprove all research activities covered by these regulations.

(b) An IRB shall require that information given to subjects as part of informed consent is in accordance with Sec. 46.116. The IRB may require that information, in addition to that specifically mentioned in Sec. 46.116, be given to the subjects when in the IRB's judgment the information would meaningfully add to the protection of the rights and welfare of subjects.

(c) An IRB shall require documentation of informed consent or may waive documentation in accordance with Sec. 46.117.

that the research presents no more than minimal risk of harm to subjects and involves no procedures for which written consent is normally required outside the research context. In cases where the documentation requirement is waived, the IRB may require the investigator to provide subjects with a written statement regarding the research.

(d) An IRB shall notify investigators and the institution in writing of its decision to approve or disapprove the proposed research activity, or of modifications required to secure IRB approval of the research activity. If the IRB decides to disapprove a research activity, it shall include in its written notification a statement of the reasons for its decision and give the investigator an opportunity to respond in person or in writing.

(e) An IRB shall conduct continuing review of research covered by these regulations at intervals appropriate to the degree of risk, but not less than once per year, and shall have authority to observe the consent process and the research.

Sec. 56.110 Expedited review procedures for certain kinds of research involving no more than minimal risk, and for minor changes in approved research.

(a) The Food and Drug Administration has established, and published in the *Federal Register*, a list of categories of research that may be reviewed by the IRB through an expedited review procedure. The list will be amended, as appropriate, through periodic republication in the *Federal Register*.

(b) An IRB may review some or all of the research appearing on the list through an expedited review procedure, if the research involves no more than minimal risk. The IRB may also use the expedited review procedure to review minor changes in previously approved research during the period for which approval is authorized. Under an expedited review procedure, the review may be carried out

HHS

[From Sec. 46.117 Documentation of informed consent.]

(c) An IRB may waive the requirement for the investigator to obtain a signed consent form for some or all subjects if it finds *either*:

(1) *That the only record linking the subject and the research would be the consent document and the principal risk would be potential harm resulting from a breach of confidentiality. Each subject will be asked whether the subject wants documentation linking the subject with* the research, and the subject's wishes will govern; or

(2) That the research presents no more than minimal risk of harm to subjects and involves no procedures for which written consent is normally required outside the research context. In cases where the documentation requirement is waived, the IRB may require the investigator to provide subjects with a written statement regarding the research.

[From Sec. 46.109] (d) An IRB shall notify investigators and the institution in writing of its decision to approve or disapprove the proposed research activity, or of modifications required to secure IRB approval of the research activity. If the IRB decides to disapprove a research activity, it shall include in its written notification a statement of the reasons for its decision and give the investigator an opportunity to respond in person or in writing.

(e) An IRB shall conduct continuing review of research covered by these regulations at intervals appropriate to the degree of risk, but not less than once per year, and shall have authority to observe the consent process and the research.

Sec. 46.110 Expedited review procedures for certain kinds of research involving no more than minimal risk, and for minor changes in approved research.

(a) The *secretary* has established, and published in the *Federal Register*, a list of categories of research that may be reviewed by the IRB through an expedited review procedure. The list will be amended, as appropriate, through periodic republication in the *Federal Register*.

(b) An IRB may review some or all of the research appearing on the list through an expedited review procedure, if the research involves no more than minimal risk. The IRB may also use the expedited review procedure to review minor changes in previously approved research during the period for which approval is authorized. Under an expedited review procedure, the review may be carried out by the IRB chairperson or by one or more experienced reviewers designated by the chairperson from among members of the IRB. In reviewing the research, the reviewers may exercise all of the authorities of the IRB except that the reviewers may not disapprove the research. A research activity may be disapproved only after review in accordance with the non-expedited procedure set forth in Sec. 56.108(b).

(c) Each IRB which uses an expedited review procedure shall adopt a method for keeping all members advised of research proposals which have been approved under the procedure.

(d) The Food and Drug Administration may restrict, suspend, or terminate an institution's or IRB's use of the expedited review procedure when necessary to protect the rights or welfare of subjects.

Sec. 56.111 Criteria for IRB approval of research.

(a) In order to approve research covered by these regulations the IRB shall determine that all of the following requirements are satisfied:

(1) Risks to subjects are minimized: (i) by using procedures which are consistent with sound research design and which do not not unnecessarily expose subjects to risk, and (ii) whenever appropriate, by using procedures already being performed on the subjects for diagnostic or treatment purposes.

(2) Risks to subjects are reasonable in relation to anticipated benefits, if any, to subjects, and the importance of the knowledge that may be expected to result. In evaluating risks and benefits, the IRB should consider only those risks and benefits that may result from the research (as distinguished from risks and benefits of therapies *that* subjects would receive even if not participating in the research). The IRB should not consider possible long-range effects of applying knowledge gained in the research (for example, the possible effects of the research on public policy) as among those research risks that fall within the purview of its responsibility.

(3) Selection of subjects is equitable. In making this assessment, the IRB should take into account the purposes of the research and the setting in which the research will be conducted.

(4) Informed consent will be sought from each prospective subject or the subject's legally authorized representative, in accordance with and to the extent required by Part 50.

(5) Informed consent will be appropriately documented, in accordance with and to the extent required by Sec. 50.27.

by the IRB chairperson or by one or more experienced reviewers designated by the chairperson from among members of the IRB. In reviewing the research, the reviewers may exercise all of the authorities of the IRB except that the reviewers may not disapprove the research. A research activity may be disapproved only after review in accordance with the non-expedited procedure set forth in Sec. 46.108(b).

(c) Each IRB which uses an expedited review procedure shall adopt a method for keeping all members advised of research proposals which have been approved under the procedure.

(d) The *secretary* may restrict, suspend, or terminate an institution's or IRB's use of the expedited review procedure when necessary to protect the rights or welfare of subjects.

Sec. 46.111 Crtieria for IRB approval of research.

(a) In order to approve research covered by these regulations the IRB shall determine that all of the following requirements are satisfied:

(1) Risks to subjects are minimized: (i) by using procedures which are consistent with sound research design and which do not unnecessarily expose subjects to risk, and (ii) whenever appropriate, by using procedures already being performed on the subjects for diagnostic or treatment purposes.

(2) Risks to subjects are reasonable in relation to anticipated benefits, if any, to subjects, and the importance of the knowledge that may *reasonably* be expected to result. In evaluating risks and benefits, the IRB should consider only those risks and benefits that may result from the research (as distinguished from risks and benefits of therapies subjects would receive even if not participating in the research). The IRB should not consider possible long-range effects of applying knowledge gained in the research (for example, the possible effects of the research on public policy) as among those research risks that fall within the purview of its responsibility.

(3) Selection of subjects is equitable. In making this assessment, the IRB should take into account the purposes of the research and the setting in which the research will be conducted.

(4) Informed consent will be sought from each prospective subject or the subject's legally authorized representative, in accordance with and to the extent required by Sec. 46.116.

(5) Informed consent will be appropriately documented, in accordance with and to the extent required by Sec. 46.117.

(6) Where appropriate, the research plan makes adequate provision for monitoring the data collected to ensure the safety of subjects.

(7) Where appropriate, there are adequate provisions to protect the privacy of subjects and to maintain the confidentiality of data.

(b) Where some or all of the subjects are likely to be vulnerable to coercion or undue influence, such as persons with acute or severe physical or mental illness, or persons who are economically or educationally disadvantaged, appropriate additional safeguards have been included in the study to protect the rights and welfare of these subjects.

Sec. 56.112 Review by institution.

Research covered by these regulations that has been approved by an IRB may be subject to further appropriate review and approval or disapproval by officials of the institution. However, those officials may not approve the research if it has not been approved by an IRB.

Sec. 56.113 Suspension or termination of IRB approval of research.

An IRB shall have authority to suspend or terminate approval of research that is not being conducted in accordance with the IRB's requirements or that has been associated with unexpected serious harm to subjects. Any suspension or termination of approval shall include a statement of the reasons for the IRB's action and shall be reported promptly to the investigator, appropriate institutional officials, and the Food and Drug Administration.

Sec. 56.114 Cooperative research.

(6) Where appropriate, the research plan makes adequate provision for monitoring the data collected to *insure* the safety of subjects.

(7) Where appropriate, there are adequate provisions to protect the privacy of subjects and to maintain the confidentiality of data.

(b) Where some or all of the subjects are likely to be vulnerable to coercion or undue influence, such as persons with acute or severe physical or mental illness, or persons who are economically or educationally disadvantaged, appropriate additional safeguards have been included in the study to protect the rights and welfare of these subjects.

Sec. 46.112 Review by institution.

Research covered by these regulations that has been approved by an IRB may be subject to further appropriate review and approval or disapproval by officials of the institution. However, those officals may not approve the research if it has not been approved by an IRB.

Sec. 46.113 Suspension or termination of IRB approval of research.

An IRB shall have authority to suspend or terminate approval of research that is not being conducted in accordance with the IRB's requirements or that has been associated with unexpected serious harm to subjects. Any suspension or termination of approval shall include a statement of the reasons for the IRB's action and shall be reported promptly to the investigator, appropriate institutional officials, and the *Secretary*.

Sec. 46.114 Cooperative research.

Cooperative research projects are those projects, normally supported through grants, contracts, or similar arrangements, which involve institutions in addition to the grantee or prime contractor (such as a contractor with the grantee, or a subcontractor with the prime contractor). In such instances, the grantee or prime contractor remains responsible to the department for safeguarding the rights and welfare of human subjects. Also, when cooperating institutions conduct some or all of the research involving some or all of these subjects, each cooperating institution shall comply with these regulations as though it received funds for its participation in the project directly from the department, except that

FDA

In complying with these regulations, institutions *involved in multi-institutional studies* may use joint review, reliance upon the review of another qualified IRB, or similar arrangements aimed at avoidance of duplication of effort.

Sec. 56.115 IRB records.

(a) An institution, or where appropriate an IRB, shall prepare and maintain adequate documentation of IRB activities, including the following:

(1) Copies of all research proposals reviewed, scientific evaluations, if any, that accompany the proposals, approved sample consent documents, progress reports submitted by investigators, and reports of injuries to subjects.

(2) Minutes of IRB meetings which shall be in sufficient detail to show attendance at the meetings; actions taken by the IRB; the vote on these actions including the number of members voting for, against, and abstaining; the basis for requiring changes in or disapproving research; and a written summary of the discussion of controverted issues and their resolution.

(3) Records of continuing review activities.

(4) Copies of all correspondence between the IRB and the investigators.

(5) A list of IRB members identified by name; earned degrees; representative capacity; indications of experience such as board certifications, licenses, etc., sufficient to describe each member's chief anticipated contributions to IRB deliberations; and any employment or other relationship between each member and the institution; for example: full-time employee, part-time employee, member of governing panel or board, stockholder, paid or unpaid consultant.

(6) Written procedures for the IRB as required by Sec. 56.108(a).

(7) Statements of significant new findings provided to subjects, as required by Sec. 50.25.

(b) The records required by this regulation shall be retained for at least 3 years after completion of the research, and the records shall be accessible for inspection and copying by authorized representatives of the Food and Drug Administration at reasonable times and in a reasonable manner.

in complying with these regulations institutions
may use joint review, reliance upon the review of another qualified
IRB, or similar arrangements aimed at avoidance of duplication of
effort.

Sec. 46.115 IRB records.

(a) An institution, or where appropriate an IRB, shall prepare and maintain adequate documentation of IRB activities, including the following:

(1) Copies of all research proposals reviewed, scientific evaluations, if any, that accompany the proposals, approved sample consent documents, progress reports submitted by investigators, and reports of injuries to subjects.

(2) Minutes of IRB meetings which shall be in sufficient detail to show attendance at the meetings; actions taken by the IRB; the vote on these actions including the number of members voting for, against, and abstaining; the basis for requiring changes in or disapproving research; and a written summary of the discussion of controverted issues and their resolution.

(3) Records of continuing review activities.

(4) Copies of all correspondence between the IRB and the investigators.

(5) A list of IRB members *as required by* Sec. 46.103(b)(3).

[From Sec. 46.103(b)(3)] A list of *the IRB* members identified by name; earned degrees; representative capacity; indications of experience such as board certifications, licenses, etc., sufficient to describe each member's chief anticipated contributions to IRB deliberations; and any employment or other relationship between each member and the institution; for example: full-time employee, part-time employee, member of governing panel or board, stockholder, paid or unpaid consultant. *Changes in IRB membership shall be reported to the secretary.*

(6) Written procedures for the IRB as required by Sec. 46.103(b)(4),

(7) Statements of significant new findings provided to subjects, as required by Sec. 46.116(b)(5).

(b) The records required by this regulation shall be retained for at least 3 years after completion of the research, and the records shall be accessible for inspection and copying by authorized representatives of the *Department* at reasonable times and in a reasonable manner.

(c) *The Food and Drug Administration may refuse to consider a clinical investigation in support of an application for a research or marketing permit if the institution or the IRB that reviewed the investigation refuses to allow an inspection under this section.*

APPENDIX 2.6

A Comparison of HHS and FDA Rules Governing Operation of Institutional Review Boards for "Expedited Reviews"

[Note: This text, comparing HHS and FDA rules, is adapted from information provided by Dr. John Petricciani of the FDA, and differences between the rules are noted in italics.]

FDA

Protection of Human Research Subjects: Clinical Investigations Which May Be Reviewed Through Expedited Review Procedure Set Forth in FDA Regulations

The agency concludes that research activities with human subjects involving no more than minimal risk and involving one or more of the following categories (carried out through standard methods), may be reviewed by an IRB through the expedited review procedure authorized in Sec. 56.110.

(1) Collection of hair and nail clippings in a non-disfiguring manner; of deciduous teeth; and of permanent teeth if patient care indicates a need for extraction.

(2) Collection of excreta and external secretions including sweat and uncannulated saliva; of placenta at delivery; and of amniotic fluid at the time of rupture of the membrane before or during labor.

(3) Recording of data from subjects who are 18 years of age or older using noninvasive procedures routinely employed in clinical practice. This category includes the use of physical sensors that are applied either to the surface of the body or at a distance and do not involve input of matter or significant amounts of energy into the subject or an invasion of the subject's privacy. It also includes such procedures as weighing, electrocardiography, electroencephalography, thermography, detection of naturally occurring radioactivity, diagnostic echography, and electroretinography. This category does not include exposure to electromagnetic radiation outside the visible range (for example, x-rays or microwaves).

(4) Collection of blood samples by venipuncture, in amounts not exceeding 450 milliliters in an eight-week period and no more often than two times per week, from subjects who are 18 years of age or older and who are in good health and not pregnant.

(5) Collection of both supra- and subgingival dental plaque and calculus, provided the procedure is not more invasive than routine prophylactic scaling of the teeth, and the process is accomplished in accordance with accepted prophylactic techniques.

(6) Voice recordings made for research purposes such as investigations of speech defects.

(7) Moderate exercise by healthy volunteers.

(8) The study of existing data, documents, records, pathological specimens, or diagnostic specimens.

HHS

Research Activities Which May Be Reviewed Through Expedited Review Procedures Set Forth in *HHS Regulations for Protection of Human Research Subjects*

Research activities involving no more than minimal risk and in which the only involvement of human subjects will be in one or more of the following categories (carried out through standard methods) may be reviewed by the Institutional Review Board through the expedited review procedure authorized in Sec. 46.110 of 45 CFR Part 46.

(1) Collection of: hair and nail clippings, in a nondisfiguring manner; deciduous teeth; and permanent teeth if patient care indicates a need for extraction.

(2) Collection of excreta and external secretions including sweat, uncannulated saliva, placenta removed at delivery, and amniotic fluid at the time of rupture of the membrane prior to or during labor.

(3) Recording of data from subjects 18 years of age or older using noninvasive procedures routinely employed in clinical practice. This includes the use of physical sensors that are applied either to the surface of the body or at a distance and do not involve input of matter or significant amounts of energy into the subject or an invasion of the subject's privacy. It also includes such procedures as weighing, *testing sensory acuity*, electrocardiography, electroencephalography, thermography, detection of naturally occurring radioactivity, diagnostic echography, and electroretinography. It does not include exposure to electromagnetic radiation outside the visible range (for example, x-rays, microwaves).

(4) Collection of blood samples by venipuncture, in amount not exceeding 450 milliliters in an eight-week period and no more often than two times per week, from subjects 18 years of age or older and who are in good health and not pregnant.

(5) Collection of both supra- and subgingival dental plaque and calculus, provided the procedure is not more invasive than routine prophylactic scaling of the teeth and the process is accomplished in accordance with accepted prophylactic techniques.

(6) Voice recordings made for research purposes such as investigations of speech defects.

(7) Moderate exercise by healthy volunteers.

(8) The study of existing data, documents, records, pathological specimens, or diagnostic specimens.

(9) *Research on individual or group behavior or characteristics of individuals, such as studies of perception, cognition, game theory, or test development,*

(9) Research on drugs or devices for which an investigational new drug exemption or an investigational device exemption is not required.

This list will be amended as appropriate and a current list will be published peridocially to the Federal Register.

where the investigator does not manipulate subjects' behavior and the research will not involve stress to subjects.

(10) Research on drugs or devices for which an investigational new drug exemption or an investigational device exemption is not required.

CHAPTER 3

Informed Consent

Although it is only one aspect of the wide field of protection in human subjects' research, informed consent is perhaps the best known feature. This is not surprising in view of the fact that it usually involves, in part, the construction of an informed consent document that has many of the characteristics of a legal contract. This document poses numerous problems for the researcher. For example, in any one particular research project, must you provide an informed consent form for the research subject? Must it be written? If so, what must it look like? Is either party legally liable if the consent procedures are not followed?

Few aspects of protection cause so much consternation and confusion for researchers and subjects alike. As we shall see in the next chapter, it is one of the least understood mechanisms within the protection process, and Institutional Review Boards continue to struggle with documentation of informed consent. In the author's experience as a member and later chairman of an IRB, informed consent was usually the area with which researchers were least familiar. Proposals to our IRB from a researcher often required reworking of their informed consent procedures before we could approve their proposals. Occasionally, this meant that researchers who were applying for federal funds had to revise their materials hastily, resubmit to the IRB, and obtain written IRB approval *before* sending their funding requests to an agency in Washington, D.C. Although goodwill and honest intentions on all sides can speed up this process, it is *not* an enviable position for a researcher to be in, especially if the researcher (as is usually the case) is competing with others for limited federal funds. Being late with your proposal is seldom an advantage.

As we shall see later in this chapter, confusion on the part of the subject in understanding a complex informed consent form also helps no one. Such confusion can breed distrust, leading to unwillingness on the part of potential research subjects to participate in worthwhile research. In the author's experience, it has not been unusual to listen to researchers who complain about potential subjects who refuse to cooperate or, once having started a project, drop out.

Whose fault is that? Because the very nature of ethical research involves *voluntary* participation, the possibility always exists that subjects will not participate, for one reason or another. The ethical and knowledgeable researcher must adequately describe all relevant aspects of a project to a potential subject and provide for maximum freedom to the subject within the necessary scientific constraints of the research.

Thus it can be seen that two basic processes must be at work for subject participation. The subject must be informed by the researcher, and the researcher must in turn obtain the consent of the subject. Both processes can take several forms in different circumstances, as we shall see.

History of Informed Consent

The development of federal regulations for informed consent in human subject research has closely paralleled the development of rules governing the operation of IRBs. For both HHS and FDA, publication of revised IRB rules have usually been accompanied, often within the same body of rules, by revised informed consent guidelines. Again, this is not surprising since one of the most important protections afforded to human research subjects is the nature of their informed consent, and IRBs closely examine these provisions before approving a research proposal.

Since the reader has already become familiar with the development of IRBs from the early 1970s through the early 1980s, some of the detail on development of informed consent will be dropped from this discussion, such as the past historical development of the ethical guidelines embodied by the Nuremberg Code and the Helsinki Code. The following highlights should provide an adequate picture of the development of informed consent elements and the surrounding procedures used to transmit those elements to a research subject.

1974

On May 30, 1974, rules by HEW were published in the *Federal Register* for protection of human subjects.[1] These rules were among the earliest actually made into regulations by HEW for protection of human subjects. Earlier, in late 1973, HEW had proposed in the *Federal Register*[2] to adapt material from their "DHEW Grants Administration Manual" about review committees (later to be

[1]*Federal Register*, Vol. 39, No. 105, Thursday, May 30, 1974, pp. 18914–18920.
[2]*Federal Register*, Vol. 38, October 9, 1973, p. 27882.

called IRBs) and make them into regulations. Following consideration by HEW of public comments to the rules proposed in 1973, the publication in 1974 of rules marked a significant beginning step in codifying agency regulations.

Essentially, the regulations of May, 1974, accomplished the following: (1) defined to whom these regulations pertained (in rather broad terms); (2) established what a review committee should consist of; (3) described how organizations could obtain *special* and *general* assurances from HEW to review their work; and (4) set down how informed consent should be obtained, using the original six elements of consent which have recently been increased to more than six, depending on the type of research involved. This information was contained in the grants policy statements of HEW from 1971 on, but the *Federal Register* publication made the information more public.

With regard to these four basic areas addressed in 1974, we have already seen in the previous chapter how IRBs have developed to their present state. In the next chapter, we shall discuss the special versus general assurance dichotomy in more detail. This leaves us with the question of to whom all these protection regulations apply, and the elements of informed consent.

It is worth repeating here that in 1974 and through the 1970s the protection regulations were often interpreted to apply to *any* projects within organizations where those organizations received *any* type of federal funding. The origins of this all-inclusive approach can be seen in the May, 1974, regulations regarding applicability[3]:

> (a) The regulations in this part are applicable to all Department of Health, Education, and Welfare grants and contracts supporting research, development, and related activities in which human subjects are involved.

The phrase "development, and related activities" covers a lot of territory! Contrast this earliest notion to the current regulations on applicability:[4]

> This subpart applies to all research involving human subjects conducted by the Department of Health and Human Services or funded in whole or in part by a Department grant, contract, cooperative agreement or fellowship.

Thus, the current rules make it clear that only research is covered by the protection rules—at least as far as the federal government is concerned—and the funds must directly support some phase of the actual research project.

Informed Consent in 1974. The fourth area originally addressed in 1974 was informed consent and is the most relevant for this chapter. The original six elements of consent, as published in 1974, are still the basic elements used today, but they have changed somewhat. If you find that these next six elements are equivalent to what you still consider to make up informed consent today,

[3]*Federal Register*, May 30, 1974, p. 18917.
[4]*Federal Register*, Vol. 46, No. 16, Monday, January 26, 1981, p. 8386.

then you will find the rest of this chapter illuminating. But before we list these six elements, just what was informed consent considered to be in 1974?

> (c) "Informed consent" means the knowing consent of an individual or his legally authorized representative, so situated as to be able to exercise free power of choice without undue inducement or any element of force, fraud, deceit, duress, or other form of constraint or coercion.[5]

Although this definition seemed straightforward enough at the time, the ensuing years saw an increasingly widespread debate between the government and the professional public on the nature of informed consent. Finally, in 1981, the current regulations dropped any definition of informed consent and simply listed the various elements. There is still cautionary language designed to *minimize* problems such as "undue inducement," but the absolute nature of former phraseology such as "free power of choice" and "without" (presumably, totally without) has been omitted. In the author's opinion, this has come about because of the difficulty of defining such absolute terms and the added problems it posed for befuddled IRBs, rather than any sentiment by government or public that forces such as "undue inducement" were ethically sound.

Elements of Consent in 1974. In general, the six elements of informed consent in use in 1974 still represent the overall nature of informed consent. But remember, these elements are now no longer correct, having been modified substantially by the final HHS and FDA regulations in January of 1981. Thus these following elements remain crucial to understanding the historical development of informed consent in the United States, but *do not use these six elements as they appear here to construct an informed consent agreement*:

> (1) A fair explanation of the procedures to be followed, and their purposes, including identification of any procedures which are experimental;
> (2) a description of any attendant discomforts and risks reasonably to be expected;
> (3) a description of any benefits reasonably to be expected;
> (4) a disclosure of any appropriate alternative procedures that might be advantageous for the subject;
> (5) an offer to answer any inquiries concerning the procedures; and
> (6) an instruction that the person is free to withdraw his consent and to discontinue participation in the project or activity at any time without prejudice to the subject.[6]

Although it would be interesting to discuss each of these six elements of consent in detail, we shall reserve that level of attention for later in this chapter when we examine the current elements. As we shall see, the current rules contain more than six elements and they give IRBs further powers to modify these elements as they see fit, powers that IRBs generally lacked in 1974.

Also codified in 1974 was the fact that there were (and still are today) two

[5]*Federal Register*, Vol. 39, No. 105, Thursday, May 30, 1974, p. 18917.
[6]Ibid.

main methods used to obtain consent: a written document, as we have heretofore portrayed consent agreement; and a *short form* wherein the consent elements are presented orally to the research subject and a brief document to that effect is signed by relevant parties. Again, this will be described in more detail when we examine the current informed consent rules.

1975

Now that the reader has a grasp on the conditions for informed consent in 1974, it must be said that it was really the rules of 1975 that dominated the field of consent for the 1970s. However, the rules published by HEW in March of 1975 actually consisted of the 1974 rules plus a few technical amendments. For example, it was the rules of 1975 that dropped the vague term *committees* and substituted for it the now familiar *Institutional Review Boards*. More than a mere name change, this action differentiated IRBs (bound by federal guidelines) from other ethics review boards operating at many institutions around the country.

Since this historical review emphasizes only regulatory highlights for informed consent, we now skip to 1978.

1978

On November 3, 1978, HEW published in the *Federal Register* a major new element to be added to the basic six elements of informed consent.[7] This seventh element dealt with informing subjects about the possibility of compensation and/or medical treatment if they suffer physical injury as a result of their participation in biomedical or behavioral research.

Work on this provision had actually been begun years earlier by a HEW task force formed shortly after their major protection regulations had come out in May, 1974. This task force was formed to develop a mechanism to compensate subjects who may be actually injured due to research participation. In January of 1977 the task force sent their report to the then-active National Commission for the Protection of Human Subjects of Biomedical and Behavioral Research. The National Commission, in turn, made its own recommendations in this area when it published its special report on IRBs in 1978 (see Chapter 2). It was then concluded by HEW to add this concern as a new separate informed consent element to make sure that subjects were informed of this aspect prior to participation.

Although the wording of this seventh element was modified along with other consent procedures in the HHS final rules of 1981, the following original

[7]*Federal Register*, Vol. 43, No. 214, Friday, November 3, 1978, p. 51559.

text of 1978 clearly shows the Department's intent to protect subjects—an intent which has not changed over the years:

> (7) With respect to biomedical or behavioral research which may result in physical injury, an explanation as to whether compensation and medical treatment is available if physical injury occurs and, if so, what it consists of or where further information may be obtained. This subparagraph will apply to research conducted abroad in collaboration with foreign governments or international organizations absent the explicit nonconcurrence of those governments or organizations.[8]

What was clear from this new element was the requirement to inform subjects about the likelihood of compensation or treatment if physical injury resulted from their research participation. However, as is usual for regulations, just how the researcher was supposed to do this (i.e., what words could be used) was not clear.

In an unusual move, HEW soon provided concerned researchers with suggested wording that could be added to informed consent documents for this new element. Why HEW did this is not apparent, but it may have been because this new element was the first major change to the basic six elements of consent in years and thus some additional advice for researchers may have seemed requisite. Certainly any suggested wording by HEW (now, HHS) for consent agreements is always welcome since, in the final analysis, it is that agency which must be satisfied with the consent documents constructed by researchers and approved by local IRBs.

The suggested wording formulated by HEW for this new seventh element was added to the *NIH (National Institutes of Health) Guide for Grants and Contracts* and sent to the 5,000 names contained on the NIH Office for Protection from Research Risks' mailing list:

> Suggested wording for consent forms: "I understand that in the event of physical injury resulting from the research procedures, (state what is available, e.g., 'medical treatment for injuries or illness is available'/'only acute/immediate/essential medical treatment (including hospitalization) is available'/'monetary compensation is available for wages lost because of injury'; or what is not available, e.g., 'financial compensation is not available, but medical treatment is provided free of charge,' etc.)."[9]

This phrasing may seem a little complicated at first, but that is because HEW needed to provide various options for individual researchers to adapt to their situation. Therefore, the phrasing contains several options (divided by slash marks) from which the researcher can choose to best fit a particular study.

The very fact that it was unusual—and still is—for a federal agency to provide suggested wording underscores an important aspect of protection regula-

[8]Ibid.

[9]*NIH (National Institutes of Health) Guide for Grants and Contracts*, Vol. 7, No. 19, December 15, 1978, pp. 33–34. Available from Office for Protection from Research Risks, National Institutes of Health, Bethesda, Maryland 20014.

tions. Almost never will a regulation contain the actual wording to be used. In our topic of protection of human research subjects, the researcher must usually: (1) obtain advice from local IRB members; (2) talk to other researchers who are experienced in subject protection issues; (3) obtain copies of research proposals previously approved by an IRB (e.g., to copy the relevant wording of consent documents); (4) examine the professional literature for examples to imitate; or (5) consult with a knowledgeable attorney.

One of the advantages of this book, we trust, is the value to be gained by examination of actual examples. With regard to informed consent, current actual wording is presented in forthcoming sections of this chapter and related examples appear in the next chapter.

Informed Consent Today

Before we examine the current consent procedures more closely, a special note: If there is any one major modification in the consent regulations published in 1981, it is that there now exist several reasons why a researcher need *not* obtain consent from a research subject—at least, that is, so far as the federal government is concerned. A researcher may still wish to obtain consent absent government decree (e.g., in accordance with professional ethical guidelines or to forestall future litigation with a clearly written document bearing the subject's signature). Obtaining consent even when not required to do so by government regulation is a choice left up to the individual researcher.

Although an entire book could be devoted to the ins and outs of informed consent, as a practical matter our primary examination of consent lies in this single chapter. Accordingly, the reader will want to investigate the full text of the regulations (appended to this chapter) for all the possible reasons why consent may not be required. Our discussion *within* this chapter will be limited to the highlighted exceptions.

This section on current consent regulations will deal with three primary topics: the basic elements of consent (now expanded beyond the seven required in 1978); a brief discussion of who must sign the consent; and a section on the two primary consent methods, the regular and the short form.

One final note before we proceed: Until noted otherwise later in this chapter, our discussion of informed consent will deal only with HHS (Department of Health and Human Services) rules on consent. Indeed, it is the HHS rules that govern the vast majority of biomedical, behavioral, social, psychological, and educational research in this country. As we noted previously, however, the rules by FDA (Food and Drug Administration) do concern certain other types of research. Consent rules for research regulated by the FDA will be discussed specifically later in this chapter.

Elements of Consent

We have already seen what elements used to be required in consent. The attentive reader will discern little change in some of the current elements, more in others as we now view the elements that were part of the final HHS rules published on January 26, 1981:[10]

> (1) A statement that the study involves research, an explanation of the purposes of the research and the expected duration of the subject's participation, a description of the procedures to be followed, and identification of any procedures which are experimental;
>
> (2) A description of any reasonably foreseeable risks or discomforts to the subject;
>
> (3) A description of any benefits to the subject or to others which may reasonably be expected from the research;
>
> (4) A disclosure of appropriate alternative procedures or courses of treatment, if any, that might be advantageous to the subject;
>
> (5) A statement describing the extent, if any, to which confidentiality of records identifying the subject will be maintained;
>
> (6) For research involving more than minimal risk, an explanation as to whether any compensation and an explanation as to whether any medical treatments are available if injury occurs and, if so, what they consist of, or where further information may be obtained;
>
> (7) An explanation of whom to contact for answers to pertinent questions about the research and research subject's rights, and whom to contact in the event of a research-related injury to the subject; and
>
> (8) A statement that participation is voluntary, refusal to participate will involve no penalty or loss of benefits to which the subject is otherwise entitled, and the subject may discontinue participation at any time without penalty or loss of benefits to which the subject is otherwise entitled.

These then are the new, current basic elements of informed consent. One interesting change from the 1978 element on physical injury was that the new wording of 1981 requires this element only for research where there is more than "minimal risk" to the subject, rather than in all informed consent documents.

Minimal Risk. We have seen previously that minimal risk is not a very serious problem for much research. For research involving questionnaires, interviews, records, or even direct observation of subjects' behavior (as in behavioral psychology, social psychology, or sociological research), minimal risk usually will not be exceeded.

The advantage of this 1981 change of the injury compensation element for researchers is clear. No longer must the investigator include an element that, at the least, raises subjects' eyebrows when they start reading about physical injury and medical treatment. In the author's experience with observational behavior research, it was not unusual for a subject, upon reading this element of consent, to ask whether we had forgotten to describe something. I mean, how is physical

[10]*Federal Register*, January 26, 1981, p. 8390.

injury possible here, huh? Fortunately, we can now drop this element if the research poses no more than minimal risk.

From the subject's viewpoint, this means less text to read in the consent document and the removal of unnecessary worry about a nonexistent danger. The regulations are quite clear, however, in insisting that the injury compensation element must be added if the possibility of injury does exist.

This one change is a clear illustration of how the 1981 regulation modifications benefited both subject and researcher.

Additional Elements of Consent. One of the problems that plagued IRBs during the 1970s was the limited and specific scope of the basic consent elements originally established by HEW. The limited scope forced IRBs to request that individual researchers consider adding items to their consent documents even though that item was not listed anywhere in HEW's published regulations. The author can recall such discussions as a member of an IRB when advice was given to researchers to add elements to their consent form. These additional elements were either better to explain various aspects of the study to the subject in certain cases or to protect the researcher by providing a comprehensive consent document that would always testify to the responsibilities of both subject and researcher.

However, an IRB is placed in an awkward position in so advising a researcher when no regulations appear to warrant such additional elements. It would not be totally illogical for the researcher to conclude that he or she was being arbitrarily singled out by an overly legalistic IRB to do more work, when in fact the IRB was trying to help.

To help alleviate this and related problems, HHS in January of 1981 provided for rather broad discretionary powers for IRBs in choosing additional elements. In the regulations, these provisions by HHS appear in the sections on informed consent (46.116—General Requirements for Informed Consent) rather than in the sections on IRBs.[11] The reader may examine these possible additional six elements in the full text as appended to this chapter. As an example, however, one of these elements is listed here:[12]

> (5) A statement that significant new findings developed during the course of the research which may relate to the subject's willingness to continue participation will be provided to the subject.

Obviously, such an element would seem most reasonable in research highly experimental such as clinical trials of experimental therapies where accumulating data may point to one therapy as clearly superior (or as very risky), which might well alter the subject's willingness to continue participating. Nevertheless, additional elements such as these appeared in HHS's 1981 regulations as one

[11]*Federal Register*, January 26, 1981, p. 8389–8390.
[12]*Federal Register*, January 26, 1981, p. 8390.

way of responding to the public comments that were received regarding special cases and the need for other consent elements.

Exceptions and Waivers of Informed Consent. In addition to presenting additional elements which may be added to informed consent, HHS also listed various situations in which an IRB now may alter or waive some or all of these consent elements. These new regulations are perhaps among the most significant of the 1981 protection regulations, for they specifically list situations in which researchers can alter the basic consent elements or even drop the consent form entirely. As has been noted before, however, for personal, ethical, or professional reasons, the researcher may still wish to employ the standard consent procedures.

It must be emphasized that nowhere in these regulations are there provisions for individual researchers to decide on their own whether or not to employ consent procedures and, if so, just which elements to use. It is always the IRB that is permitted to make such decisions, and then to advise the researcher accordingly. But we did learn in the previous chapter that there are various types of research that need never go through an IRB in the first place; and the choice of doing so can be made by the knowledgeable researcher familiar with the relevant regulations. Alternatively, the researcher may simply contact the local IRB about any proposed research to learn if it must be submitted for IRB review and, if so, what kind of informed consent might be necessary.

Although the full text of the reasons for altering or waiving entirely the consent elements is appended to this chapter,[13] these exceptions are new and important enough to highlight here:

> (c) An IRB may approve a consent procedure which does not include, or which alters, some or all of the elements of informed consent set forth above, or waive the requirement to obtain informed consent provided the IRB finds and documents that:
>
> (1) The research is to be conducted for the purpose of demonstrating or evaluating: (i) Federal, state, or local benefit or service programs which are not themselves research programs, (ii) procedures for obtaining benefits or services under these programs, or (iii) possible changes in or alternatives to these programs or procedures; and
>
> (2) The research could not practicably be carried out without the waiver or alteration.
>
> (d) An IRB may approve a consent procedure which does not include, or which alters, some or all of the elements of informed consent set forth above, or waive the requirements to obtain informed consent provided the IRB finds and documents that:
>
> (1) The research involves no more than minimal risk to the subjects;
>
> (2) The waiver or alteration will not adversely affect the rights and welfare of the subjects;
>
> (3) The research could not practicably be carried out without the waiver or alteration; and
>
> (4) Whenever appropriate, the subjects will be provided with additional pertinent information after participation.

---

[13]Ibid.

What, then, do these regulations on waivers mean? Essentially, there were two types of waivers in the regulations. First, an IRB can alter or waive consent requirements for research with large-scale government projects where data-gathering (e.g., for the census, or for welfare programs) may involve large numbers of citizens. The second type of waiver is more relevant to nongovernment studies in which not only is there no more than minimal risk present, but the researcher could not do the research without the alteration or waiver of consent requirements.

The second type of waiver is the one most researchers might examine, but the exceptions cannot be granted by an IRB simply because obtaining consent is tedious or burdensome. Only if "the research could not practicably be carried out"[14] may the IRB grant exceptions.

Remember also that *all* conditions for each situation must be met before an exception or waiver can be granted by an IRB. That is, for any given regulation, unless you see the word *or* between subcomponents of a rule, be assured that all subcomponents must be followed.

Who Must Sign the Informed Consent Form

As was true for previous versions of the regulations, it is still true that the consent document must be signed by the subject or the subject's "legally authorized representative."[15] Furthermore, the consent must be "legally effective." To be "legally effective" implies a variety of conditions, including: the mental state of the signer (not under duress or undue inducement); the signer's mental capacity to understand the information (as may be questioned with a mentally retarded individual); the readability of the document itself (i.e., is the language understandable?); and other factors. Some of these aspects of the phrase "legally effective" are discussed later in this chapter, while other features—particularly those most pertinent to special populations (e.g., the institutionalized mentally disabled)—will be discussed in subsequent chapters.

Assuming that we can obtain "legally effective consent," let us see who the "legally authorized representative" is who may sign in the subject's place if necessary;[16]

> (d) "Legally authorized representative" means an individual or judicial or other body authorized under applicable law to consent on behalf of a prospective subject to the subject's participation in the procedure(s) involved in the research.

Although one cannot blame those who worded this definition, it does not appear to offer a great deal of assistance to the researcher. The definition is

[14]Ibid.
[15]Ibid.
[16]*Federal Register*, Vol. 46, No. 16, Monday, January 26, 1981, p. 8387.

changed only slightly from the wording used in 1975.[17] According to HHS in the January, 1981, *Federal Register*,[18] there simply was not sufficient public comment on this definition to warrant any changes.

How much of a problem this is depends almost solely on the nature of the subject population. Since there are so many possible types of populations (e.g., divided by age categories, race, sex, educational achievement, income, I.Q.), it is little wonder that HHS failed to do other than recommend that the representative should be whoever is legally empowered to sign for the subject. An admittedly cynical paraphrasing of this definition might be "the legally authorized representative is whoever is legally authorized to represent the subject."

Again, it must be noted that HHS really had little choice. To understand just how complicated a problem this can be, let us take an apparently easy example: that of a male college freshman, aged 19, who participates in a male professor's research project at school. Not an unusual case and one in which the student is probably earning class credit (in place of having to write an essay paper) for participating. And let us make the research innocuous so there is no more than minimal risk. Say, for example, that the research examines response time and the student's task is to press a lever when a certain light is lit on a board in front of him in a comfortable, ventilated, well-lit laboratory room. The task takes half an hour, and our slightly bored student leaves, mildly pleased now that he does not have to write that essay paper.

Further, let us assume that the university's IRB judged that informed consent should be obtained by the professor. But is the professor safe in simply asking the student to sign the consent? At 19, the student is unlikely to be considered a minor with respect to some parts of the law, but not others. For example, he may be old enough to be considered an adult if he were to be tried in court for a felony offense and thus faces far more serious punishment than a minor. He may or may not be old enough to purchase beer, but is probably considered a minor with regard to being able to purchase liquor. Depending on the size of a purchase, he may not be able to sign a time-payment contract unless someone clearly not a minor signs conjointly and shares liability for nonpayment. He may or may not be considered of sufficient age to consent to elective major medical treatment (e.g., surgery) without a guardian's signature.

The words "may or may not" are used so frequently here because these rules about age and legal consent *vary from state to state*. Further, the consent and rights that are available to anyone at a certain age change from time to time within each state. Most commonly covered in public media are changes in the legal ages for purchasing liquor or obtaining a driver's license.

[17]*Federal Register*, Vol. 40, No. 50, Thursday, March 13, 1975, p. 11855.
[18]P. 8372.

Where, then, does this confusion leave the researcher? With three possible courses of action: (1) he can seek the advice of the local IRB or an attorney; (2) he can always obtain the signature of a parent or guardian, along with the subject's signature, whenever the subject is under an age at which reasonable doubt exists regarding the subject's ability fully to understand and consent (e.g., anyone 16 or younger); or (3) he can use his own judgment for each individual project, taking into account the overall ability of the subject to consent. He might change his own rule as to who should sign if the particular research project was more difficult or posed more risk.

In essence, a researcher should ask himself the following question: "Can the subject fully understand and consent to this research without someone else representing the subject also agreeing *and* would most persons agree with the decision I am about to make?" If the researcher can answer yes to both parts of the question, then it is unlikely that anyone else need sign the consent other than the subject. If the answer is no, then further steps must be taken and some other responsible person may be required by the IRB to be involved in the consent process and to act on behalf of the subject.

A final note of caution on this topic is warranted for situations in which parental rights have been severed by judicial action, or legal guardianship of a subject is in question. For example, the author has conducted projects in which behavioral and attitudinal data was obtained about juveniles as one way of evaluating the effectiveness of delinquency treatment programs. It was not uncommon for doubt to exist, because of unavailability of records or inaccuracies in various documents, regarding just who (i.e., parent, relative, an agency, the court) was the legal guardian of a youth. In such cases, it was deemed advisable to obtain consent or agreement from every party possible. While occasionally quite frustrating and time-consuming, such a conservative approach avoids misunderstanding later if various parties find that their memories of verbal agreements differ. Agreements which are entirely verbal do not seem prudent in most such situations.

IRB Waiver of Signature. Now that we understand how important, and confusing, it is to obtain the relevant signatures on the consent document, let us look at an exception. As the regulations were being proposed and modified during the late 1970s and early 1980s, HEW and then HHS continued to receive considerable public comment about the difficulties, in some circumstances, of getting signatures on consent documents. Earlier in this chapter, we viewed the 1981 HHS rules that granted IRBs the authority to modify some or all of the consent elements, or to waive consent requirements entirely. In a related section, HHS granted IRBs the authority to waive the requirement for *signed* consent, given that some kind of information will probably be given to subjects.

In the author's opinion, this additional provision is so close in intent to the previous regulations that perhaps all could have been published in the same section to help our understanding. Nevertheless, this is what the exception to needing *signed* consent looks like:[19]

> (c) An IRB may waive the requirement for the investigator to obtain a signed consent form for some or all subjects if it finds either:
>
> (1) That the only record linking the subject and the research would be the consent document and the principal risk would be potential harm resulting from a breach of confidentiality. Each subject will be asked whether the subject wants documentation linking the subject with the research, and the subject's wishes will govern; or
>
> (2) That the research presents no more than minimal risk of harm to subjects and involves no procedures for which written consent is normally required outside of the research context.
>
> In cases where the documentation requirement is waived, the IRB may require the investigator to provide subjects with a written statement regarding the research.

Although the language of this rule is self-explanatory, it should be remembered that what is being discussed is an *option* that an IRB *may* wish to exercise—most commonly only at the request of a researcher. The rule does *not* imply that researchers typically must ask each subject whether he or she is worried about documentation with the subject's name on it (i.e., the signed consent form). Rather, that is a step which an IRB may require a researcher to use if the researcher wants to eliminate signed consent for some or all subjects.

In the author's experience, it is usually easier simply to obtain signed consent forms. As noted previously, however, there are some types of research (especially large-scale surveys using mailed questionnaires or phone interviews) where signed forms may pose a burden. Previous sections of this volume dealing with IRBs and exemptions dealt with these and related issues in detail.

Two Types of Written Consent

Given that an investigator is going to solicit written consent, there are essentially two approaches to be taken, both of which involve the basic elements (and exceptions) examined previously. Although the most common approach has no official label, we shall refer to it as the *long form* both here and in the examples presented later in this chapter. The less common form is labelled in the regulations as the *short form*.

The Long Form Consent. First, let us note a common mistake in that the difference between the *long* and the *short* form is *not* that the long form must be written whereas the short form is simply read aloud to the subject. In reality, *both* must be written, and both may be read aloud to the subject. The differences lie in the nature of the document signed by the subject, not what the subject reads

[19]*Federal Register,* Vol. 46, No. 16, Monday, January 26, 1981, p. 8390.

or listens to. The long form contains all the elements of consent, may be read aloud, and is signed by the subject (or his representative). Thus, the signature goes on a document that lists all the elements of consent that are relevant for a particular research project. The rule for the long form is as follows:[20]

> (1) A written consent document that embodies the elements of informed consent required by Sec. 46.116. This form may be read to the subject or the subject's legally authorized representative, but in any event, the investigator shall give either the subject or the representative adequate opportunity to read it before it is signed.

Certainly, reading the consent form aloud may assist in the subject's understanding the consent form; but if understanding the information is enough of a problem to require reading aloud, there are other measures the investigator should consider. For example, if subjects appear to have difficulty understanding the form, the words used (perhaps too much research jargon) or the sentence structure (long and involved) may be altered to better suit the reading level of the subject. This topic is addressed later in this chapter when we look at a sample of problems typical of informed consent.

The Short Form Consent. The alternative short form of written consent has both advantages and disadvantages from the researcher's viewpoint. The primary advantage appears to be that the form actually signed by the subject (or representative) is short and presumably looks less legalistic or daunting than a normal consent document. Still, the subject must be given a copy of a summary (reviewed and approved by an IRB) of what the researcher says aloud about consent, so the short form itself is not the only document the subject sees or receives.

Another less obvious benefit—for both researcher and subject—is, in the author's experience, the tendency for researchers to use more common, everyday language when speaking to subjects than they use when constructing a written form. This can ease the worries of a subject who may feel more comfortable listening (and perhaps asking questions) rather than reading an imposing form that looks like a contract.

The disadvantage of the short form for the researcher is the need for the presence of a witness, and the signature of this witness in addition to the subject's signature.

The rule is as follows:[21]

> (2) A "short form" written consent document stating that the elements of informed consent required by Sec. 46.116 have been presented orally to the subject or the subject's legally authorized representative. When this method is used, there shall be a witness to the oral presentation. Also, the IRB shall approve a written summary of what is to be said to the subject or the representative. Only the short form itself is to be signed by the subject or the representative. However, the witness shall sign both the

[20]Ibid.

[21]Ibid.

short form and a copy of the summary, and the person actually obtaining consent shall sign a copy of the summary. A copy of the summary shall be given to the subject or the representative, in addition to a copy of the "short form."

Example of an Informed Consent Form

We now are ready to examine an example of informed consent. The example we will consider is the informed consent document used by the author for a study called "Looking at Life" conducted at Boys Town in 1978. This study was approved by the IRB at Boys Town and the IRB was an officially approved IRB on the HEW listing. The procedures followed by the IRB in approving this consent document and the accompanying evaluation project were the typical procedures used by IRBs around the United States in 1978.

More details on these procedures are contained in the next chapter where we will follow the "Looking at Life" proposal through the steps that eventually led to IRB approval. Thus the next chapter will serve as a suggested blueprint for investigators who desire to submit a proposal to their own IRB. The chapter will also address the question of how to form an IRB at an institution if one does not currently exist, how to get it approved by HHS, and then how to submit a proposal to the IRB and get it approved.

The "Looking at Life" Study

To understand better the consent example, a brief explanation of the study is necessary. Briefly, the project was designed originally to assess the outcome of previous Boys Town residential programs by tracking down a sample of former youths who had left Boys Town at least five years prior to 1978 and no more than 20 years prior. After being located, they would be asked in a questionnaire to answer a variety of questions about their life after Boys Town in an attempt to ascertain the relative strength of residential programs as measured by differential rates of success of youths who left at various times.

Further, the sample would be broken down into two more categories, depending on whether youths had originally left as graduates (i.e., graduated from Boys Town's high school) or for any other reason. This would provide clues regarding the relative value of a youth's remaining at Boys Town until high school graduation. The difficulties in conducting such a long-term follow-up study were numerous,[22–29] and the study was terminated before some data areas could be analyzed.

Nevertheless, the approach that was taken with regard to informed consent

[22]Maloney, D. M., Fixsen, D. L., and Maloney, K. B. *Systematic strategy for outcome evaluations.* Paper presented at the First Annual Meeting of the Evaluation Research Society, Washington, D.C., October, 1977.

was the same as would now be taken, with some exceptions. In order to understand these exceptions, and better to grasp facets of informed consent, let us examine the example of consent from "Looking at Life" as if it had been conducted following January of 1981 when HHS revised the rules.

Adequacy of Informed Consent. As we examine the forthcoming example, the reader should keep in mind two central questions as a test of how well informed consent is now understood. First, was a written long consent form even necessary? Remember that the project involved only a questionnaire to be completed by the subjects and an examination by the author's staff of the institutional files kept about subject progress when the subject had been a resident of Boys Town. No experimental treatment of any kind was involved. Hint: the reader may wish to study the actual HHS rules presented in the previous chapter regarding what types of research activities are covered by HHS rules.

Second, if the reader were to conduct a similar study today, would the example consent form (assuming the subject's written consent would be sought) be sufficient under current regulations? To aid such a decision, it would be advisable at this point for the reader to review the required elements of consent presented earlier in this chapter. Or, the actual text of the HHS rules on informed consent at the end of this chapter could be examined.

Consent Form Approved by IRB. The following text is reproduced as approved by the IRB established at Boys Town under HHS regulations.

---

[23]Collins, L. B., Warfel, D. J., Maloney, D. M., and Maloney, K. B. *The "Looking at Life" project: Finding former clients for follow-up evaluations.* Paper presented at The Neglect of Youth Conference, Boys Town, Nebraska, December, 1977.

[24]Warfel, D. J., Collins, L. B., Maloney, D. M., and Maloney, K. B. *The "Looking at Life" project: Achieving maximum response to a follow-up questionnaire.* Paper presented at The Neglect of Youth Conference, Boys Town, Nebraska, December, 1977.

[25]Maloney, D. M., Collins, L. B., Warfel, D. J., and Maloney, K. B. *The "Looking at Life" project: A management system for follow-up evaluations.* Paper presented at the Fourth Annual Western Regional Conference on Humanistic Approaches in Behavior Modification, Las Vegas, March, 1978.

[26]Collins, L. B., Warfel, D. J., Maloney, D. M., Maloney, K. B., and Fixsen, D. L. *Outcome evaluations: Finding the hard-to-locate client of child-care programs after release.* Paper presented at the Fourth Annual Meeting of the Midwestern Association of Behavior Analysis, Chicago, May, 1978.

[27]Collins, L. B., Warfel, D. J., Maloney, D. M., and Maloney, K. B. *Collection of outcome data from child-care programs: Materials and techniques.* Paper presented at the First Annual Meeting of the National Teaching-Family Association, Boys Town, Nebraska, October, 1978.

[28]Warfel, D. J., Collins, L. B., Maloney, D. M., and Maloney, D. B. *Maximizing follow-up questionnaire returns from clients of child care programs.* Paper presented at the 12th Annual Convention of the Association for the Advancement of Behavior Therapy, Chicago, November, 1978.

[29]Collins, L. B., Warfel, D. J., Maloney, D. M., and Maloney, K. B. *Outcome evaluation of behavior change programs: Relocating former clients.* Paper presented at the 12th Annual Convention of the Association for the Advancement of Behavior Therapy, Chicago, November, 1978.

* * *

Consent to Participate in Boys Town
"Looking at Life" Study Series

This form represents the voluntary consent of ______________________
(First Name)

______________________ to participate in the projects known as the
(Middle Initial) (Last Name)
Boys Town "Looking at Life" study series.

PURPOSE: I understand that the "Looking at Life" studies are performed to examine certain events in the lives of former residents of Boys Town.

PROCEDURES: I understand that my consent allows the collection, analysis, and dissemination of three main areas of information. These areas include certain information about my life *before* I entered Boys Town, *during* my stay at Boys Town, and *after* I left Boys Town.

BEFORE I came to Boys Town: I understand that this part of the study series includes information about me or my family such as: background (e.g., places lived, number of brothers and sisters); educational background (e.g., grade point averages, attendance); vocational and occupational experiences (e.g., number and types of employment); legal histories (e.g., police and court contacts); and also any related information obtained from my files at Boys Town.

DURING my stay at Boys Town: I understand that this part of the study series includes the same kinds of information as described in the previous paragraph but covering the time when I was living at Boys Town.

AFTER I left Boys Town: I understand that this part of the study series includes the same kinds of information as described in the previous two paragraphs but covering the time since I've left Boys Town. I further understand that some information for this area will require that I answer a "Looking at Life" survey, either by telephone conversation or by answering a written questionnaire.

DISCOMFORT OR RISKS: I understand that there is no physical risk or discomfort involved. I understand that the potential social or psychological risk I may experience is that my name may become publicly associated with particular facts about my life. I understand that I am protected from this risk by safeguards described in the PRIVACY PROTECTION section of this form.

PRIVACY PROTECTION: I understand that the Study Director for "Looking at Life" protects my privacy by keeping information about me in locked files. I understand that publication of study results in any form will protect my privacy and disguise me by adding together information about me along with information about others and by using some name or code number instead of my real name.

BENEFITS: I understand that the potential benefits of the studies are that they will help future boys experience programs that are most likely to help them after they leave Boys Town. It also will permit distribution of Boys Town methods around the world.

ALTERNATIVES: I understand that the alternative to my participation in the study series is to not participate.

RIGHT TO WITHDRAW: I understand that if I wish to withdraw my consent later I may freely do so without prejudice to me even after I sign this consent form. I agree that such a withdrawal will be made in writing to the Study Director for "Looking at Life."

RIGHT TO INQUIRY: I understand that I have the right to inquire at any time about the procedures described in this document. I understand that I can direct these inquiries to the Study Director.

I understand that my signature below signifies my voluntary informed consent to participate in the "Looking at Life" study series.

____________________________ ________________

(Participant signature) (Date)

For Participant under age 21, signature of guardian must also be present:

____________________________ ________________

(Guardian) (Date)

Inquiries may be sent to:

Dennis M. Maloney, Ph.D.
Study Director, "Looking at Life"
Boys Town, Nebraska 68010

* * *

Was Written Consent Necessary?

The first question to be answered as part of our test was whether the written consent would still be required if "Looking at Life" were conducted today. At this point, the knowledgeable reader realizes that this crucial question, facing all researchers at some time during the planning of a project, really consists of four questions. First, does the activity (i.e., the "Looking at Life" study in this case) come under HHS rules on protection of human subjects? Second, if it does, is informed consent required? Third, if so, must it be written? Fourth, if it must be written, what form should it take (i.e., regular long form or the short form)? These questions are the ones the author recommends to researchers when they consider what steps to take in protecting subjects' rights and in preparing a proposal for IRB review. Once again using the "Looking at Life" example, let us examine these four questions.

HHS Authority. First, does the activity come under HHS rules? Probably not. Remember, now research must be funded in whole or in part with HHS money for the rules to apply. The author received no federal funds for the "Looking at Life" project. With the 1981 rules, it no longer matters that other segments of the sponsoring institution (i.e., Boys Town) might receive HHS funds for some other projects. Further, it could be argued that the activity was not research anyway and thus required no informed consent. Although the project contained some aspects of research (e.g., randomly selected subjects, survey items from national-referenced questionnaires, computer analysis of results), it did not contain others (e.g., a control group). In essence, the project was more a straightforward evaluation of past Boys Town programs rather than a research project. The debate within professional circles over the difference between evaluation and research continues and is beyond the purview of this book. However, it is still possible that the project could be considered by some to be research, particularly in view of the current HHS definition of research:[30]

> (a) "Research" means a systematic investigation designed to develop or contribute to generalizable knowledge. Activities which meet this definition constitute "research" for purposes of these regulations, whether or not they are supported or funded under a program which is considered research for other purposes. For example, some "demonstration" and "service" programs may include research activities.

The conservative investigator would be wise to assume that an activity may be considered to be research by an IRB and proceed accordingly. In the author's opinion, it is still wise to use consent and adopt other protection procedures (i.e., name coding, locked files, etc.) to protect *both* researcher and subject. Such steps insure that adequate records are kept about any agreements between researcher and subject.

In our example, then, let us proceed as if HHS were funding it and the activity were research to follow the HHS rules to be ethically responsible to the subjects. If so, then perhaps it could still be exempt from all the rules requiring IRB review and informed consent because it was simply a survey? Not so. Although surveys funded by HHS can be exempted from the rules for various reasons (see previous chapter), "Looking at Life" would not have been exempted

[30]*Federal Register*, Vol. 46, No. 16, Monday, January 26, 1981, p. 8387.

because it met all the conditions of HHS that would require that HHS protection rules be applied:[31]

> (b) Research activities in which the only involvement of human subjects will be in one or more of the following categories are exempt . . .
>
> (3) Research involving survey or interview procedures, except where all of the following conditions exist: (i) responses are recorded in such a manner that the human subjects can be identified, directly or through identifiers linked to the subjects, (ii) the subject's responses, if they became known outside the research, could reasonably place the subject at risk of criminal or civil liability or be damaging to the subject's financial standing or employability, and (iii) the research deals with sensitive aspects of the subject's own behavior, such as illegal conduct, drug use, sexual behavior, or use of alcohol.

Because our survey questionnaire fit all the elements previously described, it would not be exempt from current HHS rules just because it was a survey.

"Looking at Life" Consent. Our second primary question to test ourselves on our knowledge so far was: if "Looking at Life" falls under HHS rules (and we are assuming it does for the reasons presented previously), then is informed consent required? In practice, the IRB would tell the researcher whether consent was required, on the basis of the description of the project given to them by the researcher. However, the conscientious researcher should anticipate the IRB's response ahead of time, if for no other reason than to avoid processing delays by submitting a consent document as part of the original proposal to the IRB.

As can be seen from the HHS informed consent rules appended to this chapter, there are situations in which an IRB may waive some *or all* of the elements of consent. The following exerpt deals with the only possible way an IRB might waive consent for our example project. We have seen this rule before, and our consent example shows its use. Does this wording allow or disallow the need to obtain informed consent for "Looking at Life"?[32]

> (d) An IRB may approve a consent procedure which does not include, or which alters, some or all of the elements of informed consent set forth above, or waive the requirements to obtain informed consent provided the IRB finds and documents that:
>
> (1) The research involves no more than minimal risk to the subjects;
>
> (2) The waiver or alteration will not adversely affect the rights and welfare of the subjects;
>
> (3) The research could not practicably be carried out without the waiver of alteration; and
>
> (4) Whenever appropriate, the subjects will be provided with additional pertinent information after participation.

Because "Looking at Life" could not fulfill all these stipulations, it is highly unlikely that an IRB would waive the requirement for consent. At a minimum, we know that the survey *could* be practicably carried out without a waiver, since

[31] *Federal Register*, January 26, 1981, p. 8386.
[32] *Federal Register*, January 26, 1981, p. 8390.

obtaining a signed, written consent involved only the inclusion of a blank consent form and a stamped return envelope with the questionnaire sent to the subjects. This increased the costs of the project (materials and personnel time) and added to the overall work of questionnaire construction (e.g., using different colored paper for the questionnaire and the consent, adding reminders about returning the consent to various parts of the questionnaire). But researchers should note that these additional steps did not make the study impractical, only more difficult to accomplish. Interest in informing the subject should outweigh mere inconvenience.

Documentation of Consent. For our third primary question, we now assume that consent is required but ask: must the consent be written? Again, an excerpt from the rules appended to this chapter will tell us:[33]

> (c) An IRB may waive the requirement for the investigator to obtain a signed consent form for some or all subjects if it finds either:
>
> (1) That the only record linking the subject and the research would be the consent document and the principal risk would be potential harm resulting from a breach of confidentiality. Each subject will be asked whether the subject wants documentation linking the subject with the research, and the subject's wishes will govern; or
>
> (2) That the research presents no more than minimal risk of harm to subjects and involves no procedures for which written consent is normally required outside of the research context.
>
> In cases where the documentation requirement is waived, the IRB may require the investigator to provide subjects with a written statement regarding the research.

This question presents a rather grey area to the investigator for projects like "Looking at Life." Our example consent would not fit the first waiver option because, in addition to the name on the consent form, identifiers appeared on the questionnaire completed by subjects. While names and other identifying information (e.g., address) were removed when completed questionnaires were returned to the author's office, a code number was entered on the answer sheets, indexed to the subject's identity through a master code-name sheet kept in a locked file.

It is not clear whether the second waiver option might apply, depending on one's judgment of whether sensitive answers on the sheets (e.g., list of arrest records) that *might* be somehow linked to the subject's real name and *might* be publicly known constituted more than minimal risk. If the reader will review the previously presented HHS definition of minimal risk,[34] it will be seen that our example survey probably does not pose more than minimal risk, so the second waiver option may have been applicable. That is, the normal risk associated with public events (e.g., that results from a test might become publicly identified with

[33]Ibid.
[34]*Federal Register*, January 26, 1981, p. 8387.

persons' names) is probably not exceeded here, depending on the sensitiveness of the questions.

Nevertheless, the author chose to use a written consent for two reasons: (1) better to explain the study to subjects who could not be met in person since they resided in various states; and (2) to allay fears of subjects who might well define risk of public knowledge of certain activities of theirs to be more than minimal risk. The Boys Town IRB agreed.

Type of Written Consent. The fourth main question for our example consent was: Given that the consent must be written, should it be the long form or the short form? Obviously, from the text of the example already examined, the short form was not used. As we noted before, implementation of the short form requires the oral reading of consent to the subject, the presence of a witness, and so forth. It was not practical or desirable to make such arrangements at long distance for each survey respondent. Further, because the age of respondents and their general IQ was not such that problems in comprehending consent would be expected, the regular written form was judged sufficient by the IRB.

Correct Consent Elements in Example

Now that the reasons have been presented on just why the consent was obtained in the described manner and approved by the IRB, let us return for a final time to the example consent itself. It may prove helpful at this point for the reader to review the basic eight elements of informed consent that were discussed earlier in this chapter.

After examining the consent form for the "Looking at Life" study, what element(s) were missing? Remember, that consent form was used in the late 1970s before the final HHS rules of January, 1981. The example consent may seem complete—and, indeed, it exceeded the requirements then—but is it? Although not appearing in the same order as contained in the rules, we find that all the elements were there, with the exception of a statement about compensation for physical injury and a related provision on whom to contact in the event of a research-related injury.

Other minor exceptions would be: (1) there was no statement made about research in the form, primarily because it could be argued that the project was closer to evaluation than research, and (2) the expected duration of the subject's participation is not spelled out, although the wording of the example's "Procedures" section seems to make it clear that participation will be limited to answering a phone interview or completing a questionnaire. As an added precaution against obtaining an invalid signature on the form, note that an added blank was situated on the form for the signature of a guardian for any subject under 21 years of age.

In summary, we can conclude that the example consent form was more than adequate for its time and, with the provision of a few more sentences, would fulfill current consent regulations. This process of studying the example also should have presented the reader with a scenario of the typical decisions about consent that must be made by an investigator who proposes to conduct research with human subjects. Because of the nature of the "Looking at Life" project, the rules of HHS would have applied, assuming HHS funds were involved.

Consent Rules for the Food and Drug Administration

As we noted before, the FDA published their final rules for protection of research subjects just one day (January 27, 1981) after HHS did so. The appendix for this chapter contains the exact text of the FDA regulations for informed consent. Also as we noted before, a side-by-side comparison of FDA and HHS rules has been provided by the federal government to assist investigators who may wonder which set of rules applies to their work or what the differences might be between the two sets of agency rules.

In keeping with the intent of this book to be a practical guide, our discussion of this section will be very brief. Interested readers may examine the complete side-by-side comparison of the two agencies' informed consent rules in the appendix to this chapter.

Although there are some administrative differences (e.g., FDA explicitly states that they may inspect the consent records maintained by the researcher), the primary difference is in exceptions or waivers. For example, the FDA rules do not permit alteration of some or all elements of informed consent in situations where the principal risk to the subject is simply a breach of confidentiality because FDA does not regulate such research.

More substantially, there is one section of the FDA rules that does not appear in the HHS rules because of the nature of FDA-regulated research. That is, while HHS permits alteration of consent only under certain situations described previously (e.g., where the research involves no more than minimal risk and could not practicably be conducted without alteration of consent requirements), FDA permits such alteration when the subject's life is in danger, as is relevant in the use of experimental drugs. All of the following must be true for alteration of consent procedures:[35]

> (1) The human subject is confronted by a life-threatening situation necessitating the use of the test article.
>
> (2) Informed consent cannot be obtained from the subject because of an inability to communicate with, or obtain legally effective consent from, the subject.
>
> (3) Time is not sufficient to obtain consent from the subject's legal representative.

[35] *Federal Register*, Vol. 46, No. 17, Tuesday, January 27, 1981, p. 8951.

(4) There is available no alternative method of approved or generally recognized therapy that provices an equal or greater likelihood of saving the life of the subject.

For additional details on these FDA rules on informed consent, see the full text of FDA rules appended to this chapter and the side-by-side comparison of HHS and FDA rules, also appended.

Informed Consent for Treatment

Although this book is designed to familiarize the reader with regulations regarding protection of human subjects in various types of research, mention must be made of treatment. Of course, it may not always be clear that the two activities are distinct and separate. Allusion is made to this even in actual regulations. It will be recalled that the HHS definition of research, presented previously, included the situation in which a project's primary activity might be service-oriented but research might still be involved. In such situations, informed consent must be sought for any subject involvement that could be considered research.

But what about the treatment itself? If the treatment were considered experimental, perhaps treatment providers would be wise to employ informed consent. If for no other reason, such a procedure can aid in the overall communication between care-giver and client, as well as produce written documentation of the agreements between the two parties. For the care-giver, such documentation can be useful later if there are ever allegations (especially from former clients) that promised treatment was not provided. At a minimum, the consent document should clearly show just what treatment was actually promised. For the subject, or client, the document may explain more about the treatment than would otherwise be provided. At a minimum, reference to the form later may remind a client on service aspects he or she had forgotten since first entering treatment.

This is, of course, a broader concept of consent than is typically referred to in research literature. It certainly is not contained in the federal regulations. Nevertheless, it may be useful for treatment providers to adopt such a procedure. This approach for consent is discussed in some detail by R. K. Schwitzgebel (1975).[36]

A Contract for Consent

Schwitzgebel contends that both parties in a treatment setting (client and service provider) would be better protected and informed if explicit contracts were drawn up prior to or at the time the client begins receiving treatment or

[36]Schwitzgebel, R. K. A contractual model for the protection of the rights of institutionalized mental patients. *American Psychologist*, 1975, August, 815–820.

services. An essential difference with this approach is that there would be more elaboration of the expectations, rewards, and potential problems for both parties than is true of a research consent agreement. For these would be contracts that would assume the closeness of a physician–patient relationship for the two parties, contracts that make explicit the goals, procedures, risks, and benefits of *treatment* rather than research. One could also expect that such a contract might necessarily be more lengthy and contain more items, because of the proposed time span of treatment and the possible events that could occur. In contrast, research projects often are single events that may not last as long and may involve fewer interactions between the researcher and the individual subject, depending on the type of research.

A similar recommendation was made by Robert Griffith,[37] who noted that consent is appropriate for both research and treatment settings. This is especially true if there are any aspects of the treatment environment or program that place clients at risk. It has been stated by an attorney that all interventions place clients at some kind of risk.[38] It may well be that the only way to guarantee that a client will never be at risk is never to administer any kind of treatment and revert to the nontherapeutic, purely custodial care more common to the nineteenth century. Rather than condemning so many people to such a fate, perhaps treatment consent should be more carefully considered by service organizations and agencies. In the medical field, treatment often requires such formal consent procedures (e.g., for surgery, hospitalization, vaccinations).

The National Teaching-Family Association

One such organization that endorses the use of consent for treatment in addition to consent for research participation is the National Teaching-Family Association.[39] The National Teaching-Family Association (NaTFA) was formed in the late 1970s by agencies and individuals (psychologists, social workers, etc.) who use the Teaching-Family Model[40] of residential care and treatment for certain client groups, primarily delinquent or emotionally disturbed adolescents. From 1967 to 1981, the model grew in use from one group home in Kansas to over 170 such homes across the country.[41] This unusual replication of a social

[37]Griffith, R. G. An adminstrative perspective on guidelines for behavior modification: The creation of a legally safe environment. *The Behavior Therapist*, 1980, 3 (1), 5–7.

[38]Martin, R. The right to receive or refuse treatment. *The Behavior Therapist*, 1980, 3 (1), 8.

[39]Maloney, D. M. *The National Teaching-Family Association* (3rd ed.). Boys Town, Nebraska: Father Flanagan's Boys' Home, January, 1982.

[40]Fixsen, D. L., Phillips, E. L., and Wolf, M. M. The Teaching-Family Model: An example of mission-oriented research. In A. C. Catania and T. A. Brigham (Eds.), *Handbook of behavior analysis: Social and instructional processes*. New York: Halsted Press, 1978.

[41]Collins, S. R., Maloney, D. M., and Collins, L. B. *1981 directory of the National Teaching-Family Association*. Boys Town, Nebraska: Father Flanagan's Boys' Home, June, 1981.

service model has come about through years of research and technology transfer to various states by the original developers of the model.[42]

Over 400 articles, manuals, textbook chapters, conference papers, films, and related items have been developed about the model[43] and further discussion of the model is beyond the scope of this book. What is most relevant for our purposes is that model proponents have endorsed the need for *consent for treatment* for residents of the homes in addition to the more typical consent for research participation. Participation of residents in research (usually observation of youth behavior or completion of written questionnaires) is voluntary so consent for research participation may or may not be given by residents. However, most Teaching-Family programs require that consent for treatment be signed, or the client (usually a youth) cannot be admitted to the program. This protects the youth who thus learns what he or she can expect from the residential program and what his or her rights are. It protects the program in that no youth unwilling to learn new skills is forced into entering residential placement.

Members of NaTFA felt strongly enough to insist that youths be informed of their rights and obligations if admitted for placement and explicit instructions for treatment consent were included in the *Standards of Ethical Conduct of the National Teaching-Family Association* (1981).[44] These standards are required reading for all NaTFA members (members so affirm when they sign membership applications) and must be followed by all relevant child care employees of any agency using the model, whether or not the staff are also NaTFA members. An example of these provisions is as follows:[45]

> 510. In addition to including the elements described in the above standards, members provide the following information in seeking informed consent for *treatment*: eligibility criteria for the program; the selection process (including selection committee make-up and function); description of each treatment element; probable length of stay (provide range); description of the training, consultation, and evaluation activities that will be conducted relative to the performance of the treatment providers.

While illustrative as an example, it should be noted that NaTFA is not unique in recommending consent for treatment to its members. Professionals in the human services field are advised to examine the ethical standards for research or treatment guidelines of their own associations for similar information. For example, as of April, 1981, the American Psychological Association was engaged

[42]Maloney, D. M., Fixsen, D. L., and Phillips, E. L. The Teaching-Family Model: Research and dissemination in a service program. *Children and Youth Services Review*, 1981, 3 (4), 343–355.
[43]Watson, E. W., Maloney, D. M., Brooks, L. E., Blase, K. B., and Collins, L. B. *Teaching-Family Bibliography* (3rd ed.). Boys Town, Nebraska: Father Flanagan's Boys' Home, December, 1980.
[44]*Standards of ethical conduct of the National Teaching-Family Association* (4th ed.). Boys Town, Nebraska: Father Flanagan's Boys' Home, December, 1981.
[45]Ibid., p. 6.

in substantially revising the APA's 1973 book on ethical principles in human subject research, a draft of which was 195 pages long.[46]

Some Problems Associated with Informed Consent

Before we leave this topic, it must be pointed out that, although knowledge of federal rules and professional association guidelines is necessary, it is seldom sufficient. As for almost any human enterprise, it looks easy on paper, but the actual implementation of informed consent can pose a variety of problems. As for many topics discussed in this book, informed consent could easily fill an entire volume of discussion and elaboration, particularly if all the possible implementation problems were dealt with in detail. In keeping with the intent of this book to serve as a practical guide, however, our discussion of consent problems will be brief and limited to two areas: resistance on the part of researchers and confusion on the part of subjects.

Researcher Resistance

A well-known spokesman for the problems and complexities of informed consent, Bradford Gray as a sociologist has brought a unique focus to the field of protection of human research subjects. As early as 1975 (while federal guidelines were beginning an era of substantial revisions), Dr. Gray published his unusual book wherein the consent process itself was the subject of his research.[47] Essentially, he studied the informed consent processes and IRB review procedures that had been followed for two rather ordinary medical research projects at a large university medical center. In essence, he found a variety of problems, including lack of adequate review on the part of an IRB and inadequate attention paid to ethical aspects of medical research by researchers who were best trained to be scientists, not protectors of their subjects.

Dr. Gray served for three years on the staff of the National Commission for the Protection of Human Subjects of Biomedical and Behavioral Research and has thus been closely involved with the development of federal rules addressed in this book. In 1978, an article[48] of his discussed some of the problems associated with informed consent, once again highlighting the limited commitment of

[46]American Psychological Association, Committee for the Protection of Human Subjects. 1200 Seventeenth Street, N.W., Washington, D.C. 20036.

[47]Gray, B. H. *Human subjects in medical experimentation: A sociological study of the conduct and regulation of clinical research.* New York: Wiley, 1975.

[48]Gray, B. H. Complexities of informed consent. *The Annals of the American Academy of Political and Social Sciences,* 1978 (May), 37–48.

professionals to the concept of informed consent. Such resistance conceivably can lead to situations in which researchers may hurry a subject through the consent process, assuming the consent document addresses all the areas it is supposed to address. It is also possible that IRBs may concentrate so much attention on the consent document itself that they ignore other aspects of the subject's participation (e.g., was undue inducement used by the researcher?).

The 1978 article by Gray concludes with the rather pessimistic view that, because the informed consent process is plagued with so many problems, IRBs should spend more time weighing actual risks and benefits of subjects and even investigating how subjects are selected (to assess the voluntariness of participation). These additional procedures, presumably, would provide more protection to subjects than simply reviewing the consent document used by researchers. It is worth noting here that federal regulations then in effect were—in this author's opinion—more confusing than current rules. Thus it is possible that clearer rules, and rules that have now been revised by researcher input in the regulatory process, may lead to better understanding by researchers and more consideration of subjects' welfare.

Whether consent was truly voluntary was also examined in an experiment by Rosen (1976), who studied subject behavior in conditions where some were given an option to refuse to participate and others were not.[49] Specifically, clients at a mental health clinic in Georgia were asked to give written consent to release medical and related information about their case to a central state system of records. Some clients were told that they had a right to refuse (voluntary consent), while others were not so informed. Of those clients who were not informed on the right to refuse, fully 100% signed the consent form, whereas only 20% of those who knew they could refuse actually consented to the release of records. Anecdotally, Rosen also noted that more females than males did not comply, and the noncompliers also tended to be more educated than compliers. In his conclusions, Rosen expressed concern that professionals may lose credibility in the future as clients come to better understand their rights—unless more explanations are given to subjects.

In this author's opinion, we should develop avenues of increasing subjects' participation more legitimate than not informing subjects of their rights—and thereby risk losing credibility for years to come.

Subject Confusion

To focus now on the subject's behavior rather than on that of the researcher, we find that subjects often may not understand the consent process at all. Two reviews of Gray's book (1975) highlighted some alarming findings.

[49]Rosen, C. E. Sign-away pressures. *Social Work*, 1976, *21* (4), 284–287.

Scheerenberger (1976)[50] noted that even adults have a hard time understanding that they are participating in research projects. He cited one of Gray's studies in which it was discovered that 39% of the subjects had never understood that they had consented to be experimental subjects. The study in question involved administration of a certain labor-inducing drug to pregnant women! If women in such a situation did not even realize that they were consenting to be experimental subjects, how much more should we wonder about the understanding of children, or of the mentally infirm. As we shall see later in this book, such concern over the protection of special populations has led to the formulation of specific rules for certain types of subjects (e.g., children, prisoners) who, for one reason or another, may have less than maximal ability to understand or to consent to research participation.

Rosenthal's review[51] of Gray's book pointed out that another disturbing finding of Gray's work was that there was evidence that informed consent frequently was not obtained even when a research review committee was involved and when researchers expressed commitment to the ideals of informed consent. This might be less of a problem today, since numerous provisions are now contained in the rules for exceptions to informed consent, as we saw earlier in this chapter. However, it is disturbing if it reflects an underlying lack of ethical concern by researchers, for it must be remembered that at that time such exceptions to informed consent were not as inclusive. Thus, despite lip service, researchers were not attending well to federal rules formulated to protect human research subjects.

If one of the causes for such a lack of attention was lack of knowledge about federal rules, then perhaps books such as this one may comprise a partial remedy. The inclusion of actual consent forms (see earlier in this chapter) and related hints (see the next chapter) may also help, for Rosenthal notes that Gray's work would have been even more helpful to researchers had it included consent examples that Gray could have recommended to readers.

Finally, there is yet another basic problem with informed consent leading to subject confusion: the readability of the consent document. We have already seen that subjects may not even realize that they are experimental subjects, and part of the problem could be how easy the form is to read and comprehend. In a summary article by Grundner (1978)[52] there is a discussion of two techniques

[50]Scheerenberger, R. C. A review of B. H. Gray's *Human subjects in medical experimentation: A sociological study of the conduct and regulation of clinical research*, 1975. In *Mental Retardation*, 1976, *14* (2), 39.

[51]Rosenthal, D. E. A review of B. H. Gray's *Human subjects in medical experimentation: A sociological study of the conduct and regulation of clinical research*, 1975. In *Social Science and Medicine*, 1975, 9 (10), 564–565.

[52]Grundner, T. M. Two formulas for determining the readability of subject consent forms. *American Psychologist*, 1978 (August), 773–775.

from the field of education that can be used to measure just how readable a consent form might be.

First, the Flesch Readability Formula (1948)[53] contains four steps that lead to a number between 0 and 100 after entering terms in a formula based on the number of syllables and sentences in the 100-word samples. Resultant scores that are low (e.g., 0–30) would be considered as very difficult and typical of professional journals. High scores (e.g., 80–90), on the other hand, are considered characteristic of pulp fiction and as easy.

Second, the Fry Readability Scale (1968)[54] has the advantage of being scored in a grade-level equivalency. The computations are generally similar, but the results are then plotted on a graph which yields an approximate grade level for the result.

Grundner notes that the Flesch Readability Formula is very sensitive to all levels of reading but is slightly more difficult to use and will not yield accurate grade equivalences beyond the seventh-grade level. The Fry system is considered more convenient and will produce grade level equivalences up to the 12th grade. Grundner recommends using both scales together,the Flesch initially and then the Fry if it is necessary to revise the consent form to bring it into a specified grade equivalent. Both systems have been used to analyze consent forms, and, of course, the reading level one chooses would be expected to vary from study to study, depending on the type of subjects.

Summary

Informed consent is perhaps the most visible aspect of the protection of human research subjects, although the written document itself is only part of the overall process of informing a subject and obtaining voluntary consent to participate. It is also perhaps one of the least understood components of subject protection and, through misuse or neglect, may eventually lead to unwillingness by an increasingly knowledgeable subject population to participate in research.

Historically, the phases of construction of a consent form, approval by an IRB, and signing of the form have gone through significant changes. By 1981, the federal rules (best exemplified by the rules by the Department of Health and Human Services) had been greatly modified regarding: (1) the makeup of the basic elements of informed consent; (2) the specific notation of additional elements of consent which may be required of an investigator by an IRB; and (3) several provisions whereby an IRB can approve changes or even whole-scale *waiver* of consent requirements for certain types of research. The early provisions

[53]Flesch, R. A new readability yardstick. *Journal of Applied Psychology*, 1948, 32 (3), 221–233.
[54]Fry, E. A readability formula that saves time. *Journal of Reading*, 1968, *11*, 513–516, 575–578.

of the 1970s regarding the regular and the short form remain valid today, providing flexibility to the researcher in choosing the method of obtaining consent. The respective advantages of these two methods suggests that, despite its name, the short form may be especially useful only in some situations.

A specific example of an informed consent form was drawn from the author's own experience and involved a survey which was approved by a federally approved IRB. The example provides an opportunity for examining the basic elements of consent as they appear in government regulations and how they can then be written for a research subject. The example also provides a case study for the reader to answer typical questions faced by a researcher today (e.g., was informed consent necessary in this or similar cases?). These questions are ones all researchers must answer, including how to get IRB approval of the consent forms and process once the researcher's proposal is complete (see next chapter).

In a contrast of the consent requirements for projects regulated by the Food and Drug Administration (FDA) and the rules of Health and Human Services (HHS), we find that they are quite similar. However, FDA does permit alteration of consent procedures when the subject's life is in danger. Even in such extreme situations, however, FDA lists several factors which must exist relative to such research.

Although the focus of this chapter and the book as a whole is on research, informed consent can be used for participation in treatment. The same contractual agreement postulated between researcher and subject can be constructed for the similar relationship between treatment provider (e.g., clinician) and the client. The example discussed in this chapter illustrates how such a relationship can be formally embodied within a code of ethics. However, it must be noted that the requirement for informed consent for treatment does not currently exist within federal regulations as it does for research.

As for any implementation of specific rules, the actual use of informed consent seldom is as easy as the published guidelines might suggest. Problems with informed consent involve the researcher's perspective and use and the subject's understanding and voluntary compliance. These problems can be solved, primarily by researchers who are sufficiently knowledgeable of the rules and who construct consent documents best suited adequately to inform subjects and lead to voluntary participation in research.

Still, even the most knowledgeable researcher must usually obtain IRB approval of consent documents and related procedures to protect subjects' rights. How to obtain such approval, and how an actual IRB handles such proposals by a researcher, are questions addressed in the next chapter.

APPENDIX 3.1

Regulations by the Department of Health and Human Services for Informed Consent

[Note: The following text is from the *Federal Register*—Vol. 46, No. 16, Monday, January 26, 1981, starting at page 8389—and contains the current HHS rules governing informed consent. This text was incorporated within overall HHS rules on protection as appended to Chapter 2 and is presented separately here to aid in reading Chapter 3.]

DEPARTMENT OF HEALTH AND HUMAN SERVICES

Office of the Secretary
45 CFR Part 46

Final Regulations Amending Basic HHS Policy for the Protection of Human Research Subjects

AGENCY: Department of Health and Human Services
ACTION: Final Rule

[Note: Deleted here are the introductory material and regulations about IRBs that were presented in Chapter 2 of this book.]

Sec. 46.116 General requirements for informed consent.

Except as provided elsewhere in this or other subparts, no investigator may involve a human being as a subject in research covered by these regulations unless the investigator has obtained the legally effective informed consent of the subject or the subject's legally authorized representative. An investigator shall seek such consent only under circumstances that provide the prospective subject or the representative sufficient opportunity to consider whether or not to participate and that minimize the possibility of coercion or undue influence. The information that is given to the subject or the representative shall be in language understandable to the subject or the representative. No informed consent, whether oral or written, may include any exculpatory language through which the subject or the representative is made to waive or appear to waive any of the subject's legal rights, or releases or appears to release the investigator, the sponsor, the institution or its agents from liability for negligence.

(a) Basic elements of informed consent. Except as provided in paragraph (c) of this section, in seeking informed consent the following information shall be provided to each subject:

(1) A statement that the study involves research, an explanation of the purposes of the research and the expected duration of the subject's participation, a description of the procedures to be followed, and identification of any procedures which are experimental;

(2) A description of any reasonably foreseeable risks or discomforts to the subject;

(3) A description of any benefits to the subject or to others which may reasonably be expected from the research;

(4) A disclosure of appropriate alternative procedures or courses of treatment, if any, that might be advantageous to the subject;

(5) A statement describing the extent, if any, to which confidentiality of records identifying the subject will be maintained;

(6) For research involving more than minimal risk, an explanation as to whether any compensation and an explanation as to whether any medical treatments are available if injury occurs and, if so, what they consist of, or where further information may be obtained;

(7) An explanation of whom to contact for answers to pertinent questions about the research and research subjects' rights, and whom to contact in the event of a research-related injury to the subject; and

(8) A statement that participation is voluntary, refusal to participate will involve no penalty or loss of benefits to which the subject is otherwise entitled, and the subject may discontinue participation at any time without penalty or loss of benefits to which the subject is otherwise entitled.

(b) Additional elements of informed consent. When appropriate, one or more of the following elements of information shall also be provided to each subject:

(1) A statement that the particular treatment or procedure may involve risks to the subject (or to the embryo or fetus, if the subject is or may become pregnant) which are currently unforeseeable;

(2) Anticipated circumstances under which the subject's participation may be terminated by the investigator without regard to the subject's consent;

(3) Any additional costs to the subject that may result from participation in the research;

(4) The consequences of a subject's decision to withdraw from the research and procedures for orderly termination of participation by the subject;

(5) A statement that significant new findings developed during the course of the research which may relate to the subject's willingness to continue participation will be provided to the subject; and

(6) The approximate number of subjects involved in the study.

(c) An IRB may approve a consent procedure which does not include, or which alters, some or all of the elements of informed consent set forth above, or waive the requirement to obtain informed consent provided the IRB finds and documents that:

(1) The research is to be conducted for the purpose of demonstrating or evaluating: (i) Federal, state, or local benefit or service programs which are not themselves research programs, (ii) procedures for obtaining benefits or services under these programs, or (iii) possible changes in or alternatives to these programs or procedures; and

(2) The research could not practicably be carried out without the waiver or alteration.

(d) An IRB may approve a consent procedure which does not include, or which alters, some or all of the elements of informed consent set forth above, or waive the requirements to obtain informed consent provided the IRB finds and documents that:

(1) The research involves no more than minimal risk to the subjects;

(2) The waiver or alteration will not adversely affect the rights and welfare of the subjects;

(3) The research could not practicably be carried out without the waiver or alteration; and

(4) Whenever appropriate, the subjects will be provided with additional pertinent information after participation.

(e) The informed consent requirements in these regulations are not intended to preempt any applicable Federal, state, or local laws which require additional information to be disclosed in order for informed consent to be legally effective.

(f) Nothing in these regulations is intended to limit the authority of a physician to provide emergency medical care, to the extent the physician is permitted to do so under applicable Federal, state, or local law.

Sec. 46.117 Documentation of informed consent.

(a) Except as provided in paragraph (c) of this section, informed consent shall be documented by the use of a written consent form approved by the IRB and signed by the subject or the subject's legally authorized representative. A copy shall be given to the person signing the form.

(b) Except as provided in paragraph (c) of this section, the consent form may be either of the following:

(1) A written consent document that embodies the elements of informed consent required by Sec. 46.116. This form may be read to the subject or the subject's legally authorized representative, but in any event, the investigator shall give either the subject or the representative adequate opportunity to read it before it is signed; or

(2) A "short form" written consent document stating that the elements of informed consent required by Sec. 46.116 have been presented orally to the subject or the subject's legally authorized representative. When this method is used, there shall be a witness to the oral presentation. Also, the IRB shall approve a written summary of what is to be said to the subject or the representative. Only the short form itself is to be signed by the subject or the representative. However, the witness shall sign both the short form and a copy of the summary. A copy of the summary shall be given to the subject or the representative, in addition to a copy of the "short form."

(c) An IRB may waive the requirement for the investigator to obtain a signed consent form for some or all subjects if it finds either:

(1) That the only record linking the subject and the research would be the consent document and the principal risk would be potential harm resulting from a breach of confidentiality. Each subject will be asked whether the subject wants documentation linking the subject with the research, and the subject's wishes will govern; or

(2) That the research presents no more than minimal risk of harm to subjects and involves no procedures for which written consent is normally required outside of the research context.

In cases where the documentation requirement is waived, the IRB may require the investigator to provide subjects with a written statement regarding the research.

Sec. 46.118 Applications and proposals lacking definite plans for involvement of human subjects.

Certain types of applications for grants, cooperative agreements, or contracts are submitted to the Department with the knowledge that subjects may be involved within the period of funding, but definite plans would not normally be set forth in the application or proposal. These include activities such as institutional type grants (including bloc grants) where selection of specific projects is the institution's responsibility; research training grants where the activities involving subjects remain to be selected; and projects in which human subjects' involvement will depend upon completion of instruments, prior animal studies, or purification of compounds. These applications need not be reviewed by an IRB before an award may be made. However, except for research described in Sec. 46.101(b), no human subjects may be involved in any project supported by these awards until the project has been reviewed and approved by the IRB, as provided in these regulations, and certification submitted to the Department.

Sec. 46.119 Research undertaken without the intention of involving human subjects.

In the event research (conducted or funded by the Department) is undertaken without the intention of involving human subjects, but it is later proposed to use human subjects in the research, the research shall first be reviewed and approved by an IRB, as provided in these regulations, a certification submitted to the Department, and final approval given to the proposed change by the Department.

Sec. 46.120 Evaluation and disposition of applications and proposals.

(a) The Secretary will evaluate all applications and proposals involving human subjects submitted to the Department through such officers and employees of the Department and such experts and consultants as the Secretary determines to be appropriate. This evaluation will take into consideration the risks to the subjects, the adequacy of protection against these risks, the potential benefits of the proposed research to the subjects and others, and the importance of the knowledge to be gained.

(b) On the basis of this evaluation, the Secretary may approve or disapprove the application or proposal, or enter into negotiations to develop an approvable one.

Sec. 46.121 Investigational new drug or device 30-day delay requirement.

When an institution is required to prepare or submit a certification with an application or proposal under these regulations, and the application or proposal involves an investigational new drug (within the meaning of 21 U.S.C. 355(i) or 357(d)) or a significant risk device (as defined in 21 CFR 812.3(m)), the institution shall identify the drug or device in the certification. The institution shall also state whether the 30-day interval required for investigational new drugs by 21 CFR 312.1(a) and for significant risk devices by 21 CFR 812.30 has elapsed, or whether the Food and Drug Administration has waived that requirement. If the 30-day interval has expired, the institution shall state whether the Food and Drug Administration has requested that the sponsor continue to withhold or restrict the use of the drug or device in human subjects. If the 30-day interval has not expired, and a waiver has not been received, the institution shall send a statement to the Department upon expiration of the interval. The Department will not consider a certification acceptable until the institution has submitted a statement that the 30-day interval has elapsed, and the Food and Drug Administration has not requested it to limit the use of the drug or device, or that the Food and Drug Administration has waived the 30-day interval.

Sec. 46.122 Use of federal funds.

Federal funds administered by the Department may not be expended for research involving human subjects unless the requirements of these regulations, including all subparts of these regulations, have been satisfied.

Sec. 46.123 Early termination of research funding; evaluation of subsequent applications and proposals.

(a) The Secretary may require that Department funding for any project be terminated or suspended in the manner prescribed in applicable program requirements, when the Secretary finds an institution has materially failed to comply with the terms of these regulations.

(b) In making decisions about funding applications or proposals covered by these regulations the Secretary may take into account, in addition to all other eligibility requirements and program criteria, factors such as whether the applicant has been subject to a termination or suspension under paragraph (a) of this section and whether the applicant or the person who would direct the scientific and technical aspects of an activity has in the judgment of the Secretary materially failed to discharge responsibility for the protection of the rights and welfare of human subjects (whether or not Department funds were involved).

Sec. 46.124 Conditions.

With respect to any research project or any class of research projects the Secretary may impose additional conditions prior to or at the time of funding when in the Secretary's judgment additional conditions are necessary for the protection of human subjects.

APPENDIX 3.2

Regulations by the Food and Drug Administration for Informed Consent

[Note: The following regulations are from the *Federal Register*—Vol. 46, No. 17, Tuesday, January 27, 1981, starting at page 8951—and address how FDA views the process and form of informed consent. These rules were not included in the appended material on the FDA for Chapter 2 since those rules at the end of the chapter dealt with FDA rules for IRBs, not informed consent.]

DEPARTMENT OF HEALTH AND HUMAN SERVICES

Food and Drug Administration
21 CFR Parts 50, 71, 171, 180, 310,
312, 314, 320, 330, 361, 430, 431, 601,
630, 812, 813, 1003, 1010
(Docket No. 78N–0400)

Protection of Human Subjects; Informed Consent

AGENCY: Food and Drug Administration
ACTION: Final Rule

[Note: Deleted here are introductory information and FDA rules on the operation of IRBs. Also, the rules will conclude with the sections on informed consent, since this author has not included the varied technical changes in other FDA rules on color additives, food additives, related areas as contained in Parts 71, 171, 180, 310, 312, 314, 320, 330, 361, 430, 431, 601, 630, 812, 813, 1003, and 1010 of Title 21 of the Code of Federal Regulations. Interested readers will wish to consult the cited source in the *Federal Register* for this supplementary information.]

b. By adding new Subpart B to read as follows.

Subpart B—Informed Consent of Human Subjects

Subpart B—Informed Consent of Human Subjects

Sec. 50.20 General requirements for informed consent.

Except as provided in Sec. 50.23, no investigator may involve a human being as a subject in research covered by these regulations unless the investigator has obtained the legally effective informed consent of the subject or the subject's legally authorized representative. An investigator shall seek such consent only under circumstances that provide the prospective subject or the representative sufficient opportunity to consider whether or not to participate and that minimize the possibility of coercion or undue influence. The information that is given to the subject or the representative shall be in language understandable to the subject or the representative. No informed consent, whether oral or written, may include any exculpatory language through which the subject or the representative is made to waive or appear to waive any of the subject's legal rights, or releases or appears to release the investigator, the sponsor, the institution, or its agents from liability for negligence.

Sec. 50.21 Effective date.

The requirements for informed consent set out in this part apply to all human subjects entering a clinical investigation that commences on or after July 27, 1981.

Sec. 50.23 Exception from general requirements.

(a) The obtaining of informed consent shall be deemed feasible unless, before use of the test article (except as provided in paragraph (b) of this section), both the investigator and a physician who is not otherwise participating in the clinical investigation certify in writing all of the following:

(1) The human subject is confronted by a life-threatening situation necessitating the use of the test article.

(2) Informed consent cannot be obtained from the subject because of an inability to communicate with, or obtain legally effective consent from, the subject.

(3) Time is not sufficient to obtain consent from the subject's legal representative.

(4) There is available no alternative method of approved or generally recognized therapy that provides an equal or greater likelihood of saving the life of the subject.

(b) If immediate use of the test article is, in the investigator's opinion, required to preserve the life of the subject, and time is not sufficient to obtain the independent determination required in paragraph (a) of this section in advance of using the test article, the determinations of the clinical investigator shall be made and, within 5 working days after the use of the article, be reviewed and evaluated in writing by a physician who is not participating in the clinical investigation.

(c) The documentation required in paragraph (a) or (b) of this section shall be submitted to the IRB within 5 working days after the use of the test article.

Sec. 50.25 Elements of informed consent.

(a) Basic elements of informed consent. In seeking informed consent, the following information shall be provided to each subject:

(1) A statement that the study involves research, an explanation of the purposes of the research and the expected duration of the subject's participation, a description of the procedures to be followed, and identification of any procedures which are experimental.

(2) A description of any reasonably foreseeable risks or discomforts to the subject.

(3) A description of any benefits to the subject or to others which may reasonably be expected from the research.

(4) A disclosure of appropriate alternative procedures or courses of treatment, if any, that might be advantageous to the subject.

(5) A statement describing the extent, if any, to which confidentiality of records identifying the subject will be maintained and that notes the possibility that the Food and Drug Administration may inspect the records.

(6) For research involving more than minimal risk, an explanation as to whether any compensation and an explanation as to whether any medical treatments are available if injury occurs and, if so, what they consist of, or where further information may be obtained.

(7) An explanation of whom to contact for answers to pertinent questions about the research and research subjects' rights, and whom to contact in the event of a research-related injury to the subject.

(8) A statement that participation is voluntary, that refusal to participate will involve no penalty or loss of benefits to which the subject is otherwise entitled, and that the subject may discontinue participation at any time without penalty or loss of benefits to which the subject is otherwise entitled.

(b) Additional elements of informed consent. When appropriate, one or more of the following elements of information shall also be provided to each subject:

(1) A statement that the particular treatment or procedure may involve risks to the subject (or to the embryo or fetus, if the subject is or may become pregnant) which are currently unforeseeable.

(2) Anticipated circumstances under which the subject's participation may be terminated by the investigator without regard to the subject's consent.

(3) Any additional costs to the subject that may result from participation in the research.

(4) The consequences of a subject's decision to withdraw from the research and procedures for orderly termination of participation by the subject.

(5) A statement that significant new findings developed during the course of the research which may relate to the subject's willingness to continue participation will be provided to the subject.

(6) The approximate number of subjects involved in the study.

(c) The informed consent requirements in these regulations are not intended to preempt any applicable Federal, State, or local laws which require additional information to be disclosed for informed consent to be legally effective.

(d) Nothing in these regulations is intended to limit the authority of a physician to provide emergency medical care to the extent the physician is permitted to do so under applicable Federal, State, or local law.

Sec. 50.27 Documentation of informed consent.

(a) Except as provided in Sec. 56.109(c), informed consent shall be documented by the use of a written consent form approved by the IRB and signed by the subject or the subject's legally authorized representative. A copy shall be given to the person signing the form.

(b) Except as provided in Sec. 56.109(c), the consent form may be either of the following:

(1) A written consent document that embodies the elements of informed consent required by Sec. 50.25. This form may be read to the subject or the subject's legally authorized representative, but, in any event, the investigator shall give either the subject or the representative adequate opportunity to read it before it is signed.

(2) A "short form" written consent document stating that the elements of informed consent required by Sec. 50.25 have been presented orally to the subject or the subject's legally authorized representative. When this method is used, there shall be a witness to the oral presentation. Also, the IRB shall approve a written summary of what is to be said to the subject or the representative. Only the short form itself is to be signed by the subject or the representative. However, the witness shall sign both the short form and a copy of the summary, and the person actually obtaining the consent shall sign a copy of the summary. A copy of the summary shall be given to the subject or the representative in addition to a copy of the short form.

APPENDIX 3.3

A Comparison of HHS and FDA Rules Governing Informed Consent

[Note: The following text contains a side-by-side comparison of FDA and HHS rules and is adapted from information provided by Dr. John Petricciani of the Food and Drug Administration. These comparisons were not included in the similar comparisons on IRBs that were appended to Chapter 2 of this book and represent new material. Differences in the rules are noted in italics.]

FDA

Sec. 50.20 General requirements for informed consent.

Except as provided in Sec. 50.23, no investigator may involve a human being as a subject in research covered by these regulations unless the investigator has obtained the legally effective informed consent of the subject or the subject's legally authorized representative. An investigator shall seek such consent only under circumstances that provide the prospective subject or the representative sufficient opportunity to consider whether or not to participate and that minimize the possibility of coercion or undue influence. The information that is given to the subject or the representative shall be in language understandable to the subject or the representative. No informed consent, whether oral or written, may include any exculpatory language through which the subject or the representative is made to waive or appear to waive any of the subject's legal rights, or releases or appears to release the investigator, the sponsor, the institution, or its agents from liability for negligence.

Sec. 50.23 Exception from general requirements.

(a) The obtaining of informed consent shall be deemed feasible unless, before use of the test article (except as provided in paragraph (b) of this section), both the investigator and a physician who is not otherwise participating in the clinical investigation certify in writing all of the following:

(1) The human subject is confronted by a life-threatening situation necessitating the use of the test article.

(2) Informed consent cannot be obtained from the subject because of an inability to communicate with, or obtain legally effective consent from, the subject.

(3) Time is not sufficient to obtain consent from the subject's legal representative.

(4) There is available no alternative method of approved or generally recognized therapy that provides an equal or greater likelihood of saving the life of the subject.

(b) If immediate use of the test article is, in the investigator's opinion, required to preserve the life of the subject, and time is not sufficient to obtain the independent determination required in paragraph (b) of this section in advance of using the test article, the determinations of the clinical investigator shall be made and, within 5 working days after the use of the article, be reviewed and evaluated in writing by a physician who is not participating in the clinical investigation.

HHS

Sec. 46.116 General requirements for informed consent.

Except as provided *elsewhere in this or other subparts*, no investigator may involve a human being as a subject in research covered by these regulations unless the investigator has obtained the legally effective informed consent of the subject or the subject's legally authorized representative. An investigator shall seek such consent only under circumstances that provide the prospective subject or the representative sufficient opportunity to consider whether or not to participate and that minimize the possibility of coercion or undue influence. The information that is given to the subject or the representative shall be in language understandable to the subject or the representative. No informed consent, whether oral or written, may include any exculpatory language through which the subject or the representative is made to waive or appear to waive any of the subject's legal rights, or releases or appears to release the investigator, the sponsor, the institution, or its agents from liability for negligence.

FDA

(c) The documentation required in paragraph (a) or (b) of this section shall be submitted to the IRB within 5 working days after the use of the test article.

Sec. 50.25 Elements of informed consent.

(a) Basic elements of informed consent.

In seeking informed consent, the following information shall be provided to each subject:

(1) A statement that the study involves research, an explanation of the purposes of the research and the expected duration of the subject's participation, a description of the procedures to be followed, and identification of any procedures which are experimental.

(2) A description of any reasonably foreseeable risks or discomforts to the subject.

(3) A description of any benefits to the subject or to others which may reasonably be expected from the research.

(4) A disclosure of appropriate alternative procedures or courses of treatment, if any, that might be advantageous to the subject.

(5) A statement describing the extent, if any, to which confidentiality of records identifying the subject will be maintained *and that notes the possibility that the Food and Drug Administration may inspect the records.*

(6) For research involving more than minimal risk, an explanation as to whether any compensation and an explanation as to whether any medical treatments are available if injury occurs and, if so, what they consist of, or where further information may be obtained.

(7) An explanation of whom to contact for answers to pertinent questions about the research and research subjects' rights, and whom to contact in the event of a research-related injury to the subject.

(8) A statement that participation is voluntary, *that* refusal to participate will involve no penalty or loss of benefits to which the subject is otherwise entitled, and *that* the subject may discontinue participation at any time without penalty or loss of benefits to which the subject is otherwise entitled.

(b) Additional elements of informed consent. When appropriate, one or more of the following elements of information shall also be provided to each subject:

HHS

Sec. 46.116 General requirements for informed consent [continued].

(a) Basic elements of informed consent. *Except as provided in paragraph (c) of this section,*
in seeking informed consent, the following information shall be provided to each subject:

(1) A statement that the study involves research, an explanation of the purposes of the research and the expected duration of the subject's participation, a description of the procedures to be followed, and identification of any procedures which are experimental;

(2) A description of any reasonably foreseeable risks or discomforts to the subject;

(3) A description of any benefits to the subject or to others which may reasonably be expected from the research;

(4) A disclosure of appropriate alternative procedures or courses of treatment, if any, that might be advantageous in the subject;

(5) A statement describing the extent, if any, to which confidentiality of records identifying the subject will be maintained;

(6) For research involving more than minimal risk, an explanation as to whether any compensation and an explanation as to whether any medical treatments are available if injury occurs and, if so, what they consist of, or where further information may be obtained;

(7) An explanation of whom to contact for answers to pertinent questions about the research and research subjects' rights, and whom to contact in the event of a research-related injury to the subject; *and*

(8) A statement that participation is voluntary, refusal to participate will involve no penalty or loss of benefits to which the subject is otherwise entitled, and the subject may discontinue participation at any time without penalty or loss of benefits to which the subject is otherwise entitled.

(b) Additional elements of informed consent. When appropriate, one or more of the following elements of information shall also be provided to each subject:

FDA

(1) A statement that the particular treatment or procedure may involve risks to the subject (or to the embryo or fetus, if the subject is or may become pregnant) which are currently unforeseeable.

(2) Anticipated circumstances under which the subject's participation may be terminated by the investigator without regard to the subject's consent.

(3) Any additional costs to the subject that may result from participation in the research.

(4) The consequences of a subject's decision to withdraw from the research and procedures for orderly termination of participation by the subject.

(5) A statement that significant new findings developed during the course of the research which may relate to the subject's willingness to continue participation will be provided to the subject.

(6) The approximate number of subjects involved in the study.

HHS

(1) A statement that the particular treatment or procedure may involve risks to the subject (or to the embryo or fetus, if the subject is or may become pregnant) which are currently unforeseeable;

(2) Anticipated circumstances under which the subject's participation may be terminated by the investigator without regard to the subject's consent;

(3) Any additional costs to the subject that may result from participation in the research;

(4) The consequences of a subject's decision to withdraw from the research and procedures for orderly termination of participation by the subject;

(5) A statement that significant new findings developed during the course of the research which may relate to the subject's willingness to continue participation will be provided to the subject; *and*

(6) The approximate number of subjects involved in the study.

(c) *An IRB may approve a consent procedure which does not include, or which alters, some or all of the elements of informed consent set forth above, or waive the requirement to obtain informed consent provided the IRB finds and documents that:*

(1) The research is to be conducted for the purpose of demonstrating or evaluating: (i) federal, state, or local benefit or service programs which are not themselves research programs, (ii) procedures for obtaining benefits or services under these programs, or (iii) possible changes in or alternatives to these programs or procedures; and

(2) The research could not practicably be carried out without the waiver or alteration.

(d) *An IRB may approve a consent procedure which does not include, or which alters, some or all of the elements of informed consent set forth above, or waive the requirements to obtain informed consent provided the IRB finds and documents that:*

(1) The research involves no more than minimal risk to the subjects;

(2) The waiver or alteration will not adversely affect the rights and welfare of the subjects;

(3) The research could not practicably be carried out without the waiver or alteration; and

(4) Whenever appropriate, the subjects will be pro-

(c) The informed consent requirements in these regulations are not intended to preempt any applicable Federal, State, or local laws which require additional information to be disclosed for informed consent to be legally effective.

(d) Nothing in these regulations is intended to limit the authority of a physician to provide emergency medical care to the extent the physician is permitted to do so under applicable Federal, State, or local law.

Sec. 50.27 Documentation of informed consent.

(a) Except as provided in Sec. 56.109(c), informed consent shall be documented by the use of a written consent form approved by the IRB and signed by the subject or the subject's legally authorized representative. A copy shall be given to the person signing the form.

(b) Except as provided in Sec. 56.109(c), the consent form may be either of the following:

(1) A written consent document that embodies the elements of informed consent required by Sec. 50.25. This form may be read to the subject or the subject's legally authorized representative, but, in any event, the investigator shall give either the subject or the representative adequate opportunity to read it before it is signed.

(2) A "short form" written consent document stating that the elements of informed consent required by Sec. 50.25 have been presented orally to the subject or the subject's legally authorized representative. When this method is used, there shall be a witness to the oral presentation. Also, the IRB shall approve a written summary of what is to be said to the subject or the representative. Only the short form itself is to be signed by the subject or the representative. However, the witness shall sign both the short form and a copy of the summary, and the person actually obtaining *the* consent shall sign a copy of the summary. A copy of the summary shall be given to the subject or the representative in addition to a copy of the short form.

Sec. 56.109 IRB review of research.

vided with additional pertinent information after partcipation.

(e) The informed consent requirements in these regulations are not intended to preempt any applicable federal, state, or local laws which require additional information to be disclosed *in order* for informed consent to be legally effective.

(f) Nothing in these regulations is intended to limit the authority of a physician to provide emergency medical care to the extent the physician is permitted to do so under applicable federal, state, or local law.

Sec. 46.117 Documentation of informed consent.

(a) Except as provided in paragraph (c) of this section, informed consent shall be documented by the use of a written consent form approved by the IRB and signed by the subject or the subject's legally authorized representative. A copy shall be given to the person signing the form.

(b) Except as provided in paragraph (c) of this section, the consent form may be either of the following:

(1) A written consent document that embodies the elements of informed consent required by Sec. 46.116. This form may be read to the subject or the subject's legally authorized representative, but in any event, the investigator shall give either the subject or the representative adequate opportunity to read it before it is signed; *or*

(2) A "short form" written consent document stating that the elements of informed consent required by Sec. 46.116 have been presented orally to the subject or the subject's legally authorized representative. When this method is used, there shall be a witness to the oral presentation. Also, the IRB shall approve a written summary of what is to be said to the subject or the representative. Only the short form itself is to be signed by the subject or the representative. However, the witness shall sign both the short form and a copy of the summary, and the person actually obtaining consent shall sign a copy of the summary. A copy of the summary shall be given to the subject or the representative in addition to a copy of the "short form."

Sec. 46.117 Documentation of informed consent [continued].

(c) An IRB may waive the requirement for the investigator to obtain a signed consent form for some or all subjects if it finds either:

FDA

(c) ....that the research presents no more than minimal risk of harm to subjects and involves no procedures for which written consent is normally required outside the research context. In cases where the documentation requirement is waived, the IRB may require the investigator to provide subjects with a written statement regarding the research.

(1) *That the only record linking the subject and the research would be the consent document and the principal risk would be potential harm resulting from a breach of confidentiality. Each subject will be asked whether the subject wants documentation linking the subject with the research, and the subject's wishes will govern; or*

(2) That the research presents no more than minimial risk of harm to subjects and involves no procedures for which written consent is normally required outside of the research context. In cases where the documentation requirement is waived, the IRB may require the investigator to provide subjects with a written statement regarding the research.

CHAPTER 4

How to Get Research Approved by an Institutional Review Board

We have seen in previous chapters how Institutional Review Boards have come to wield the power they currently have and we have examined some of the issues which they examine. For example, we have seen that procedures for informed consent appear to be one of the most common areas of a research proposal to receive the closest scrutiny by IRBs. Thus one entire chapter of this volume was devoted to the intricacies of constructing a consent form so that a researcher can ethically protect subjects. An assumption maintained throughout these discussions has been that the well-prepared researcher who, among other activities, constructs a good consent form will have little difficulty obtaining IRB approval for the research. Such approval is often then followed by the researcher's sending various materials on to a funding agency to receive funds necessary to support the research. Obtaining IRB approval of the researcher's subject-protection procedures is therefore usually the last step before obtaining research funds.

The experienced researcher, of course, usually has long been in contact with the funding agency before this final step. Securing the interest and support of funding agencies for research projects requires numerous professional activities beyond the scope of this book; but, typically, no matter how creative or vital the research concept, approval of an IRB will be needed before any funds are provided. Indeed, it is the official policy of many agencies not even to consider an initial application for research funds unless the application is accompanied by an IRB approval concerning protection of the subjects-to-be. Since the chronology of steps required varies between agencies and funding sources, the hopeful researcher is advised to contact any intended funding source for specific instructions.

It is obvious, then, that a researcher must be familiar with the operations of whatever local IRB has jurisdiction over the researcher's anticipated project. It may well come about that IRB approval will be almost automatic, except in those complicated situations discussed previously in which an IRB may waive any or all of the detailed provisions for protecting subjects (e.g., see waiver of informed consent as noted earlier in this book). Still, the researcher usually cannot be certain of how difficult (or easy) it will be to obtain IRB approval until the process of submitting the research proposal to the local IRB has been completed.

This chapter addresses three aspects of the relationship between researcher and IRB. First, we will examine how an IRB is established and approved by the federal government. This section will be especially useful for research groups planning to establish a new IRB, an activity not unusual in either new research settings or in expanding organizations. Second, once an IRB has been established, we will examine how it operates. Adopting both viewpoints of IRB and researcher, we will go through the steps required by an IRB by using an example based upon the author's experience as a member, and later as the chairman, of a federally approved IRB. Finally, we will examine the kinds of experiences reported by others in dealing with IRBs. By drawing upon the survey results of a major study of IRBs in the late 1970s, we will gain insight into what one may typically expect of the researcher–IRB relationship.

Establishing an Institutional Review Board

This section is included here rather than in the earlier chapter on IRBs because, now that the reader is familiar with the national development and history of IRBs, it is time to focus more specifically on how an IRB operates—especially if the reader desires IRB approval for a research project. The sequence of events necessary to form an individual IRB makes clear the two-way responsibility of an IRB; namely, to the researcher and the researcher's subjects on one hand, and to the federal government on the other as a guarantee that federal regulations are indeed being followed. The process of establishing an IRB can provide a valuable viewpoint to the researcher when it comes time to propose or even defend a proposal before an IRB.

Obtaining Federal Approval of an IRB

To illustrate how an IRB comes into existence, we will follow the steps taken at Boys Town when its IRB was first formed. The author was involved in these steps with other colleagues in the late 1970s when it became clear that, because of the increasing number of research projects at or sponsored by Boys Town, working through IRBs at other institutions was hindering researchers' applications to federal agencies for funds. Although it was not always clear that the activities were indeed research (especially since this was during the time when the official definition of *research* was developing), it was clear that IRB approval was required by funding agencies. This was true even for staff training and demonstration projects. Boys Town's IRB has greatly reduced its activities since the late 1970s, primarily because Boys Town has reexamined its responsibilities and increased activities in nonresearch areas (e.g., staff training and treatment of youth rather than research), but its IRB was formally approved by DHEW after completing the required steps. Therefore, its procedures in establishing an IRB still serve as a useful example for other organizations intending to establish an IRB.

To form an IRB, Boys Town went through two phases. First, it developed an internal policy related to protection of research subjects. Second, it engaged in correspondence with DHEW to obtain necessary materials to apply to become an IRB. The entire process took approximately 18 months.

Internal Policy. As for most organizations, Boys Town formulates policy and procedures for various aspects of organizational management (financial accountability, personnel guidelines, etc.). The *policy* for any such area contains the general aspects and the *procedures* describe the detailed means by which a policy is to be carried out. In the spring of 1976, the author worked with others at Boys Town to draft a policy and procedure regarding organizational commitment to protection of subjects. These statements would clarify management responsibilities within the organization and serve as documentation of principles to be forwarded later to DHEW as part of applying to establish an IRB.

A retyped version of this policy and procedure follows. Please note that, since this is not a photostatic copy of the original, the signatures which appeared on the original documents cannot be reproduced. Other documents, in this chapter and elsewhere in the book, similarly have been retyped for publication purposes and do not bear reproductions of original signatures. The letters are all matters of public record and contain no confidential information about either Boys Town or HEW. Indeed, the documents and the work that went into them signify the importance of Boys Town's commitment to protecting youth rights and the leadership of its Executive Director, Rev. Robert Hupp.

* * *

BOYS TOWN POLICY

TITLE: Human Rights Review Committee

EFFECTIVE DATE: June 14, 1976 CODE: 59400

PURPOSE

It is the purpose of this policy to establish a Human Rights Review Committee to insure that the rights of all participants in any research, development, or related activity at Boys Town are protected.

POLICY

The Home will take all necessary precautions to insure that all individuals in their care or employ are protected from undue risk as a result of participation in any research, development, or related activity at the home.

The Human Rights Review Committee will be established to insure that all relevant procedures will be reviewed and approved so that the best interests of the participants involved are assured. The Committee will have the authority and responsibility to approve or disapprove those procedures which relate to the protection of human rights. (See note.)

The Committee will establish: (1) procedures to follow in its initial and continuing review of relevant applications, proposals and activities; (2) procedures to provide advice and counsel to staff with regard to the Committee's actions; and (3) procedures to insure prompt reporting to the Committee of proposed changes in an activity and of unanticipated problems involving risk to participants or others.

The Committee will make recommendations to the Director of the Home.

NOTE: Code of Federal Regulations Title 45, subtitle A—part 46.

* * *

Before we proceed, it should be noted that the Human Rights Review Committee (HRRC) mentioned in the policy is the committee that came to be considered as the IRB for Boys Town. The official designation of an "Institutional Review Board" was deemed insufficiently descriptive at Boys Town, so HRRC was adopted in its stead. Federal guidelines do not require that the IRB be, in fact, called an IRB. What is crucial is the following of required steps, not what the group is called. Thus the reader should keep in mind that whenever the Boys Town's Human Rights Review Committee is mentioned, the label is synonymous with IRB.

Internal Procedure. Following adoption of the general policy on protection of subjects in June of 1976, work then began at Boys Town to develop specific procedures to implement the policy. Again, these internal administrative guidelines would be used in support of Boys Town's application to the Department of Health, Education, and Welfare to set up an IRB at Boys Town. To expedite this, and to ensure that organizational guidelines would appropriately match IRB guidelines, the published federal rules on IRBs were used as a model in constructing Boys Town's procedures for its Human Rights Review Committee.

The following text was approved on February 23, 1977, by Boys Town as the *procedure* for operating the HRRC.

* * *

BOYS TOWN PROCEDURE

TITLE: Human Rights Review Committee CODE: 59400–A

EFFECTIVE DATE: February 23, 1977

PURPOSE

It is the policy of the Department of Health, Education and Welfare that "no activity involving human subjects to be supported by HEW grants or contracts shall be undertaken unless a committee of the organization has reviewed and approved such activity and the organization has submitted to HEW a certification of such review and approval . . ." (*Federal Register,* Vol. 39, No. 105, p. 18917, 1974). The primary purpose of this procedure is to provide guidelines for the selection of the Committee. In addition, procedures for processing research proposals through the Human Rights Review Committee is included. The responsibility of the HRRC is to safeguard the rights and welfare of subjects at risk, whether supported by HEW or by Boys Town. The scope of the committee is limited to those procedures that are used to insure the rights of subjects at risk in research projects, i.e., informed consent, confidentiality, degree of risk and possible benefits.

PROCEDURE

I. Committee Composition and Selection

A. *Composition*—The Committee will consist of voting members from the following areas of the Home and the Omaha Community:

1. Three (3) members from the Department of Youth Care; a Community Director, a member of the Division of Education and a Family Living Teacher.
2. Two (2) members from the Boys Town Institute.
3. Two (2) members from the Boys Town Center for the Study of Youth Development.
4. A member of the Spiritual Affairs Department will be appointed by the Executive Director of the Home.
5. A member from the Greater Omaha Metropolitan Area, who is not affiliated in any manner with Boys Town, will be appointed by the Executive Director of the Home.
6. The Chairperson will be appointed by the Executive Director of the Home. The Chairperson is a non-voting member of the Committee.

The Chairperson may vote when there is a tie vote of the Committee.

B. *Selection*—It is the responsibility of the Department Director to select and appoint the representative from his/her area. The candidates should have had experience with research projects and be cognizant of the critical issues involving the protection of human rights engaged in research projects.

1. The Greater Omaha Metropolitan Area representative will be nominated by the Committee and forwarded to the Executive Director for his approval.
2. The Executive Director will also appoint the Chairperson of the Human Rights Committee. The Chairperson will report directly to the Executive Director.
3. All appointments are for a one-year period, including the Chairperson. Members may be reappointed by the Department Director or the Executive Director in the case of the Chairperson. Reappointment is not restricted to a specific number of years. The appointments will begin on January 1 and end on December 31 of each year.

II. Human Rights Review Committee Presentation Format

A. *Submission of Project to Committee*—The following steps must be completed before the HRRC will consider reviewing a proposed research project:
1. Department Directors must approve all proposed research projects before they are brought to the Committee for review.
2. Project author delivers ten (10) copies of project proposal to the HRRC Chairperson.
3. The HRRC Chairperson distributes copies of the proposal to HRRC members so that members may read the proposal prior to its review at the next HRRC meeting.
4. The Project Investigator may be invited to attend the HRRC meeting when his/her proposal is to be reviewed to clarify any questions the Committee may have.

B. *Human Rights Review Committee Process*

1. A quorum of six (6) HRRC members must be present for meetings in which projects are reviewed. (1981 regulations have since required that the lay member must be present.)
2. The HRRC members will determine if the project proposal delin-

eates procedures sufficient to protect the rights of its participants according to HEW requirements in the following areas:
   a. Informed consent of participants.
   b. Privacy and confidentiality of data collected.
3. The HRRC members will vote to approve or not approve the project proposal's procedures to protect the rights of its participants (hereafter, "Procedures"). A simple majority vote will be carried.
   a. If the majority of members vote to approve the Procedures, the proposal's cover sheet will be signed by the HRRC Chairperson indicating approval and copies will be sent to both the Investigator and the Investigator's Department Director.
   b. If the majority members vote not to approve the Procedures, the Investigator will be so informed and given specific recommendations for changes in the Procedures that would meet the Committee's requirements. The Committee and the Project Investigator would set a date for the proposal to be resubmitted. This process may be repeated as frequently as the Investigator desires until the Procedures are approved.

III. Human Rights Review Committee Procedural Handbook

A procedural handbook is available from the Chairperson of the HRRC.

IV. Timing for Review

The Chairperson will convene the Committee within twenty-four to forty-eight hours of a request for review to insure that the grant is procured in an expeditious manner.

* * *

Correspondence with Federal Agency in Washington

After establishing its own internal policy and procedures to set up an IRB, Boys Town then initiated correspondence with officials within the Department of Health, Education, and Welfare in Washington, D.C. Then, as now, such correspondence and applications for establishing an IRB were handled by the Office for Protection from Research Risks, Office of the Director, NIH (now within HHS).

The series of correspondence between Boys Town and HEW ultimately resulted in a federally-approved IRB being established at Boys Town (Boys Town's HRRC). There were three primary steps in this process: (1) HEW's notification to Boys Town that an IRB should be established to review research grants; (2) Boys Town's response to HEW with supporting documents; and (3)

HEW's formal approval of Boys Town's HRRC as an operational Institutional Review Board.

Need for General Assurance. HEW first contacted Boys Town regarding a need to establish an IRB in connection with a particular grant application that had been forwarded by a Boys Town staff member to HEW for funding. To expedite the review of such grants, at least as far as protection of human subjects was concerned, HEW recommended that an IRB be formed at Boys Town. Anticipation of such a request, and the desirability of having its own IRB, is what had led to the previous development at Boys Town of its own Human Rights Review Committee and related policies and procedures.

The ongoing mechanism of such review was referred to as a *general assurance* by HEW at that time. The number eventually assigned by HEW to Boys Town's IRB was its general assurance number and is the official designation used by all IRBs when referring to an IRB as a separate entity within any organization. These numbers serve as a shorthand method of referring to any individual IRB and can be found listed for the appropriate institutions in the appendix to this chapter.[1]

On October 7, 1977, Dr. J. R. Marches (Office for Protection from Research Risks) notified the Rev. Robert P. Hupp (Executive Director, Boys Town) that establishment of a general assurance mechanism by forming an IRB would be necessary. The following letter is self-explanatory and conveys what any institution can expect to provide should it desire to form its own IRB. The focus here is not on the one particular research project that serves as the example, but rather on the steps required by federal officials to speed up reviews by forming local IRBs rather than relying totally on federal reviews of every research project supported by HEW. There was no question of any problems with research at the institute where Boys Town helps children with speech or hearing defects, or of Boys Town's procedures, but rather it was time for Boys Town to form its own IRB to speed up processing of support requests.

* * *

DEPARTMENT OF HEALTH, EDUCATION, AND WELFARE

Public Health Service
National Institutes of Health
Bethesda, Maryland 20014

October 7, 1977

Boys Town
Boys Town, Nebraska 68010

[1]*61st Cumulative List of Institutions which have Established General Assurances of Compliance with*

In order to satisfy the Department's requirements as a grantee, it will be necessary to establish an ongoing mechanism (General Assurance) at your institution, with a duly appointed review panel to regularly review sub-projects or avenues of research that emerge during the course of the grant.

To facilitate this task, please find a three-part sample enclosed: Part 1 is essentially a paraphrasal of the regulations to which one subscribes—simply transpose to Boys Town letterhead stationery and affix your signature as the institutional official; Part 2 is the description of the procedures that will be used to implement the regulations; Part 3 is the roster of persons who will serve on the review panel (Institutional Review Board).

For your convenience, a copy of the *Code of Federal Regulations* is also enclosed. It might be helpful to point out areas for quick reference, for example:

1. Membership composition
2. Review coverage
3. Informed consent

In reviewing the three-part model, it should be noted that the principal ingredients include:

1. A statement of compliance (Part 1) signed by the institutional official.
2. Implementation of the regulations (Part 2)
    a. Citation or set of ethical principles
    b. Hierarchy of administrative staff and sequence for processing proposals
    c. Membership composition of institutional board reflecting medical and behavioral expertise as well as one or more members not affiliated with institution
    d. Procedure for conducting review, recording of minutes, and regularity of meetings (quarterly, monthly, etc.)
    e. Definition of quorum as a majority of the total membership
    f. Final authority of review board on re-review of appeals or resubmissions
    g. Explanation of informed consent procedures
    h. Distribution and accessibility of the assurance plan to project directors, subjects, and other personnel of the Department.
3. Institutional Review Board Roster

HEW *Regulations on Protection of Human Subjects* (February, 1980). Bethesda, Maryland: DHEW, Public Health Service, National Institutes of Health.

If we can be of help, please let us know (301) 496–7005.

Sincerely,

/s/ J. R. Marches

J. R. MARCHES, PH.D.
Office for Protection
from Research Risks
Office of the Director

* * *

[Note: The reader should be aware that assurance format, like any other aspect of federal regulations, changes from time to time. It is possible that the examples used by the government and presented in this chapter may change by the time this work is published. Persons wishing to form an IRB should contact NIH Office for Protection from Research Risks for current examples and forms.]

* * *

PART 1

EXAMPLE

INSTITUTIONAL ASSURANCE OF COMPLIANCE

This is an *example*, not a form. *Please* prepare the Institutional Assurance of Compliance on your *institution's letterhead with the original signature of authorized official*. Mail to Office for Protection from Research Risks, OD, National Institutes of Health, Bethesda, Maryland 20205. Note: It is required that documentation of Assurance Implementation prescribing institutional practices and procedures for protection of human subjects accompany this Assurance of Compliance.

(PLEASE USE INSTITUTIONAL LETTERHEAD)

ASSURANCE OF COMPLIANCE WITH DHEW REGULATIONS
ON PROTECTION OF HUMAN SUBJECTS

1. The (full name of institution) will comply with the Department of Health, Education, and Welfare regulations on Protection of Human Subjects (45 CFR 45 as amended), accordingly:

2. This Institution has established and will maintain an Institutional Review Board competent to review projects and activities that involve human subjects. The Board shall determine for each activity as planned and conducted whether subjects will be placed at risk and if risk is involved, whether:

> The risks to the subject are so outweighed by the sum of the benefit to the subject and the importance of the knowledge to be gained as to warrant a decision to allow the subject to accept these risks;
>
> The rights and welfare of any such subjects will be adequately protected;
>
> Legally effective informed consent will be obtained by adequate and appropriate methods in accordance with the provision of the regulation;
>
> The conduct of the activity will be reviewed at timely intervals.

3. This Institution will provide for Board reviews to be conducted with objectivity and in a manner to ensure the independent judgment of the members. Members will be excluded from review of projects or activities in which they have an active role or conflict of interest.

4. This Institution will encourage continuing constructive communication between the Board and the activity director as a means of safeguarding the rights and welfare of the subjects.

5. This Institution will have available the facilities and professional attention required for subjects who may suffer physical, psychological, or other injury as a result of participation in an activity.

6. This Institution acknowledges that it will bear full responsibility for the proper performance of all work and services involving human subjects under any grant or contract covered by this assurance, including continuing compliance with pertinent state or local laws, particularly those concerned with informed consent.

7. This Institution will maintain appropriate and informative records of the Board's review of applications and activities, of documentation of informed consent, and of other documentation that may pertain to the selection, participation, and protection of subjects and to the review of circumstances that adversely affect the rights or welfare of individual subjects.

8. This Institution will at least annually reassure itself through appropriate administrative overview that its practices and procedures designed for the protection of the rights and welfare of human subjects are being effectively applied and are consistent with the regulation and with the implementation of this assurance as accepted by the Department of Health, Education, and Welfare.

9. This general assurance of compliance applies to The (full name of the institution as used in applying for grants and contracts) .

(Signature of authorized official)

(Name-typewritten)

(Titled)

(Date)

(Telephone)

* * *

[Note: Part 2 of this letter is omitted, since it dealt with a different example of how an institution internally monitored its own IRB. What was later sent back to HEW by Boys Town was the policy and procedure for its HRRC and a description of related administrative steps that eventually became part of Boys Town's HRRC Application Kit for use by project directors at Boys Town. This kit was developed by the author and is described in more detail later in this chapter. Part 3 of the application follows.]

* * *

PART 3

INSTITUTIONAL REVIEW BOARD ROSTER

[Instructions: Please use this format. *Do not* send curriculum vitae, bibliographies, or detailed information other than that indicated.]

Name of Institution ______________________________
(Typewritten)

General Assurance Number

G ______________________

DATE: ______________

Institutional Officer (other than chairperson to whom correspondence should be addressed)

(Name – typewritten)

(Title – typewritten)

REGULAR MEMBER	ALTERNATE MEMBER
Name (typewritten): ______ Earned Degrees: ______ Position or Occupation: ______ Board certification, licenses, etc.: ______ Relationship to institution: ______ Representative capacity: ______	Designated Alternate (if any)
Name (typewritten): ______ Earned Degrees: ______ Position or Occupation: ______ Board certification, licenses, etc.: ______ Relationship to institution: ______ Representative capacity: ______	Designated Alternate (if any)
Name (typewritten): ______ Earned Degrees: ______ Position or Occupation: ______ Board certification, licenses, etc.: ______ Relationship to institution: ______ Representative capacity: ______	Designated Alternate (if any)
Name (typewritten): ______ Earned Degrees: ______ Position or Occupation: ______ Board certification, licenses, etc.: ______ Relationship to institution: ______ Representative capacity: ______	Designated Alternate (if any)

(Repeat as Needed)

NOTES

(a) The IRB must consist of not less than *five* members identified as above. above.
(b) Alternate members (if any) should be similarly identified and keyed to a particular member(s).
(c) The "Representative capacity" line should be used to indicate whether the individual serves the IRB as an investigator, administrator, practitioner, lawyer, ethicist, or in other capacities such as community representative.
(d) An IRB member may serve in more than one representative capacity.

(e) The Chairman of the IRB must be identified.
(f) IRBs reviewing activities involving the use of investigational new drugs (INDs) must include at least two individuals licensed to administer drugs and one person who is not so licensed. Review for other purposes should utilize IRBs of equal or greater breadth.
(g) No IRB may be composed entirely of members from the same professional group or entirely of members employed either full time or part time by the institution.
(h) When updating an existing committee, please submit a new roster, listing full membership as it currently stands.

[Note: See current IRB rules in this book for a more complete listing of new requirements.]

* * *

Institution Response to Form IRB. In response to the HEW request, Boys Town next forwarded various materials in support of receiving approval to form an IRB and being assigned a general assurance number. On October 25, 1977, a Boys Town staff member returned documents to HEW containing Boys Town's "Assurance of Compliance" with protection guidelines for human subjects and a roster of the first members of the IRB-to-be. The following versions of this correspondence are self-explanatory, but the reader should note that several of the provisions in the letters referred to federal regulations which have since been changed. An organization that might seek to establish its own IRB today would, of course, use terminology and protection rules currently approved by the Department of Health and Human Services as contained, for example, in this book.

* * *

J. R. Marches, Ph.D.
Office for Protection From
Research Risks
Office of the Director
Public Health Service
National Institutes of Health
Bethesda, Maryland 20014

October 25, 1977

Dear Dr. Marches:

Enclosed are the materials you stated were necessary to issue Boys Town a General Assurance number for Human Subject protection. I hope everything you need is here.

If I can supply any more information or help in any way feel free to call.

Thanks for your assistance in this matter.

Sincerely,

/s/(Boys Town Staff Member)
Enclosures

ASSURANCE OF COMPLIANCE WITH DHEW REGULATIONS ON PROTECTION OF HUMAN SUBJECTS

1. *Boys Town* will comply with the Department of Health, Education, and Welfare regulations on Protection of Human Subjects (45 CFR 46 as amended), accordingly:

2. This Institution has established and will maintain an individual Review Board competent to review projects and activities that involve human subjects. The Board shall determine for each activity as planned and conducted whether subjects will be placed at risk and if risk is involved, whether:

> The risks to the subject are so outweighed by the sum of the benefit to the subject and the importance of the knowledge to be gained as to warrant a decision to allow the subject to accept these risks;
>
> The rights and welfare of any such subjects will be adequately protected;
>
> Legally effective informed consent will be obtained by adequate and appropriate methods in accordance with the provision of the regulation;
>
> The conduct of the activity will be reviewed at timely intervals.

3. This Institution will provide for Board reviews to be conducted with objectivity and in a manner to ensure the exercise of independent judgment of the members. Members will be excluded from review of projects or activities in which they have an active role or conflict of interest.

4. This Institution will encourage continuing constructive communication between the Board and the activity director as a means of safeguarding the rights and welfare of the subjects.

5. This Institution will have available the facilities and professional attention required for subjects who may suffer physical, psychological, or other injury as a result of participation in an activity.

6. This Institution acknowledges that it will bear full responsibility for the proper performance of all work and services involving human subjects under any grant or contract covered by this assurance, including continuing compliance with pertinent state or local laws, particularly those concerned with informed consent.

7. This Institution will maintain appropriate and informative records of the Board's review of applications and activities, of documentation of informed consent, and of other documentation that may pertain to the selection, participation, and protection of subjects and to the review of circumstances that adversely affect the rights or welfare of individual subjects.

8. This Institution will at least annually reassure itself through appropriate administrative overview that its practices and procedures designed for the protection of the rights and welfare of human subjects are being effectively applied and are consistent with the regulation and with the implementation of this assurance as accepted by the Department of Health, Education, and Welfare.

9. This general assurance of compliance applies to *Boys Town*.

(Boys Town official)

____
(Signature of authorized official)

____
(Name – typewritten)

____
(Titled)

____
(Date)

____
(Telephone number)

INSTITUTIONAL REVIEW BOARD ROSTER

Name of Institution: Boys Town

General Assurance Number

G ____________________

DATE: ________________

Institutional Officer (other than chairperson to whom correspondence should be addressed)

____
(Name – typewritten)

____
(Title – typewritten)

REGULAR MEMBER

Name (typewritten): David D. Coughlin, Chairman
Earned Degrees: M.A., Ph.D.
Position or Occupation: Assistant Director of Youth Care
Board certification, licenses, etc.:
Relationship to institution: Employee
Representative capacity: Administrator

Name (typewritten): Pamela Daly
Earned Degrees: M.A., Ph.D.
Position or Occupation: Co-Director, Staff Education
Board Certification, licenses, etc.:
Relationship to institution: Employee
Representative capacity: Practitioner

Name (typewritten): James Garbarino
Earned Degrees: M.A., Ph.D.
Position or Occupation: Research Associate
Board certification, licenses, etc.:
Relationship to institution: Employee
Representative capacity: Investigator

Name (typewritten): Father James Kelly
Earned Degrees: A.A., B.A., M.S.
Position or Occupation: Catholic Pastor
Board certification, licenses, etc.:
Relationship to institution: Employee
Representative capacity: Ethicist

Name (typewritten): Edward LaCrosse
Earned Degrees: M.A., Ed.D.
Position or Occupation: Coordinator of Extramural Proj. and Resource Dev.
Board certification, licenses, etc.:
Relationship to institution: Employee
Representative capacity: Administrator

Name (typewritten): Dennis Maloney
Earned Degrees: M.A., Ph.D.
Position or Occupation: Co-Director of Teaching–Family Program
Board certification, licenses, etc.:

Relationship to institution: Employee
Representative capacity: Investigator

Name (typewritten): James Peck
Earned Degrees: M.A., Ph.D.
Position or Occupation: Coordinator of Audiological Services
Board certificaton, licenses, etc.:
Relationship to institution: Employee
Representative capacity: Practitioner

Name (typewritten): William Thornton
Earned Degrees: M.A., Ed.D.
Position or Occupation: Coordinator of Student and Program Evaluation
Board certification, licenses, etc.:
Relationship to institution: Employee
Representative capacity: Investigator

Name (typewritten): John Barksdale III
Earned Degrees: B.S., M.S.
Position or Occupation: Family Living Teacher
Board certification, licenses, etc.:
Relationship to institution: Employee
Representative capacity: Practitioner

Name (typewritten): Shelton Hendricks
Earned Degrees: M.A., Ph.D.
Position or Occupation: Professor, University of Nebraska at Omaha
Board certification, licenses, etc.:
Relationship to institution: None
Representative capacity: Community Representative/Investigator

* * *

Approval of IRB. The third and final step in the sequence of obtaining approval of Boys Town's IRB took place on November 29, 1977. At that time, Dr. D. T. Chalkley of the HEW Office for Protection from Research Risks approved Boys Town's assurance procedures (operation of its Human Rights Review Committee as an IRB) and established G3112 as Boys Town's General Assurance number.

* * *

DEPARTMENT OF HEALTH, EDUCATION AND WELFARE

Public Health Service
National Institutes of Health
Bethesda, Maryland 20014

November 29, 1977

General Assurance G3112

Director
Boys Town
Boys Town, Nebraska 68010

The Department of Health, Education, and Welfare has approved a general institutional assurance dated October 25, 1977, as amended November 18, 1977, submitted by Boys Town in compliance with the requirements for the protection of human subjects prescribed in Part 46 of Title 45 of the Code of Federal Regulations (45 CFR 46).

The implementing procedures submitted in support of the assurance are accepted subject to the issuance of additional or superseding regulations, and further negotiation as may be considered desirable. It is essential that proposed changes in membership of the Institutional Review Board, and changes in institutional policies, or in procedures for protection of the rights and welfare of human subjects be reported to this office.

This assurance supersedes previous assurances and is effective immediately for all applications hereafter submitted for new and renewal of DHEW support of activities involving human subjects, and will become effective within the ensuing year for current activities as each such activity is the subject of continuing interim review by the Institutional Review Board and prior to application for continuation of support for the activity.

Approval of the assurance will remain in effect until superseded or terminated by either your institution or the Department following appropriate advance notification.

We are prepared to assist you in any way we can in your institutional efforts to provide for the protection of human subjects. Please call us on Area Code 301, 496–7005 if you feel we can be of service to you.

Sincerely,

/s/D. T. Chalkley

D. T. Chalkley, Ph.D.
Director, Office for Protection from Research Risks

* * *

Institutional Review Boards around the United States

Before we leave the topic of how to set up an IRB, a brief note is warranted. Although Boys Town did indeed go through such a step, its own IRB is now less active since research is no longer a major part of Boys Town programs. Nevertheless, the steps taken by Boys Town and experienced by the author as formation of an IRB was sought serve as appropriate examples for other institutions. Many organizations have followed such steps in forming their own IRBs.

Several hundred IRBs already exist in the United States, and they are listed in the *61st Cumulative List* (1980) that is contained in the appendix to this chapter. This reference should be especially useful for individuals or organizations considering formation of their own IRB. Since such a step can take months to accomplish, a researcher may be better off locating an IRB in the list closest to his or her locale to seek advice or potential IRB approval of a project. On the other hand, if research is to be an ongoing activity and several projects may be seeking federal funding support, it may be advisable to consider formation of a new IRB closer to hand. This may especially be true if an IRB does exist close by but is not approved by HHS to review a particular *type* of research. Sufficient information is contained in the *61st Cumulative List* in the appendix for this chapter to enable the reader to judge the review authority of all IRBs in the United States as of this publication.

Special Assurance and General Assurance

Up until now, we have used only the phrase *general assurance* in discussing the formation of an IRB. Essentially, the terms refer to an ongoing assurance by an IRB to the federal government that appropriate review will be conducted by the IRB of any research proposal that is sent to a federal funding agency. Once a general assurance number is assigned by the Department of Health and Human Services to an IRB, the IRB then conducts the review of research proposals and the government generally will simply follow the recommendations of the IRB unless some doubt exists about the adequacy of the researcher's protection of subjects.

However, this almost total reliance on a general assurance mechanism is relatively recent. In the 1970s there was an additional mechanism used by HEW, referred to as a *special assurance*. The reader should note that this special assurance mechanism is no longer part of federal regulations, although the author has heard researchers mistakenly refer to it still as an alternative to working through an IRB. In fact, the term *general assurance* is no longer used (since there is no *special assurance* with which to contrast it), but the process of obtaining assurance remains similar to what used to fall under the general assurance label.

Special Assurance

Since special assurance is no longer part of HHS rules, we will only deal with the topic briefly. Researchers who become familiar with its history can readily see that it no longer is a readily obtainable alternative to working through an established IRB.

On May 30, 1974, HEW published its early rules on protection of human subjects.[2] We have already discussed most of the sections of these early rules with regard to informed consent, IRBs, and so forth. Also contained in these rules were provisions for a general assurance and a special assurance. A special assurance was defined as:

> (b) Special assurances. A special assurance will, as a rule, describe those review and implementation procedures applicable to a single activity or project. A special assurance will not be solicited or accepted from an organization which has on file with DHEW an approved general assurance.[3]

Basically, the special assurance procedure required the researcher and his or her organization to describe in detail how subjects would be protected, what kind of consent procedure would be used, how a local committee was going to review the research, and so on for each separate project. If this material was approved by HEW, then they granted a special assurance number for that project only. If another investigator wished to seek funding for a project, he or she would go through all the same steps again and HEW would then review that project.

Obviously, this approach increased the detailed review steps for HEW staff and presumably lengthened the amount of time necessary for review by requiring back-and-forth correspondence between Washington and a researcher's organization. But it should be noted that there were fewer IRBs then in the country, so HEW had little choice if it was to ensure that protection procedures were reviewed by someone.

General Assurance

In 1974, general assurance steps were listed along with those for a special assurance.

> (a) General assurances. A general assurance describes the review and implementation procedures applicable to all DHEW-supported activities conducted by an orga-

[2]*Federal Register*, Vol. 39, No. 105, Thursday, May 30, 1974, pp. 18914–18920.
[3]Ibid., p. 18918.

> nization regardless of the number, location, or types of its components or field activities. General assurances will be required from organizations having a significant number of concurrent DHEW-supported projects or activities involving human subjects.[4]

Thus, if an institution were to have several projects requiring some kind of review for subject protection, HEW did not wish to go repeatedly through a special assurance procedure for each project.

Although some details on the general assurance mechanism changed from 1974 to 1981, the essential nature of the process and its reliance on IRBs to conduct the actual reviews have not significantly changed. The 1981 rules[5] on general assurance are also provided separately in the appendix. In the 1981 rules, the reader will note that no longer is there any listing of special assurance since HHS's reliance for protection review has totally shifted to IRBs. However, the Department of Health and Human Services still retains the final review authority to approve or disapprove the recommendations of an IRB when a project is forwarded to HHS for possible funding.

Applying to an IRB for Approval

Now that we are familiar with the steps needed to form an IRB, let us look more closely at how they operate. We have already learned in previous chapters that specific federal regulations exist governing such factors as how many members must comprise an IRB, what type of informed consent can be approved (or waived) by an IRB, and other procedures regarding how an IRB documents its review steps in case HHS wishes to audit its procedures. Earlier in this chapter we saw how an IRB can be established.

But how does the individual investigator interact with an IRB to get his or her research proposal approved? In addition to needing feedback on the appropriateness and adequacy of subject protection, the anxious researcher typically is already worrying about federal agency submission deadlines when the proposal goes to the IRB for review. What if the IRB takes so long to review the proposal, or requires so many revisions (e.g., in the consent procedures) that the researcher can not get the final approved version to Washington in time for consideration of funding support by the relevant agency?

Although the author has never experienced or heard of an IRB's deliberately delaying review of a proposal, problems in meeting deadlines certainly do arise. It would appear to be the responsibility of the researcher to submit sufficient information to an IRB for their review. Preferably, the proposal will contain all the information necessary for IRB members (usually researchers themselves) to

[4]Ibid., p. 18917–18918.
[5]*Federal Register*, Vol. 46, No. 16, Monday, January 26, 1981, pp. 8366–8392.

conduct a thorough review of the steps to be taken by the researcher to protect subjects' rights. At the same time, the IRB should provide enough information to the researcher so that it is clear *ahead of time* just what type of information will be required for IRB review of the research.

The author was involved in developing an application kit[6] at Boys Town, in consultation with other HRRC members, so that project directors would know what, when, and how to submit a proposal to our IRB. This kit was then distributed and made available to staff who were submitting or considering submitting proposals to Boys Town's IRB. The following sections are taken from this kit, consisting of five parts: (1) a memo to investigators describing IRB review steps; (2) detailed instructions on preparing a proposal for IRB review; (3) a roster of IRB members; (4) a schedule of IRB meetings; and (5) an appendix of various sample IRB and informed consent forms.

This kit naturally contains some information particular only to Boys Town, but persons at other organizations could substitute relevant material. This kit was developed toward the end of the author's tenure as chairman of the IRB in direct response to a variety of problems that arose in the absence of such a kit. In the author's experience, and in the light of comments by a subsequent IRB chairman, such a kit (or anything that achieves the same results) can be an invaluable aid to researchers and IRB members alike. As we will see later in this chapter, friction between researchers and IRBs is not at all unusual, and it may be that administrative tools such as this kit can alleviate difficulties. The President's Commission, in cooperation with NIH and FDA, is working on a looseleaf binder to contain IRB rules and a section for a particular IRB's member roster, meeting schedule, and so forth. It was projected to be available to the public in 1983.

Memo Describing IRB Review Steps

This memo served as the first part of the kit and described to investigators how to submit a proposal to the IRB. The reader should keep in mind that any reference within the kit to federal regulations was based on regulations in effect *at that time*. Any changes in the rules (especially those of January, 1981) would need to be reflected in any current kit.

* * *

HUMAN RIGHTS REVIEW COMMITTEE

TO: Persons submitting a project for review by the Humans Rights Review Committee (HRRC)

[6] *Application Kit* (Human Rights Review Committee). Boys Town, Nebraska: Father Flanagan's Boys' Home, January, 1979.

FROM: Dennis Maloney, Chairman (1978) Human Rights Review Committee of Boys Town

RE: Review Procedure

DATE: January 22, 1979

The following excerpt from Boys Town's Policy #59400 establishes the need for the HRRC:

> It is the purpose of this policy to establish a Human Rights Review Committee to insure that the rights of all participants in any research, development, or related activity at Boys Town are protected. Thus, persons directing any research related projects should submit an application to the Human Rights Review Committee.

Persons needing to submit a project to the Human Rights Review Committee must follow the established review procedures described below:

STEP 1. Obtain an "Application for Review" form (HRRC-1) from either the HRRC Secretary or from a representative from your Boys Town Department who is an HRRC member. Current HRRC members are listed on the "HRRC Committee Roster" included in this HRRC APPLICATION KIT. *Do not use* the example forms included in this KIT. Forms can be obtained from your HRRC representative or from the HRRC Secretary.

STEP 2. Submit twelve (12) copies of the completed application form to your Department representative HRRC member. If this is not possible, send your 12 copies directly to the HRRC Secretary. Applicants should note that their application may be directly returned for revisions by either their Department representative or by the HRRC Secretary *before* review of your application by the HRRC. This typically occurs when the application is incorrectly completed or submitted in such a fashion as to prevent adequate and timely review by the HRRC. Please refer to the "Instructions for Completing a HRRC Application" section in this HRRC APPLICATION KIT for further details.

STEP 3. As the applicant, you will then receive a "Notice of Receipt of Application" (HRRC-4) to signify that the HRRC is processing your application.

STEP 4. Your application will be discussed and reviewed at the next regularly scheduled HRRC meeting. Application review meetings will be normally scheduled to occur in the first week of each month (see HRRC Meetings Schedule for exact days). Inasmuch as scheduled meetings are cancelled if no applications are pending, applications should be submitted at least 10 days prior to the scheduled meeting. This permits adequate time for the chairperson to confirm the meeting. Applicants submitting grants to extra-mural funding agencies and needing the review process expedited in order to meet funding application deadlines should so notify the chairperson of the HRRC.

STEP 5. Following the review of your application, you will receive a "Notice of Review Decision" (HRRC-3). This notice will indicate either: (a) approval of your application; (b) approval, pending revisions of your application; or (c) a vote of nonapproval.

If your application is approved, no further action is necessary on your part unless your approved project changes subsequently in such a manner that protection of research subjects is affected. If such changes occur, notice of such changes must be forwarded by the applicant to the HRRC. Additionally, all approved projects are requested to file brief progress statements with the HRRC within one year of the original HRRC approval. It is the responsibility of the applicant to file an "Annual Progress Report" (HRRC-6) within 12 months of HRRC approval of a project. An example copy of the "Annual Progress Report" is included in this HRRC APPLICATION KIT.

If your application is approved with revisions, it is necessary for you to resubmit the application, usually by starting again with Step #1 of these procedures.

The final disposition of your application also will be communicated to your Department Director.

SPECIAL NOTES

It is emphasized that the HRRC was established to protect the rights and welfare of participants in research, development, or related activities supported by Boys Town. The primary role of the HRRC is to ensure that adequate safeguards are instituted by principal investigators to protect such participants and that appropriate procedures are followed. Applicants are urged to consult with their Departmental representatives or other HRRC members on matters of participants' rights.

* * *

Detailed Instructions for Preparing IRB Proposal

The second part of the kit in our example is the section containing detailed instructions to investigators regarding how to submit proposals for review by Boys

Town's IRB (HRRC). The following is taken from the HRRC Application Kit. Note the extent of the detail (i.e., the level of instructions to the applicants) that was used to minimize the need for any additional communication between the HRRC and the individual investigator. Certain portions of the instructions refer to different parts of various HRRC forms that are presented later in this chapter.

* * *

HUMAN RIGHTS REVIEW COMMITTEE

Instructions for Completing HRRC Application (HRRC-1)

PART A - APPLICANT INFORMATION

- Complete all sections of Part A.
- For "Date," enter the date you complete the application.
- For "Title of Project," enter the official title of the proposed project. If a grant is involved, enter the title as it appears on the grant application.
- For "Brief Description of Project, enter a *brief* description of the activity. Do not use more space than allowed on the application form.
- Do not append any supplementary information to the application to further explain items for Part A. If the HRRC requires such information, you will be contacted.

PART B - PROTECTION OF SUBJECTS

- Do not use more space than allowed for any question in Part B.
- It is not necessary to fill in each line of each question if shorter responses will still provide the HRRC with adequate information.

1. *Statement of the problem to be investigated:* Enter a clear and concise description of the problem being addressed by the project or activity. It is appropriate here to state the hypothesis to be tested or the general questions to be answered.
2. *Description of the method, procedures, and subjects involved:* Enter a concise description of the research methods and procedures to be used in the project. This may include statements about where the study will take place, what kind of data will be collected, and what types of measurement will be used (e.g., questionnaires vs. direct observations vs. clinical exams vs. medical records). This section must contain a description of the types of subjects who will be involved (e.g., age, sex).
3. *Review of literature describing extent to which any portion of the project is "experimental" in nature and comment on any "experimental treatment" that may be involved:* Enter references that relate the project to previous work. In-

clude citations of appropriate references but it is not necessary to include full references nor is an appended bibliography necessary. A statement must be included describing the presence or absence of any "experimental treatment."
4. *Description of potential risks and potential benefits for subjects:* Enter a description of both risks and benefits, if any. Please note that these risks and benefits (if any) must also be explained to subjects on the Informed Consent form, but the language used in this section of the HRRC Application may differ from the language used on the Informed Consent form.

- Risks. "Subject at risk" has been defined by the Department of Health, Education, and Welfare (DHEW) in the Code of Federal Regulations, Title 45, Subtitle A, Part 64, Subpart 46.3(b) for the Protection of Human Subjects, 3/13/75:

> Subjects at risk means any individual who may be exposed to the possibility of injury, including physical, psychological, or social injury, as a consequence of participation as a subject in any research, development, or related activity which departs from the application of those established and accepted methods necessary to meet his needs, or which increases the ordinary risks of daily life, including the recognized risks inherent in a chosen occupation or field of service.

Any risk to a subject which is included in the above definition must be described in this section for the HRRC and described in some manner to each subject.

- Benefits. Two types of benefits have been established by the DHEW as possible outcomes of research in the Code of Federal Regulations, Title 45, Subtitle A, Part 46, Subpart 46.2(b) (1) for the Protection of Human Subjects, 3/13/75. These two possible types of benefits for subjects are:

> (specific) ". . . benefit to the subject . . ."
>
> (general) ". . . and the importance of the knowledge to be gained . . ."

Although specific benefit to the subject is preferable, it is not relevant for many types of approved research and thus the general type of benefit is also permissible. The type of benefit involved must be described in this section and on the Informed Consent form. If specific, this section of the HRRC Application form must include information regarding benefits the subjects can expect to receive, directly or indirectly, as a result of participation in the research. If general, the importance of the knowledge to be gained must be described.

5. *Description of steps taken to ensure rights and welfare of subjects:* Enter a description of how the previously described risks will be minimized or avoided entirely. This may include how the research data will be kept confidential, how

subjects' privacy will be maintained, how any possible subject embarrassment or discomfort will be alleviated, etc. For projects involving any form of deception, this section must contain a description of how subjects are to be debriefed following their participation in the project.

Please note that this information does *not* have to be included in the Informed Consent form. However, applicants are encouraged to include a summary of these steps anyway in their Informed Consent forms for the benefit of subjects. It is often advisable to inform subjects how they will be protected from any risks described in the "Risks" portion of the Informed Consent form.

6. *Description of methods used to obtain "informed consent" of subjects found to be "at risk"*: Enter a brief description of the procedures to be used in obtaining subject consent. This will be either in the form of (a) a written consent document, signed by each subject or his authorized representative, embodying all of the elements of informed consent, (b) a "short form" written consent document indicating that the basic elements of informed consent have been presented orally to the subject or his authorized representative, or (c) a modification of the above procedures approved by the Board (HRRC). In each case, however, the procedures will follow the requirements set forth in Subtitle A of Title 45 of the Code of Federal Regulations, Part 46, Subpart 46.10 on the Protection of Human Subjects:

> "The actual procedure utilized in obtaining legally effective informed consent and the basis for Institutional Review Board (HRRC) determinations that the procedures are adequate and appropriate shall be fully documented. The documentation of consent will employ one of the following three forms:
>
> "(a) Provision of a written consent document embodying all of the basic elements of informed consent. This may be read to the subject or to his legally authorized representative, but in any event he or his legally authorized representative must be given adequate opportunity to read it. This document is to be signed by the subject or his legally authorized representative. Sample copies of the consent form as approved by the Board are to be retained in his records."
>
> "(b) Provision of a 'short form' written consent document indicating that the basic elements of informed consent have been presented orally to the subject or his legally authorized representative. Written summaries of what is to be said to the patient are to be approved by the Board. The short form is to be signed by the subject or his legally authorized representative and by an auditor witness to the oral presentation and to the subject's signature. A copy of the approved summary, annotated to show any addi-

tions, is to be signed by the persons officially obtaining the consent and by the auditor witness. Sample copies of the consent form and of the summaries as approved by the Board are to be retained in its records."

"(c) Modifications of either of the primary procedures outlined in paragraphs (a) and (b) of this section. Granting of permission to use modified procedures imposes additional responsibility upon the Board and the institution to establish: (1) that the risk to any subject is minimal, (2) that use of either of the primary procedures for obtaining informed consent would surely invalidate objectives of considerable immediate importance, and (3) that any reasonable alternative means for attaining these objectives would be less advantageous to the subjects. The Board's reasons for permitting the use of modified procedures must be individually and specifically documented in the minutes and in reports of Board actions to the files of the institution. All such modifications should be regularly reconsidered as a function of continuing review and as required for annual review, with documentation of reaffirmation, revision, or discontinuation, as appropriate." (Code of Federal Regulations: 45 CFR 46:110, revised as of January 11, 1978.)

The "long form," the "short form," and the subjects' "legally authorized representative" are defined as follows:

- Long Form ". . . (a) Provision of a written consent document embodying all the basic elements of informed consent. This may be read to the subject or to his legally authorized representative, but in any event he or his legally authorized representative *must be given adequate opportunity to read it* (italics added). This document is to be signed by the subject or his legally authorized representative . . ." (Code of Federal Regulations, Title 45, Subtitle A, Part 46, Subpart 46.10 (a) for the Protection of Human Subjects, 3/13/75).
- Short Form ". . . (b) Provision of a 'short form' written consent document indicating that the basic elements of informed consent have been presented orally to the subject or his legally authorized representative. Written summaries of what is to be said to the patient are to be approved by the Board. Short form is to be signed by the subject or his legally authorized representative and by an auditor witness to the oral representation and to the subject's signature. A copy of the approved summary, annotated to show any additions, is to be signed by the persons officially obtaining the consent and by the auditor witness . . ." (Code of Federal Regulations, Title 45, Subtitle A, Part 46, Subpart 46.10(b) for the Protection of Human Subjects, 3/13/75).
- Legally Authorized Representative ". . . 'legally authorized representative' means an individual or judicial or other body authorized under applicable law

to consent on behalf of a prospective subject to such subject's participation in the particular activity or procedure . . ." (Code of Federal Regulations, Title 45, Subtitle A, Part 46, Subpart 46.3(h) for the Protection of Human Subjects, 3/13/75).

Parental permission is required for all minor children unless the youth is emancipated. A minor child's consent also may have to be obtained (see appendix).

– Regardless of the type of informed consent used, it must contain *seven elements* as determined by the DHEW. These elements are described as follows and are taken directly from the Code of Federal Regulations, Title 45, Subtitle A, Part 46, Subpart 46.3(c) (1–6) for the Protection of Human Subjects (3/13/75):

(c) "Informed consent" means knowing consent of an individual or his legally authorized representative, so situated as to be able to exercise free power of choice without undue inducement or any element of force, deceit, duress, or other form of constraint or coercion. These basic elements of information necessary to such consent include:

1. A fair explanation of the procedures to be followed, and their purposes, including identification of any procedures which are experimental;
2. A description of any attendant discomforts and risks reasonably to be expected;
3. A description of any benefits reasonably to be expected;
4. A disclosure of any appropriate alternative procedures that might be advantageous for the subject;
5. An offer to answer any inquiries concerning the procedures;
6. An instruction that the person is free to withdraw his consent and to discontinue participation in the project or activity at any time without prejudice to the subject; and
7. Effective 1/2/79, a seventh element of Informed Consent was added (43 FR, 11/3/78): 'with respect to biomedical or behavioral research which may result in physical injury, an explanation as to whether compensation and medical treatment is available if physical injury occurs and, if so, what it consists of or where further information may be obtained. This subparagraph will apply to research conducted abroad in collaboration with foreign governments or international organizations absent the explicit nonconcurrence of those governments or organizations."

The National Institute of Health Guide for Grants and Contracts (Vol. 7, No. 19, 12/15/78) recommended the following possible wordings to satisfy element 7 for Informed Consent:

Suggested wording for consent forms: "I understand that in the event of physical injury resulting from the research procedures, (state what is available, e.g., 'medical treatment for injuries or illness available'/'only acute/immediate/essential medical treatment (including hospitalization) is available'/'monetary compensation is available for wages lost because of injury') or what is not available, e.g., 'financial compensation is not available, but medical treatment is provided free of charge,' etc."

- Final Note on Informed Consent: The HRRC typically does *not* approve of "negative return" informed consent procedures in which investigators send a letter to subjects asking them to return a form only if they *do not* want to participate. The assumption with such a procedure is that if a subject does not return a form, he/she has agreed to participate. The HRRC typically disapproves such procedures for the following reasons:

    1. It assumes that the form was delivered to the subject and that no form was returned due to subject consent . . . ignoring the fact that the subject may not have received the form in the first place;
    2. It usually does not provide any opportunity for subjects to ask questions of the investigator before deciding *not* to return the form;
    3. There is no written record of agreement signed by the subject indicating the precise nature of the agreement;
    4. There is no proof that the investigator presented the seven required elements of informed consent to the subject;
    5. It is not easy for a subject to later decline participation in the project unless the investigator has made provisions for sending in a "no" on a form *after* the study has begun.

7. *Special Considerations:* In this section of the HRRC Application, enter a description of any unusual circumstances regarding the project relevant for HRRC review. For example, if the project is already underway and the "proposed" Informed Consent forms have already been used, inform HRRC of this fact in this section. If there are agreements with other Institutional Review Boards—usually with grants where several organizations are involved—describe the situation here.

- Sign the application form.
- Attach the Informed Consent form(s) to the application.
- If the project involves questionnaire(s), please attach one copy of each questionnaire to the HRRC Application form. If questionnaire(s) *will be* used but are not yet developed, the investigator is responsible for sending a copy of the questionnaire to the HRRC later *before* actually using the questionnaire.

* * *

IRB Roster

The third portion of the IRB Application Kit was a roster of the names of the IRB members. This roster served two purposes: (1) it was available to forward to HEW as part of their general assurance process; and (2) it alerted investigators to the names of colleagues on the IRB whom they could call upon for more information. From a purely practical standpoint, it is to be expected that many organizations will prefer that representatives from their major administrative sections serve on their IRB. Thus these persons will serve as primary liaisons for members of the various sections who may submit a proposal to the organization's IRB. The following roster was included in the Boys Town IRB Application Kit.

* * *

HUMAN RIGHTS REVIEW COMMITTEE

HRRC Membership Roster (Date: January 12, 1979)

As of the above date, the following persons serve on the HRRC:

Member	Boys Town Department
Dennis Culver	Youth Care
Jim Garbarino, Ph.D.	Center
Betty Johnson	Youth Care (Family-Teacher)
Shelton Hendricks, Ph.D.	University of Nebraska at Omaha
Father James Kelly	Spiritual Affairs
Richard Lippman, Ph.D.	Institute
David Cyr (Chairman)	Institute
Vaughn Call, Ph.D.	Center
Jim Peck, Ph.D.	Institute
Cary Bell, Ph.D.	Youth Care (Education)
Jackie Bousha (Secretary)	Institute

* * *

Schedule of IRB Meetings

The fourth section of the kit contained a schedule of prearranged IRB meetings so that investigators could more easily plan when to submit proposals to the IRB to later meet funding deadlines for various federal agencies. The following is the list from the Boys Town HRRC Application Kit.

* * *

HUMAN RIGHTS REVIEW COMMITTEE

HRRC Meetings Schedule – 1979

February 15	1:30–3:00 P.M.
March 8	1:30–3:00 P.M.
April 12	1:30–3:00 P.M.
May 10	1:30–3:00 P.M.
June 14	1:30–3:00 P.M.
July 12	1:30–3:00 P.M.
August 9	1:30–3:00 P.M.
September 13	1:30–3:00 P.M.
October 11	1:30–3:00 P.M.
November 8	1:30–3:00 P.M.
December 13	1:30–3:00 P.M.

As of February 14, 1979, HRRC meetings are held in the Research Seminar Room at the Boys Town Institute (555 No. 30th, Omaha, NE). Meetings are open to HRRC members only unless nonmembers have been specifically requested to attend by the committee.

Notices of upcoming HRRC meetings typically are published in *The Boys Towner* at least one week in advance.

* * *

Sample Forms

The final section of the kit contained samples of forms used to submit information to the IRB and different types of consent forms that could be used by the investigators with their subjects. The following examples of forms from the HRRC kit consist of two parts: (a) forms used as part of IRB processing of a research proposal and (b) examples of various consent forms geared toward different types of subjects.

IRB Form ("Application for Review"). The first form that an investigator would see would be the application form with which the research to be reviewed by the IRB would be described. The blank form is self-explanatory and was designed to alert the IRB to matters of subject protection primarily, whereas information on research design and related matters was considered secondary but necessary. This form, and the others subsequent to it, have been retyped for this book.

* * *

Boys Town, Boys Town, Nebraska 68010

HUMAN RIGHTS REVIEW COMMITTEE

For HRRC Use Only:

Application No.: ______________

Date Received: ______________

(If Revision, indicate with R)

Application for Review

PART A - APPLICANT INFORMATION

Principal investigator: ______________ Date: ______________

Principal investigator's title: ______________

Principal investigator's
business address: ______________
Organization

____
Street

Phone: ______________ ______________
City State Zip code

Home address: ______________
Street

Phone: ______________ ______________
City State Zip code

Title of project: ______________

Brief description of project: ______________

____

____

PART B - PROTECTION OF SUBJECTS

1. Statement of the problem to be investigated: ______________

____

2. Description of the method, procedures, and subjects involved:

3. Review of literature describing extent to which any portion of the project is "experimental" in nature and comment on any "experimental treatment" that may be involved: ______

4. Description of potential risks and potential benefits for subjects: ______

5. Description of steps taken to ensure rights and welfare of subjects: ______

6. Special considerations: ______________________________

(Signature of principal investigator)

Copy of *actual* informed consent form(s) must be attached to this application.

HRRC-1 (Rev. 1/79)

* * *

IRB Form ("Notice of Receipt of Application"). Early in the development of this particular IRB, it had been noted that some researchers (particularly those with approaching funding guidelines) became concerned after submitting an application about the timeliness of the review process. As it turned out, a substantial portion of this concern resulted from researchers's uncertainty as to when the process of review commenced after submission of an application. In addition, the IRB wished to inform the researcher quickly if any revisions in the application were immediately apparent. Hence, the "Notice of Receipt of Application" was developed.

* * *

Boys Town, Boys Town, Nebraska 68010

HUMAN RIGHTS REVIEW COMMITTEE

For HRRC use only:

Application No.: ________________

Date Received: ________________

(If Revision, indicate with R)

Notice of Receipt of Application

TO: Principal Investigators

FROM: Dennis Maloney, Chairman

RE: Receipt of Application

DATE:

Thank you for submitting your application to the Human Rights Review Committee (HRRC). To review your application adequately, further information

is not required _____. is required as follows: _________

____

____

____

If further information is needed, it must be received by the HRRC chairman as soon as possible. You may wish to discuss your application with your Boys Town Department HRRC representative. Additionally, feel free to call on the HRRC chairman to discuss any related issues at any time during the review process. Please reference your application number in any correspondence with the HRRC.

Thank you.

For HRRC use only:

Principal Investigator: ______________________________

Title of Project: ______________________________

HRRC-4 (Rev. 5/78)

* * *

IRB Form ("Notice of Review Decision"). Following IRB deliberation, this form officially notified the researcher of the result of the review. Of course, most researchers will hear of the result on an informal basis, but such a form documents the IRB's decision both for the applicant and for the IRB files which are subject to HHS examination. Nationally, there has been concern expressed that IRB files do not always document a decision; such documentation can be very useful if future questions about a project ever arise. Simply relying on the memory of IRB members (who may review dozens of projects a year) is not advisable.

* * *

Boys Town, Boys Town, Nebraska 68010

HUMAN RIGHTS REVIEW COMMITTEE

For HRRC use only:

Application No.: ______________

Date Received: ______________

(If Revision, indicate with R)

Notice of Review Decision

TO: Principal Investigators

FROM: Dennis Maloney, Chairman

RE: Review of Application to HRRC

DATE:

Your application, as described below, has been reviewed by the HRRC. The decision has been made to:

Approve ____ Approve with revision(s) ____ Not approve ____
your application.

Any questions regarding this decision should be directed to the HRRC chairman. Please reference your application number in any correspondence with the HRRC.

Thank you.

For HRRC use only:

Principal Investigator: ______________________________

Title of Project: ______________________________

Date of Approval: ______________________________

HRRC-3 (Rev. 1/79)

* * *

IRB Form ("Annual Progress Report"). In compliance with HHS regulations, the kit also contained a blank form to be used by the researcher once a year if the research was still going on.

* * *

Boys Town, Boys Town, Nebraska 68010

HUMAN RIGHTS REVIEW COMMITTEE

For HRRC use only:

Application No.: ____________

Date Received: ____________

(If Revision, indicate with R)

Annual Progress Report

PART A - APPLICANT INFORMATION

Principal investigator: ______________________ Date: __________

Business address: ______________________________________
Organization

____
Street

Phone: __________ ______________________________________
City State Zip code

Title of project (as approved by HRRC): ______________________

____

Brief description of project (as approved by HRRC): __________

____

____

PART B - PROTECTION OF SUBJECTS

1. Project conducted in any way different from HRRC approved application as far as protection of subjects are concerned?

   Yes ____ No ____

2. If yes, explain (attach additional pages if needed): __________

____

____

____

____

____
(Signature of Principal Investigator)

HRRC-6 (Rev. 1/79)

* * *

Informed Consent Example ("'Long Form' for Adult or Mature Youth"). This form, along with subsequent consent examples, was included in the kit to aid researchers who were not readily familiar with consent forms. Although these consent forms might need to be revised slightly to conform to 1981 consent rules in some instances, they remain essentially valid and serve as useful examples. The elements of consent are clearly and completely specified. The consent forms were written for different types of research.

Note that the example is listed as being for an adult or mature youth. In the opinion of the author and of the IRB, the language in this form should have been understood by most older adolescents or adults, whereas later forms presented in this chapter were designed for use with younger children. However, the wording used in the forms was based on reactions of subjects reported to the IRB before the application kit had been developed. Although this information certainly proved useful in developing the different types of consent forms, current researchers may produce superior forms (in terms of readability) by using the formulas discussed in the previous chapter for aiming a form at a particular age or education level audience.

Note also with this form that, although the federal rules are vague on this point, the form requests the signature of a guardian if the subject is under 21. As we have seen before, in the absence of a clear definition of when parental consent is also needed, the safest procedure is to request it anyway.

* * *

EXAMPLE ONLY

"Long Form" Consent for Adult or Mature Youth

Youth Place, Anywhere, USA 00001

INFORMED CONSENT TO PARTICIPATE IN YOUTH PLACE'S "FOLLOW-UP STUDY" RESEARCH PROJECT

This form represents the voluntary consent of ______________________
(First name)

______________________ to participate in the project known as the
(Middle initial) (Last name)
"Follow-Up Study."

PROCEDURES: I understand that my consent allows the collection, analysis, and dissemination of information about my life after I left Youth Place. I understand that this includes information about me or my family since I left Youth

Place such as: living environment (e.g., places lived); educational experiences (e.g., grade point averages, attendance); vocational and occupational experiences (e.g., number and types of employment); and legal histories (e.g., police and court contacts).

DISCOMFORTS OR RISKS: I understand that there is no physical risk or discomfort involved. I understand that the potential social or psychological risk or discomfort I may experience is that my name may become publicly associated with particular facts about my life. I understand that I am protected from this risk by safeguards described in the PRIVACY PROTECTION section of this form.

PRIVACY PROTECTION: I understand that the Study Director for the "Follow-Up Study" protects my privacy by keeping information about me in locked files. I understand that publication of study results in any form will protect my privacy and disguise me by adding together information about me along with information about others and by using some name or code number instead of my real name.

AVAILABILITY OF MEDICAL TREATMENT AND COMPENSATION: I understand that in the event of physical injury resulting from the research procedures, emergency medical treatment will be provided but financial compensation is not available.

BENEFITS: I understand that the potential benefits of the study are that it will provide information to help future youths experience programs that are most likely to help them after they leave Youth Place. It also will permit distribution of successful Youth Place program methods around the world.

ALTERNATIVES (Freedom of Choice): I understand that the alternative to my participation in the study series is to not participate.

RIGHT TO WITHDRAW: I understand that if I wish to withdraw my consent later I may do so without penalty to me even after I sign this consent form. I agree that such a withdrawal will be made in writing to the Study Director for the "Follow-Up Study."

RIGHT TO INQUIRE: I understand that I have the right to inquire at any time about the procedures described in this document. I understand that I can direct these inquiries to the Study Director.

I understand that my signature below signifies my voluntary informed consent to participate in the "Follow-Up Study."

________________________________ ________________

(Participant signature) (Date)

For Participant under age 21, signature of guardian must be present.

______________________________ _______________

(Guardian signature) (Date)

Inquiries may be sent to:

John Doe, Ph.D.
Study Director, "Follow-Up Study"
Youth Place
Anywhere, USA 00001
Phone: (999) 000–0000

* * *

Informed Consent Example ("'Long Form' Consent for Child"). In this form, the wording has been simplified to be more personal and the two sections on risks and confidentiality have been combined. The form is thus shorter and presumably easier to read and comprehend so that a younger child would have no difficulty with it. The assumption inherent in this form is that the subject is a young adolescent or older preadolescent and is capable of reading and consenting. Later forms will be the short forms where information is read aloud to the subject and thus the child's reading level is less important.

* * *

EXAMPLE ONLY

"Long Form" Consent for Child

Youth Place, Anywhere, USA 00001

INFORMED CONSENT TO TAKE PART IN
YOUTH PLACE'S "NEIGHBORHOOD INTERVIEWS"

This form represents the agreement of ______________________
(First name)

______________________________ to take part in the research project
(Middle initial) (Last name)

known as "Neighborhood Interviews."

PROCEDURES: I understand that if I take part in this research, I will answer questions during an interview about what I think of my neighborhood. Questions will include what I think of neighborhood schools, churches, social agencies and what I think of the effect these things have on my family. I understand that my answers will be written down and studied by the people doing this research.

POTENTIAL RISKS OR DISCOMFORTS: I understand that if I take part in this study there will be no physical danger or even discomfort other than answering questions for about an hour for the interview. I also know that the researchers will make every effort to keep my name secret. If the researchers ever tell anyone in any way what I said during the interview, they will use only code numbers or false names and never my real name. My answers will be kept in locked files at Youth Place.

AVAILABILITY OF MEDICAL TREATMENT AND MONEY: I understand that, in the event of physical injury resulting from the research procedures, emergency medical treatment will be provided but extra money is not available.

BENEFITS: I understand that it is expected that my answers—and the answers of others—may help improve neighborhoods and make life better for families.

ALTERNATIVES (Freedom of Choice): I understand that I do not have to take part in this study if I don't want to.

WITHDRAWAL OF CONSENT: I understand that I can stop taking part in the study later if I want to—even after I sign this agreement—and the researchers will agree to let me stop. If I want out later, I will tell the researchers in writing that I want to quit.

RIGHT TO INQUIRE: I understand that if I have questions about this study, I can write or call the Study Director listed at the end of this form.

I understand that my signature means I understand and agree to what this form says.

______________________________ ______________
(Participant signature) (Date)

For participant under age 21, signature of guardian must be present.

______________________________ ______________
(Guardian signature) (Date)

Inquiries may be made to:

John Doe, Ph.D.
Study Director, "Neighborhood Interviews"
Youth Place
Anywhere, USA 00001
Phone (999) 000–0000

* * *

Informed Consent Example ("'Short Form' Consent for Child"). The reader may recall our previous discussion of the short-form approach to informed consent. In contrasting its advantages and disadvantages, one immediate disadvantage appeared to be the paperwork involved, despite the title. This is, in fact, the case. Forms for the short-form approach are required to indicate later what was read to the subject (Part 1) and the short form for the subject to sign (Part 2). In addition, a witness must be present for this approach and a space left on the forms for the witness to sign.

* * *

EXAMPLE ONLY

"Short Form" Consent For Child
(*Script to be read aloud to child*)

PART 1

Youth Place, Anywhere, USA 00001

SCRIPT FOR CONSENT OF YOUTH TO TAKE PART IN THE "STUDY HOUR" RESEARCH

This form represents the voluntary consent of ______________________
(First name)
______________________________________ to participate in the "Study
(Middle initial) (Last name)
Hour" research project.

SCRIPT

PROCEDURES: When you sign your consent form, it means that you let us look at and show other people different types of information that we get about you and your family. We need this information so we can see how well "Youth Place" teaches your parents to help you with your schoolwork. The kind of information we'll collect includes what you do during study hour in your home and what kinds of grades you get in school. We'll get this information by coming to your home and just watching you and your parents during study hour and sometimes writing down what we see. We'll get your grades from your teachers at school.

DISCOMFORTS OR RISKS AND PRIVACY PROTECTION: There is no physical harm involved in the program, but your name and information about you could become known by other people. We'll do our best to make sure this doesn't happen by locking up all the information we have about you and by using a made-up name or number for you instead of your real name.

AVAILABILITY OF MEDICAL TREATMENT AND MONEY: If there is any kind of physical injury resulting from the research procedures, emergency medical treatment will be provided for you but extra money is not available.

BENEFITS: The information we get about you could help us work better with families about school issues.

ALTERNATIVES (Freedom of Choice) AND RIGHT TO WITHDRAW: You do not have to sign your consent form and take part in the research unless you want to. Even if you do want to and you sign the agreement, you can later get out of the program if you want to by telling us in writing that you want to stop. If you do do this later, we'll honor your request and we won't object in any way.

RIGHT TO INQUIRE: You can ask us anytime about what it is we're doing or about any other parts of the project you don't understand.

My signature below signifies that this informed consent document was read to the research participant whose name appears at the top of this form.

______________________________ ______________

(Witness signature) (Date)

Inquiries may be sent to:

Ms. Jane Doe, ACSW
Study Director, "Study Hour"
Youth Place
Anywhere, USA 00001
Phone: (999) 999–9999

EXAMPLE ONLY

"Short Form" Consent
For Child

PART 2

Youth Place, Anywhere, USA O0001

CONSENT FOR YOUTH TO TAKE PART IN THE "STUDY HOUR" RESEARCH

I have had read to me the information consent form titled "Script for Consent to Youth to Take Part in the 'Study Hour' Research." I understand that my signature below means I understood and agreed to all that was read to me and agree to take part in the "Study Hour" research.

_________________________________ _______________
(Participant signature) (Date)

_________________________________ _______________
(Guardian signature) (Date)

_________________________________ _______________
(Witness signature) (Date)

* * *

Informed Consent Form ("Parental Permission for Child Participation"). Finally, we have an example of a parental permission form indicating permission for a child to participate. This form represents an extension of the previous short form approach with a child, except in this case the parent was not present at the reading of the consent form. The assumption here is that the youth's consent will still be sought and the complete information will still be read aloud to the child. In those instances where the subject's consent cannot be obtained (e.g., because of severe mental deficiency), the guardian or parent's consent would be sought directly as a proxy for the subject's consent. See later chapters in this book for more discussions of such special cases.

* * *

EXAMPLE ONLY

"Parental Permission for
Child Participation" Form

Youth Place, Anywhere, USA 00001

I have received the information you sent regarding the "Study Hour" research. I understand that my signature below means that I give permission to the Study Director to invite the participation of my child in this research.

____
(Child's name)

____________________________________ ________________
(Parent or guardian signature) (Date)

Inquiries may be sent to:

Ms. Jane Doe, ACSW
Study Director, "Study Hour" Project
Youth Place
Anywhere, USA 00001
Phone: (999) 999–9999

* * *

THE RESEARCHER AND THE IRB: A COMPLICATED RELATIONSHIP

We have now examined the primary federal rules for informed consent and IRBs and have seen that historical developments on a national level have greatly modified subject protection procedures since the 1970s. The example of an IRB application kit just described presents a straightforward view to researchers of how to submit proposals to an IRB and how to do so efficiently. Timely completion of IRB forms like those presented, and inclusion of adequate consent forms, should usually ensure that IRB review will be prompt and the researcher can then seek funds for the project. A properly educated researcher can significantly minimize any delays due to proposal or consent revisions as he or she works through an IRB.

But is that the way it works? Unfortunately, there is evidence that the process is not so simple. Readers who have either served as part of an IRB or submitted a proposal for IRB review may draw upon personal experience to assess the IRB-researcher relationship, and we also can look to a national survey involving both "sides" of the issue. The word *sides* is used cautiously, for it would be inaccurate to portray the relationship as an antagonistic one, ideally and often it is a cooperative relationship. Nevertheless, the personal and professional investment most researchers place in their research can lead to anxiety and intense feelings when the project comes up for IRB review. Such emotions can quickly move from petty bickering to serious complications between colleagues who would otherwise work to their mutual benefit.

Because of the importance of this relationship, a national survey was conducted under the auspices of the original National Commission for the Protection of Human Subjects of Biomedical and Behavioral Research. Directed by Cooke and Tannenbaum[7] of the Institute for Social Research at the University of Michigan in Ann Arbor, along with Bradford Gray of the Commission, the study produced numerous findings on the operations of IRBs in the 1970s. This data was reviewed by the National Commission and successor bodies in forming the rules on IRBs and informed consent that we use today.

A *National IRB Survey*

The survey is perhaps the most significant in-depth examination of how IRBs operate in the United States. The detail and the sheer amount of data collected are beyond the scope of this volume and interested readers should consult the original source as cited in the notes at the end of this chapter. As one who has been on both sides of the IRB—researcher relationship, the author found the survey results fascinating and enlightening. Examination of the survey would have been even more useful had the author had the opportunity to review the national survey results *before* ever taking any proposal to any IRB.

It is in the interest of just such preparation that material from the survey will be summarized in subsequent sections of this chapter. The survey was so widespread that it examined aspects of the IRB—researcher relationship far beyond the scope of any one IRB or the experience of any one researcher. The survey results can thus be viewed as a source of valuable hints on what to expect and how to operate, from both the researcher's viewpoint and that of an IRB member.

IRB Survey Design

Funded by the National Commission under Contract No. N01–HU–6–2110, the national survey involved IRBs at 61 institutions where research was

[7]Cooke, R. A., Tannenbaum, A. S., & Gray, B. A Survey of Institutional Review Boards and Research Involving Human Subjects. In the *Appendix to Report and Recommendations, Institutional Review Boards* by the National Commission for the Protection of Human Subjects of Biomedical and Behavioral Research. Washington, D.C.: U.S. Government Printing Office (Stock No. 052–003–00556–5), 1977, pp. 1–310.

submitted to the IRBs between July 1, 1974, and June 30, 1975. Thus survey respondents answered questions about research that had actually been submitted to an IRB, the research had already begun, and subjects had already participated.

A total of approximately 3,800 persons were interviewed with face-to-face interviews with study personnel and/or phone interviews and/or self-administered questionnaires, depending on whether the respondent was a principal investigator (researcher), a member of an IRB or key person associated with the IRB, a subject, or a subject's proxy. Of the total sample of respondents, about 2,000 were researchers, 800 were IRB members or associated key persons, and 1,000 were subjects or subject's proxies.

Because of the sensitivity of issues involved, the survey team went to considerable effort to ensure confidentiality and privacy of the respondent's opinions of researchers, IRBs, the review of research, and the consent process for subjects. The report itself contains copies of the various letters and agreements made by the survey team with the different types of respondents.

Of the institutions involved, 18% were universities, 59% were medical schools, and 15% were hospitals. The remainder consisted of institutions for the mentally infirm, biomedical research institutions, and related organizations. An interesting feature of the survey was the use of the Flesch scale noted earlier in this book to score consent forms as the survey team attempted to assess the readability of consent forms used at the various institutions.

Survey Results

This section will address some general findings and some results regarding the performance of the IRBs. Once again, the interested reader is urged to examine the report itself to get a true appreciation of the scope of data obtained in the survey. The survey resulted in the availability of a wide variety of information regarding the opinions and experiences of all major parties to any research project.

General Results. As it turned out, 60% of the research projects surveyed were primarily biomedical (reflecting the proportion of medical schools in the sample), 30% were behavioral, and the rest were secondary analyses. According to researchers, actual harm to subjects as a result of participation was very rare, and 24% of the time their IRBs had required changes in the proposed consent procedures. Where consent was obtained, it was *written* consent 78% of the time. In 27% of the studies examined, proxy consent was obtained rather than from the subject, with age of the subject given as the main reason for the proxy approach. Of the institutions where proxy consent was used, hospitals had the highest proportion, with 38% of their research involving proxy consent.

Overall, consent forms were used 80% of the time but the survey team judged the forms as generally difficult to read. This difficulty was judged to be due not to the use of medical or technical terms but to complex sentence structure and general reading difficulty. The more complete forms obtained from IRBs (i.e., they had all the consent elements) were no harder or easier to read than the less complete ones. Further, no evidence was found that IRB's affected the readability of proposed consent forms although IRBs did ask for changes in the forms.

As we noted previously, the notion of readability is so important because it lies at the very heart of the concept of true informed consent. That is, if the subject cannot read the consent form with sufficient comprehension, has consent truly been given? The question becomes of more than academic interest in research in which the subject is placed at risk by participating or in which any type of actual injury occurs.

Finally, it was found that although most IRB members were scientists, fully 30% of the members did not classify themselves as either behavioral or biomedical scientists. Instead, they were administrators, lawyers, nurses, members of the clergy, and so on. When compared to biomedical scientists, behavioral scientists and other IRB members placed more emphasis on informed consent procedures and issues of privacy and confidentiality. In contrast, the biomedical researchers reported that they were more often interested in contributions to scientific knowledge.

IRB Performance. Perhaps most relevant to the researcher is what he or she can typically expect to experience when submitting a proposal to an IRB. Although the results of the survey bear on IRB performance from the mid-1970s, there does not appear to be, in the author's opinion, an appreciable difference in how IRBs operate today. Hence the following highlights should be of interest. These highlights are from the portions of the survey where each IRB was scored by its own members and by researchers on how comprehensively IRBs reviewed proposals, what procedures were used to monitor progress of approved research, what kinds of changes were required by IRBs in proposals better to protect subjects, whether IRBs approved readable and complete consent forms, how positively IRBs were viewed, and whether IRBs were judged to have been doing a good job.

In general, the survey found that IRBs that scored high on one dimension did not necessarily score high on others. For example, no correlation was found between the evaluation of an IRB by its members with the evaluation by researchers who submitted proposals to the IRB. For statistical purposes, any correlation that was reported in the survey to be about or greater than $r = .23$ was significant at the .05 level. The following highlights represent findings on the IRB–researcher relationship.

(1) One of the highest correlations reported ($r = .43$) was the positive correlation between the frequency with which IRBs required changes in the researcher's consent documents and the completeness of the researcher's consent

forms. The most complete forms tended to be submitted to and approved by IRBs that made the most modifications.

(2) There was a negative correlation ($r = -.35$) in one area in which researchers were less likely to say that review procedures actually protected subjects, whereas IRBs frequently requested additional information from researchers regarding their proposals.

(3) The nature of the research in the proposals appeared to be an especially important correlate of performance. IRBs whose work load included a large proportion of biomedical research tended to rank high on many criteria of performance. They made slightly more modifications regarding consent ($r = .23$) and risk ($r = .29$) in proposals, more often monitored ongoing projects ($r = .35$), and their discussions of proposals were relatively comprehensive ($r = .24$). But they were also more likely to approve research where there was no provision for confidentiality of the data ($r = -.46$) and to approve less readable ($r = -.38$) through more complete ($r = .40$) consent forms.

(4) IRBs were found to vary widely in the amount of time and effort they put into the review process. Some IRBs spent less than 10 "member-hours" per proposal, while others spent as much as 250 member-hours per project. IRBs that spent relatively more time tended to make changes in consent procedures ($r = .30$) and with regard to reducing risk or discomfort to subjects ($r = .48$).

(5) IRBs that more often assigned proposals to individuals or to subcommittees made a greater variety of changes in researcher proposals ($r = .27$). IRBs that approved proposals which were subjected to a second review made more changes concerning risk ($r = .36$) and scientific design ($r = .40$) than did other IRBs, but they were also less likely to monitor the approved research later ($r = -.51$).

(6) IRBs that had members with a bachelor's degree or less education required a higher variety of changes in researcher proposals ($r = .28$). Also, such IRBs tended to produce consent forms that were more complete ($r = .29$) but no more readable.

(7) IRBs that were defined as "participative" (i.e., members shared influence among themselves) tended to be smaller ($r = .52$) and to have a lower work load ($r = .38$) than less participative IRBs. However, participative IRBs did not appear to be more active than nonparticipative boards. Interestingly, researchers who submitted proposals to participative boards were not so likely to have a high opinion of them ($r = -.37$).

(8) Divisive disagreements did not often occur among IRB members, but where it did happen nonmembers were more likely to say that the review process impedes research ($r = -.24$) and does not protect subjects ($r = .29$) and that IRBs may not learn about any actual harm that later befalls a subject anyway ($r = .34$). Also in such situations where the board experienced internal conflict, researchers were less likely to have a favorable attitude toward the IRB ($r = -.40$) or to feel that IRBs protect the rights and welfare of subjects ($r = -.26$). Finally,

such IRBs made slightly more frequent changes in consent procedures ($r = .31$) and a greater variety of changes in proposals ($r = .26$).

(9) About 25% of the researchers who were involved in the survey had been or were on an IRB at the time of the survey. Such researchers were less likely to have their proposals changed by an IRB, but this could have been because of their greater experience either with IRBs or with conducting research projects in general.

(10) The survey did *not* find that children, women, minority group members, or low-income subjects were any more likely to participate in projects that were above average in risk. At the same time, such subjects (some of whom have been protected by special federal regulations) were also no less likely to participate in projects that were intended to benefit subjects.

(11) The survey obtained interesting responses from subjects, but for several reasons the survey authors cautioned us that they were unable to have a totally representative sample of research subjects or proxies. In those situations in which subject responses were obtained, the survey team reported that although most subjects recalled giving consent, 10% said they did not understand they were consenting to *research*; 98% reported that they had voluntarily consented; 67% said they had benefited directly from the research; 13% said they had experienced unexpected difficulties; and 70% said they would be very willing to participate in a similar study again.

(12) Encouragingly for the field, most researchers and IRB members viewed the review process favorably, but a substantial minority of researchers felt that the review is an unwarranted intrusion on a researcher's autonomy, that IRBs get involved in inappropriate areas, that IRBs make judgments they are not qualified to make, and that IRBs impede research.

(13) The problem cited most often by IRB members was administrative, namely, the difficulty of getting members together for meetings! Problems such as the need for rapid action because of deadlines imposed by funding agencies, the lack of precise federal guidelines, and the amount of IRB time spent in reviewing research involving little risk were mentioned by 25–33% of IRB member survey respondents.

(14) Finally, both subjects and proxies offered a variety of other comments from their perspective, including the need for more information about the research (19%) and for more care and courtesy on the part of researchers toward the subjects (11%).

Summary

In this chapter we have examined several aspects of the IRB–researcher relationship with the primary purpose of providing hints to the reader on how to submit a successful research proposal to an IRB.

First, we saw how an IRB is established at an institution and approved by a federal agency. The formation of an IRB clearly establishes the IRB as the local intermediary, so far as protection of research subjects is concerned, between the federal government on one hand and the individual researcher on the other. Readers planning to submit a research project to a federal agency or simply desiring official approval of subject protection steps can either submit their project to an established local IRB or form a new IRB. Formation of a new IRB is most relevant in those cases where an existing IRB has not been approved by the government to review a particular type of research or where an IRB simply is not geographically close enough to the researcher. The appendix to this chapter contains a listing of currently approved IRBs in the United States, and the chapter contains examples of actual correspondence from one case of forming an IRB.

The second aspect of the IRB–researcher relationship that was addressed concerned the practical "how to" information a researcher should know to submit a proposal for IRB review. Again, examples were taken from an approved IRB showing the various forms to be completed by a researcher when seeking IRB approval. As a further practical aid, examples of different types of consent forms were presented from the application kit used by the IRB for researchers' proposals. These consent forms were designed for different types of subjects, and elements of the forms were based on material discussed in previous chapters. This aspect of the IRB–researcher relationship thus presented examples of forms to complete and IRB review procedures to enable the reader to initiate most expeditiously review and approval of his or her own research by an IRB.

Finally, the chapter contained various results obtained in a national survey of IRBs conducted by Cooke, Tannenbaum, and Gray[8] under a contract with the original National Commission for the Protection of Human Subjects of Biomedical and Behavioral Research. Because of the diversity and scope of the survey data, only highlights were presented and the reader was referred to the original study for more detail. In general, we saw that IRBs typically required revisions of the informed consent portion of a researcher's proposal and that there were areas in which IRB members and researchers saw room for improvement. From the subjects' standpoint, survey results strongly hinted at aspects of the researcher–subject relationship that could be improved. These survey results, besides influencing current federal rules, suggested techniques that researchers should consider, both to increase subject satisfaction (and, presumably, their willingness to participate again) and to increase chances of successful IRB review.

Supporting information on these topics is contained in the appendix to this chapter.

[8]Ibid.

APPENDIX 4.1

61st Cumulative List of Institutions Which Have Established General Assurances of Compliance with HEW (Now HHS) Regulations on Protection of Human Subjects

(February 1, 1980)

[Note: The following pages contain a list of federally approved IRBs in the United States. The IRBs are organized alphabetically by state, with the institution's name, city, and general assurance number given. Any type of restrictions placed on particular IRBs are explained in footnotes. These restrictions generally govern what type of research an IRB has *not* been approved by HHS to review.]

General Assurance Number	Institution	City
	ALABAMA	
G0198	Alabama, University of, Birmingham	Birmingham
G0610	Alabama, University of	University

(continued)

Restriction code
XM—No Medical or IND Studies
XB —No Behavioral Studies
XO—Other Restriction (call 301-496-7005)
* —Final Acceptance of Revisions Pending

General Assurance Number	Institution	City
G2522	South Alabama, University of, including the Medical Center	Mobile
G0055	Tuskegee Institute	Tuskegee Inst.
	ARIZONA	
G0799XM	Arizona State University	Tempe
G0026	Arizona, University of	Tucson
G1287XB	St. Joseph's Hospital and Medical Center, including Barrow Neurological Institute	Phoenix
	ARKANSAS	
*G0336	Arkansas, University of, Medical Center	Little Rock
	CALIFORNIA	
G0784XM	American Institutes for Research	Palo Alto
G0144	California Institute of Technology	Pasadena
G0261	California State Health and Welfare Agency Department of Health Services Department of Mental Health Department of Developmental Services Department of Social Services Department of Alcohol and Drug Abuse Office of Statewide Health Planning and Development	Sacramento
G2354XM	California State University, Dominguez Hills	Carson
G0278	California, University of, Berkeley	Berkeley
G0195	California, University of, Davis	Davis
G0380	California, University of, Irvine	Irvine
G0238	California, University of, Los Angeles	Los Angeles
G0193	California, University of, Riverside	Riverside
G0389	California, University of, San Diego	La Jolla
G0155	California, University of, San Francisco	San Francisco
G0282	California, University of, Santa Barbara	Santa Barbara
G0205XM	California, University of, Santa Cruz	Santa Cruz
G0234	Cedars-Sinai Medical Center	Los Angeles
*G2564	Charles R. Drew Postgraduate Medical School	Los Angeles
G0135	Childrens Hospital of Los Angeles	Los Angeles
G0287XB	Children's Hospital of San Francisco	San Francisco
G0009	City of Hope National Medical Center	Duarte

General Assurance Number	Institution	City
G0432	Institute for Medical Research	San Jose
G2169XM	Institute for Research in Social Behavior	Oakland
*G2400	Institute of Medical Sciences	San Francisco
*G2367	John Wesley County Hospital Attending Staff Association	Los Angeles
G0214	Kaiser Foundation Research Institute, a Division of Kaiser Foundation Hospitals, Kaiser Foundation Hospitals in Northern California, Southern California, Oregon, and Hawaii	Oakland
G0278	Lawrence Berkeley Laboratory, University of California	Berkeley
G0226	Loma Linda University	Loma Linda
G0226	Loma Linda University (La Sierra Campus)	Riverside
G0359XB	Los Angeles County Harbor General Hospital, Professional Staff Association of	Torrance
*G0213	Los Angeles County Hospital—University of Southern California Medical Center, Professional Staff Association of	Los Angeles
*G2564	Martin Luther King, Jr. General Hospital and Its Staff Association Foundation	Los Angeles
*G0367XO	Mental Research Institute	Palo Alto
G0145	Mount Zion Hospital and Medical Center	San Francisco
G2961	Northern California Cancer Program	Palo Alto
*G2400	Pacific, University of, School of Dentistry	San Francisco
*G2400	Pacific, University of, School of Medical Sciences	San Francisco
G0082XB	Palo Alto Medical Research Foundation	Palo Alto
*G2400	Presbyterian Hospital of the Pacific Medical Center, Inc.	San Francisco
G0067	Rancho Los Amigos Hospital and Its Professional Staff Association	Downey
G1416	Rand Corporation	Santa Monica
*G1096	Salk Institute	San Diego
G0420	San Diego State University	San Diego
G0420	San Diego State University Foundation	San Diego
G0729XM	San Francisco State University	San Francisco
G1472XM	Scientific Analysis Corporation (including the Institute for Scientific Analysis and Performance Sites in Berkeley, Los Angeles, and San Diego)	San Francisco

(continued)

General Assurance Number	Institution	City
G0108	Scripps Clinic and Research Foundation	La Jolla
*G0122	Southern California, University of	Los Angeles
G0378	SRI International	Menlo Park
G0011	Stanford University	Stanford
G0010	USPHS Hospital, San Francisco	San Francisco
G1620	Veterans Administration Hospital—Long Beach	Long Beach
G1796	Veterans Administration Hospital—Sepulveda	Sepulveda
G1658XB	Veterans Administration Wadsworth Medical Center	Los Angeles
G1591XM	Wright Institute	Berkeley
	COLORADO	
G0845	Children's Hospital of Denver	Denver
G2823XB	Colorado Regional Cancer Center, Inc.	Denver
G1144	Colorado State University	Fort Collins
G0455XM	Colorado, University of	Boulder
G0268	Colorado, University of, Medical Center	Denver
G0277XM	Denver, University of (Colorado Seminary)	Denver
G0267	National Asthma Center	Denver
G0068	National Jewish Hospital and Research Center	Denver
	CONNECTICUT	
G0672XB	Connecticut State Department of Health	Hartford
G0039	Connecticut, University of	Storrs
G1966	Connecticut, University of, Health Center	Farmington Hartford
G1533	Connecticut Valley Hospital	Middletown
G0331XB	Hartford Hospital	Hartford
G0007	Institute of Living	Hartford
G0003	John B. Pierce Foundation of Connecticut, Inc.	New Haven
G0821	Saint Francis Hospital	Hartford
G2895	Waterbury Hospital Health Center	Waterbury
*G0243	Yale University	New Haven
	DELAWARE	
G0148	Delaware Division of Mental Health of the Department of Health and Social Services	New Castle

General Assurance Number	Institution	City
G0013	Delaware, University of	Newark
G2480	Wilmington Medical Center, Inc.	Wilmington

DISTRICT OF COLUMBIA

General Assurance Number	Institution	City
G0784XM	American Institutes for Research	Washington
G1398XB	American National Red Cross	Washington
G1674XM	Bureau of Social Science Research, Inc.	Washington
G0365XM	Catholic University of America	Washington
G0042	Children's Hospital of D.C., Research Foundation of	Washington
G1249	Gallaudet College	Washington
G0294	George Washington University	Washington
G0228	Georgetown University	Washington
G0841	Howard University	Washington
G1989	National Academy of Sciences, including National Academy of Engineering, Institute of Medicine, and National Research Council	Washington
G1036XM	National Children's Center, Inc.	Washington
G3045XO	Veterans Administration Central Office NCI-VA Coop Cancer Therapy Projects Only	Washington
G0180XB	Washington Hospital Center	Washington

FLORIDA

General Assurance Number	Institution	City
G2905	Florida State Department of Health and Rehabilitation Services	Tallahassee
G0119	Florida State University	Tallahassee
G0382	Florida, University of	Gainesville
G0840XB	Miami Heart Institute	Miami Beach
G0301	Miami, University of	Coral Gables
G0301	Miami, University of	Miami
G0275	Mt. Sinai Medical Center of Greater Miami	Miami Beach
G2321	Papanicolaou Cancer Research Institute	Miami
G1065	South Florida, University of	Fort Myers
G1065	South Florida, University of	St. Petersburg
G1065	South Florida, University of	Sarasota
G1065	South Florida, University of	Tampa

(continued)

General Assurance Number	Institution	City
	GEORGIA	
*G0138	Emory University Division of Basic Health Sciences	Atlanta
G0622	Emory University Graduate School of Arts and Sciences	Atlanta
*G0138	Emory University Nell Hodgson Woodruff School of Nursing	Atlanta
G0595	Emory University School of Dentistry	Atlanta
*G0138	Emory University School of Medicine	Atlanta
G0197	Georgia Dept of Human Resources, All Units	Atlanta
G0520	Georgia Institute of Technology, including its Engineering Experiment Station, and its Georgia Tech Research Institute	Atlanta
G0151	Georgia, Medical College of	Augusta
G0248	Georgia, University of, and its Research Foundation	Athens
*G1467	Grady Memorial Hospital	Atlanta
	HAWAII	
G0104	Hawaii, University of (including the Entire System and the Research Corporation)	Honolulu
G0214	Kaiser Foundation Hospital	Honolulu
	IDAHO	
G0218XM	Idaho, University of	Moscow
*G1869	Mountain States Tumor Institute	Boise
	ILLINOIS	
G0395	American Dental Association and ADA Health Foundation	Chicago
G0332XM	Anna Mental Health and Developmental Center	Anna
G1626	Chicago, University of (Division of Biological Sciences and Pritzker School of Medicine, Division of Social Sciences, School of Social Service Administration, and Related Units)	Chicago
G0114	Children's Memorial Hospital	Chicago

General Assurance Number	Institution	City
G0430	Downey Veterans Administration Hospital	Downey
G0182	Evanston Hospital	Evanston
G0430	Health Sciences, University of/Chicago Medical School	Chicago
G0053	Hektoen Institute for Medical Research	Chicago
G2959	Illinois Cancer Council	Chicago
G1561	Illinois State Psychiatric Institute	Chicago
G0345	Illinois, University of, at Chicago Circle	Chicago
G0291	Illinois, University of, Medical Center Campus, including Rockford School of Medicine, and Peoria School of Medicine	Chicago
G0102	Illinois, University of, and all components at Urbana-Champaign	Urbana
G1373XM	Lincoln Developmental Center	Lincoln
G2369	Loyola University Medical Center	Maywood
G0095	Michael Reese Hospital and Medical Center	Chicago
G2763	Mount Sinai Hospital Medical Center of Chicago	Chicago
G0518XM	Northern Illinois University	DeKalb
G0190	Northwestern Memorial Hospital	Chicago
G0190	Northwestern University	Chicago
G0190	Northwestern University	Evanston
G0190	Rehabilitation Institute of Chicago	Chicago
G0175	Rush-Presbyterian-St Luke's Medical Center	Chicago
G0309	Southern Illinois University	Carbondale
G2606	Southern Illinois University	Edwardsville
G0309	Southern Illinois University School of Medicine	Springfield
G0190	VA Lakeside Hospital of Chicago	Chicago
G1970XM	Western Illinois University	Macomb
	INDIANA	
G0314XM	DePauw University	Greencastle
G0945	Fort Wayne State Hosp and Training Center	Ft. Wayne
G0185	Indiana University and Indiana University Foundation	Bloomington Indianapolis
G0185	Indiana University (Law School)	Indianapolis
G0185	Indiana University-Purdue University (Medical Facility)	Indianapolis
G0185	Indiana University Regional Campuses at Fort	

(*continued*)

General Assurance Number	Institution	City
	Wayne, Gary, Kokomo, New Albany, Richmond, and South Bend	
G1397	Indianapolis Center for Advanced Research	Indianapolis
G0883	Larue D. Carter Memorial Hospital	Indianapolis
G1697	Marion County Association for Retarded Citizens, Inc.	Indianapolis
G1246XB	Methodist Hospital of Indiana, Inc.	Indianapolis
G1119XM	Notre Dame, University of	Notre Dame
G0131	Purdue University and Purdue Research Foundation	West Lafayette
	IOWA	
*G0343XO	Iowa State University of Science and Technology	Ames
G0209	Iowa, University of	Iowa City
*G0856XO	Woodward State Hospital—School	Woodward
	KANSAS	
G0038	Kansas State University	Manhattan
*G0157XO	Kansas, University of—Lawrence Campus	Lawrence
G0229	Kansas, University of, Medical Center College of Health Sciences and Hospital	Kansas City
G0081	Menninger Foundation	Topeka
	KENTUCKY	
G0223	Kentucky, University of, and the Research Foundation	Lexington
G0527	Louisville, University of, and Louisville, University of, Foundation, Inc.; except Schools of Medicine and Dentistry	Louisville
G0023	Louisville, University of, Schools of Medicine and Dentistry	Louisville
G0223	Veterans Administration Hospital	Lexington
	LOUISIANA	
G0442	Alton Ochsner Medical Foundation	New Orleans
*G0262XO	Louisiana State University and A & M College System	Baton Rouge
*G2428XO	Louisiana State University in Baton Rouge	Baton Rouge

General Assurance Number	Institution	City
*G2485	Louisiana State University Medical Center	New Orleans
G0137XM	Pinecrest State School	Pineville
G0018	Tulane Medical Center	New Orleans
G2386	Tulane University except Medical Center	New Orleans
G0010	USPHS Hospital (National Leprosarium)	Carville
G0010	USPHS Hospital, New Orleans	New Orleans
	MAINE	
G3007	Eastern Maine Medical Center	Bangor
G0513	Maine Medical Center	Portland
	MARYLAND	
G0402	Baltimore City Hospitals	Baltimore
G0010	Division of Hospitals and Clinics, BMS, DHEW (includes 9 USPHS Hospitals and 26 Outpatient Clinics)	West Hyattsville
G3109XB	Frederick Cancer Research Center	Frederick
G0186	Friends Medical Science Research Center, Inc.	Baltimore
G0640	Institute for Behavioral Research, Inc.	Silver Spring
G2857	John F. Kennedy Institute for Handicapped Children	Baltimore
G0352	Johns Hopkins University Faculty of Arts and Sciences	Baltimore
G0274	Johns Hopkins University School of Hygiene and Public Health	Baltimore
G0174	Johns Hopkins University School of Medicine and Johns Hopkins Hospital	Baltimore
G0326	Maryland, University of, at Baltimore	Baltimore
G0296	Maryland, University of, at College Park	College Park
G0400	Sinai Hospital of Baltimore, Inc.	Baltimore
G0010	USPHS Hospital, Baltimore	Baltimore
	MASSACHUSETTS	
G3270	Affiliated Hospitals Center, Inc. (Formerly listed as the Boston Hospital for Women, Peter Bent Brigham Hospital, and Robert Bent Brigham Hospital)	Boston

(*continued*)

General Assurance Number	Institution	City
G0784XM	American Institutes for Research	Cambridge
G0286	Beth Israel Hospital	Boston
G0554	Boston City Hospital	Boston
G0554	Boston Department of Health and Hospitals	Boston
G0554	Boston Division of Community Health Services	Boston
G0783	Boston Mental Health Foundation, Inc.	Boston
G0783	Boston State Hospital	Boston
G0322XM	Boston University, Charles River Campus	Boston
G0550	Boston University Medical Center	Boston
G0550	Boston University School of Graduate Dentistry	Boston
G0550	Boston University School of Medicine	Boston
G0132XM	Brandeis University	Waltham
G2396XB	Center for Blood Research, Inc.	Boston
G0406	Children's Hospital Medical Center	Boston
G0117XM	Clark University	Worcester
G0098	Danvers State Hospital	Hathorne
G2042	Eunice Kennedy Shriver Center	Waltham
G0410XB	Eye Research Institute of Retina Foundation	Boston
G0412XB	Forsyth Dental Center	Boston
G2397	Harvard Medical School and School of Dental Medicine	Boston
G2238	Harvard University, Faculty of Arts and Sciences	Cambridge
G2494XM	Harvard University, Graduate School of Education	Cambridge
G2398	Harvard University, School of Public Health	Boston
G2573	Harvard University, University Health Services	Cambridge
G0791	Joslin Diabetes Foundation, Inc.	Boston
G0087XB	Lahey Clinic Foundation, Inc.	Boston
G0449XB	Lemuel Shattuck Hospital	Jamaica Plain
G0554	Long Island Chronic Disease Hospital	Boston Harbor
G2304XO	Mass Developmental Disabilities Council	Boston
G0212XB	Massachusetts Eye and Ear Infirmary	Boston
G0167	Massachusetts General Hospital	Boston
G0097	Massachusetts Institute of Technology	Cambridge
G0597	Massachusetts Mental Health Center	Boston
G0597	Massachusetts Mental Health Research Corp	Boston
G0147	Massachusetts, University of—Amherst	Amherst

General Assurance Number	Institution	City
G2736	Massachusetts, University of, Medical School	Worcester
G0554	Mattapan Chronic Disease Hospital	Mattapan
G0576	McLean Hospital	Belmont
G3123	Mount Auburn Hospital	Cambridge
G0494XB	New England Deaconess Hospital	Boston
G0133	New England Medical Center Hospital	Boston
G3064	North Charles Mental Health Research & Training Foundation, Inc.	Cambridge
G0297	Northeastern University	Boston
G3018XB	Pondville Hospital	Walpole (Norfolk)
G2515XM	Research Institute for Educational Problems	Cambridge
G0771	Sidney Farber Cancer Institute, Inc.	Boston
G0351XB	St. Margaret's Hospital for Women	Boston
G0554	Trustees of Health & Hospitals, City of Boston, Inc.	Boston
G2107	Tufts-New England Medical Center	Boston
G2379	Tufts University—Boston	Boston
*G0266XO	Tufts University—Medford	Medford
G0550	University Hospital, Inc.	Boston
G0010	USPHS Hospital, Boston	Boston
G2042	Walter E. Fernald State School	Belmont
G0085	Worcester Foundation for Experimental Biology	Shrewsbury

MICHIGAN

General Assurance Number	Institution	City
G0363	Child Research Center of Michigan	Detroit
G0363	Children's Hospital of Michigan	Detroit
G0497XM	Detroit, University of	Detroit
G0357	Edsel B. Ford Institute for Medical Research	Detroit
G0357	Henry Ford Hospital	Detroit
G0101	Lafayette Clinic	Detroit
G0656XB	Michigan Cancer Foundation	Detroit
G0615	Michigan Department of Public Health	Lansing
G0384	Michigan State University	East Lansing
G0295	Michigan, University of	Ann Arbor
G0295	Michigan, University of	Dearborn
G0295	Michigan, University of	Flint
G1009XM	Oakland University	Rochester
G0008	Sinai Hospital of Detroit	Detroit

(*continued*)

General Assurance Number	Institution	City
G2483	Veterans Administration Medical Center—Ann Arbor	Ann Arbor
G0354	Wayne State University	Detroit
G0720	Western Michigan University	Kalamazoo
	MINNESOTA	
G0065	Hennepin County Medical Center	Minneapolis
G0041	Mayo Foundation	Rochester
G0065	Minneapolis Medical Research Foundation, Inc.	Minneapolis
G0463	Minnesota, University of	Austin
G0463	Minnesota, University of	Duluth
G0463	Minnesota, University of	Minneapolis
G0463	Minnesota, University of	Morris
G0463	Minnesota, University of	St Paul
G0463	Minnesota, University of, Technical College	Crookston
G0463	Minnesota, University of, Technical College	Waseca
	MISSISSIPPI	
G0561XM	Mississippi, University of (Main Campus)	University
G0366	Mississippi, University of, Medical Center	Jackson
G0366	Veterans Administration Center	Jackson
	MISSOURI	
G2791XM	American Nurses' Association, Inc.	Kansas City
G0298XB	Cancer Research Center	Columbia
G0298XB	Ellis Fischel State Cancer Hospital	Columbia
G0424	Jewish Hospital of St Louis	St Louis
G0320XB	Kirksville College of Osteopathic Medicine	Kirksville
G2936XB	Missouri Cancer Programs	Columbia
G0292	Missouri, University of	Columbia
G0292	Missouri, University of	Kansas City
G0292	Missouri, University of	Rolla
G0292	Missouri, University of	St Louis
G0356	St Louis University	St Louis
*G2056	Veterans Administration Medical Center—Kansas City	Kansas City

General Assurance Number	Institution	City
G0279XM	Washington University (excluding Schools of Medicine and Dentistry)	St Louis
G2351	Washington University School of Medicine	St Louis
	NEBRASKA	
G1120	Creighton University	Omaha
G3112	Father Flanagan's Boys' Home	Boys Town
G0407	Nebraska, University of, including the Medical Center	Omaha
G0407	Nebraska, University of, including the State University of Nebraska	Lincoln
	NEVADA	
G0762	Nevada, University of, Reno Campus (includes all components of the University, Reno Campus)	Reno
	NEW HAMPSHIRE	
G0427XM	Dartmouth College Faculty of Arts & Sciences	Hanover
G0227	Dartmouth Medical School	Hanover
	NEW JERSEY	
G0349	College of Medicine and Dentistry of New Jersey	Newark
G0349	CMDNJ—New Jersey Dental School	Newark
G0349	CMDNJ—New Jersey Medical School	Newark
G0349	CMDNJ—New Jersey School of Osteopathic Medicine	Piscataway
G0349	CMDNJ—Rutgers Medical School	Piscataway
G0981XM	Educational Testing Service	Princeton
G0492XB	Institute for Medical Research	Camden
G0837	Newark Beth Israel Medical Center	Newark
G0570XM	Princeton University	Princeton
G0349	Rutgers Medical School	Piscataway
G0300	Rutgers, The State University of New Jersey	Camden
G0300	Rutgers, The State University of New Jersey	New Brunswick
G0300	Rutgers, The State University of New Jersey	Newark

(*continued*)

General Assurance Number	Institution	City
	NEW MEXICO	
*G0472	New Mexico, University of	Albuquerque
*G0353	New Mexico, University of, Health Sciences Group—School of Medicine, Colleges of Pharmacy and Nursing	Albuquerque
	NEW YORK	
G0217	Albany Medical College of Union University	Albany
G2129	American Health Foundation	New York
G1064XB	Associated Universities, Inc., including Brookhaven National Laboratory	Upton
G2116XM	Barnard College	New York
G0341	Beth Israel Medical Center	New York
G1831XM	Center for Policy Research, Inc.	New York
G2277XM	Child Welfare League of America, Inc.	New York
G0381	Children's Hospital of Buffalo	Buffalo
G0437XM	Columbia University, except Health Sciences Division	New York
G0034	Columbia University Health Sciences and Presbyterian Hospital	New York
G0063XM	Cornell University, except Medical College	Ithaca
G0245	Cornell University Medical College	New York
G0109XO	Eastman Dental Center	Rochester
G2912	Genesee Hospital	Rochester
G2887XB	Health Research, Inc., Albany Division	Albany
G2904	Health Research, Inc., Roswell Park Division	Buffalo
G2887XB	Helen Hayes Hospital	West Haverstraw
G2475XB	Highland Hospital of Rochester	Rochester
G0621	Hospital for Joint Diseases and Medical Center, including the Research Institute for Skeletomuscular Diseases	New York
G0047	Hospital for Special Surgery	New York
G2920	ICD Rehabilitation and Research Center	New York
G1005XO	Jewish Board of Family and Children's Services, Inc.	New York
G0369	Kingsbrook Jewish Medical Center and Isaac Albert Research Institute	Brooklyn
G2375	Long Island Jewish-Hillside Medical Center	New Hyde Park
G0397	Maimonides Medical Center	Brooklyn
G0086XB	Mary Imogene Bassett Hospital	Cooperstown

General Assurance Number	Institution	City
G0383XB	Memorial Hospital for Cancer and Allied Diseases	New York
G0059	Montefiore Hospital and Medical Center	New York
G0139	Mount Sinai School of Medicine of CUNY	New York
G2518XB	Nassau County Medical Center, including Meadowbrook Medical Education and Research Foundation, Inc.	East Meadow
G0173XM	New School for Social Research	New York
G0526XB	New York Blood Center, Inc.	New York
G1370	New York City Department of Health	New York
G0765XM	New York, City University of	New York
G0765XM	New York, City University of, Baruch College	New York
G0765XM	New York, City University of, Bronx Community College	Bronx
G0765XM	New York, City University of, Brooklyn College	Brooklyn
G0765XM	New York, City University of, City College	New York
G0765XM	New York, City University of, Graduate School and University Center	New York
G0765XM	New York, City University of, Herbert H. Lehman College	Bronx
G0765XM	New York, City University of, Hunter College	New York
G0765XM	New York, City University of, John Jay College of Criminal Justice	New York
G0765XM	New York, City University of, Kingsborough Community College	Brooklyn
G0765XM	New York, City University of, Manhattan Community College	New York
G0765XM	New York, City University of, Medgar Evers College	Brooklyn
G0765XM	New York, City University of, New York City Community College	Brooklyn
G0765XM	New York, City University of, Queens College	Flushing
G0765XM	New York, City University of, Queensborough Community College	New York
G0765XM	New York, City University of, Research Foundation	New York
G0765XM	New York, City University of, College of Staten Island	Staten Island
G0765XM	New York, City University of, York College	Jamaica

(*continued*)

General Assurance Number	Institution	City
G2887XB	New York Department of Health and Regional Offices	Albany
G0208	New York Medical College—Flower and Fifth Avenue Hospitals, Metropolitan Hospital, Bird S. Coler Memorial Hospital, and Lincoln Medical and Mental Health Center, Bronx	Valhalla
G0374XM	New York, State University of, at Albany	Albany
G0508	New York, State University of, at Binghamton	Binghamton
G0163	New York, State University of, at Buffalo	Buffalo
G0596	New York, State University of, at Stony Brook	Stony Brook
G0474	New York, State University of, College at Brockport	Brockport
G0506XM	New York, State University of, College at Buffalo	Buffalo
G0259XM	New York, State University of, College at Cortland	Cortland
G0386XM	New York, State University of, College at Oswego	Oswego
G0457	New York, State University of, College of Arts and Science at Geneseo	Geneseo
G0220	New York, State University of, College of Environmental Science and Forestry	Syracuse
G0305	New York, State University of, Downstate Medical Center	Brooklyn
G0127	New York, State University of, Upstate Medical Center	Syracuse
G0390XO	New York University, exclusive of the Medical Center	New York
G0405	New York University Medical Center	New York
G0788	North Shore University Hospital	Manhasset
G2887XB	Oxford Home for Veterans and their Dependents	Oxford
G0034	Presbyterian Hospital (Columbia University Health Sciences)	New York
G2247	Public Health Research Institute of the City of New York, Inc.	New York
G2375	Queens Hospital Center	Jamaica
G1001	Rensselaer Polytechnic Institute	Troy
G0624	Research Foundation for Mental Hygiene, Inc. (NYS Office of Mental Health, NYS	Albany

General Assurance Number	Institution	City
	Office of Mental Retardation and Developmental Disabilities, NYS Division of Alcoholism and Substance Abuse—all operating components of the NYS Department of Mental Hygiene) including	
	Capital District Psychiatric Center	Albany
	Long Island Research Institute	Central Islip
	NYS Institute for Basic Research in Mental Retardation	Staten Island
	New York State Psychiatric Institute	New York
	Research Institute on Alcoholism	Buffalo
	Rockland Research Institute	Orangeburg
	Special Projects Research Unit	Albany
G0172	Rochester, University of	Rochester
G0230	Rockefeller University	New York
G2036	Rockland County Health & Social Services Complex	Pomona
G0491	Roosevelt Hospital	New York
G2904	Roswell Park Memorial Institute	Buffalo
G0536XB	Sloan-Kettering Institute for Cancer Research	New York
G0076	St Luke's Hospital Center	New York
G2972XB	St Mary's Hospital	Rochester
G0376	St Vincent's Hospital and Medical Center of New York	New York
G0280XM	Syracuse University	Syracuse
G0373XM	Teachers College, Columbia University	New York
G2438XB	Trudeau Institute	Saranac Lake
G0010	USPHS Hospital, Staten Island	New York
G2045	Veterans Administration Hospital—Northport	Northport
G0002	Yeshiva University, Albert Einstein College of Medicine	Bronx

NORTH CAROLINA

General Assurance Number	Institution	City
G0191	Duke University Medical Center and Duke University exclusive of the Medical Center	Durham
G2242	North Carolina A & T State University	Greensboro
G0285	North Carolina State University at Raleigh	Raleigh
G0177	North Carolina, University of, at Chapel Hill	Chapel Hill

(continued)

General Assurance Number	Institution	City
G0242	Wake Forest University, Bowman Gray School of Medicine	Winston-Salem
G1087	Wake Forest University, except School of Medicine	Winston-Salem
G0078XM	Western Carolina Center	Morganton
G0191	Whitehead Medical Research Institute	Durham
	OHIO	
G3143XB	Akron General Medical Center	Akron
G0485	Akron, University of	Akron
G0683	Battelle Columbus Laboratories of the Battelle Memorial Institute	Columbus
G2843XB	Cancer Center, Inc.	Cleveland
G1167	Case Western Reserve University	Cleveland
G0088	Children's Hospital Medical Center	Cincinnati
G0249	Children's Hospital Research Foundation, Inc.	Columbus
G0667XM	Cincinnati, University of, except Medical Center	Cincinnati
G0337	Cincinnati, University of, Medical Center, including Colleges of Medicine, Nursing, and Health and Pharmacy	Cincinnati
G0502XB	Cleveland Clinic Foundation and Cleveland Clinic Educational Foundation	Cleveland
G0171	Cleveland Psychiatric Institute	Cleveland
G2678	Cox Heart Institute	Dayton
G2678	Fels Research Institute	Yellow Springs
G1936XB	Jewish Hospital of Cincinnati, including the School of Nursing, Children's Psychiatric Center, and May Institute for Medical Research	Cincinnati
G0542	Kent State University	Kent
G1724	Medical College of Ohio at Toledo	Toledo
G0269XB	Mount Sinai Hospital of Cleveland	Cleveland
G0121	Ohio Agricultural Research & Development Center	Wooster
G0121	Ohio State University and Research Foundation	Columbus
G0662	St Luke's Hospital Association (United Methodist Church)	Cleveland

General Assurance Number	Institution	City
G0350XB	University Hospitals of Cleveland	Cleveland
G2678	Wright State University (all components)	Dayton
G2907XM	Youngstown State University	Youngstown
	OKLAHOMA	
G0327	Oklahoma Medical Research Foundation, Inc.	Oklahoma City
*G2402XO	Oklahoma, University of	Norman
G0241	Oklahoma, University of, Health Sciences Center	Oklahoma City
	OREGON	
G0134	Good Samaritan Hospital and Medical Center	Portland
G0214	Kaiser Foundation Hospital	Portland
G0043XB	Medical Research Foundation of Oregon and Oregon Regional Primate Research Center	Portland
G2399XM	Northwest Regional Educational Laboratory	Portland
G0505XM	Oregon Research Institute	Eugene
G0111XM	Oregon State University	Corvallis
G0288XM	Oregon, University of	Eugene
G2931	Oregon, University of, Health Sciences Center, including Schools of Medicine, Dentistry, and Nursing	Portland
	PENNSYLVANIA	
G0125	Albert Einstein Medical Center, Northern Division	Philadelphia
G0952XB	Allegheny Health, Education and Research Corporation—Allegheny-Singer Research Corporation and Allegheny General Hospital	Pittsburgh
G0784XM	American Institutes for Research	Pittsburgh
G0338	Arthur P. Noyes Research Foundation	Norristown
G0436XM	Bucknell University	Lewisburg
G1218	Carnegie-Mellon University	Pittsburgh
G0129	Children's Hospital of Philadelphia (Joseph Stokes, Jr., Research Institution of)	Philadelphia

(*continued*)

General Assurance Number	Institution	City
G0360	Children's Hospital of Pittsburgh	Pittsburgh
G0045	Fox Chase Cancer Center (Institute for Cancer Research and American Oncologic Hospital)	Philadelphia
G1556	Franklin Institute	Philadelphia
G0594	Geisinger Medical Center, Institute for Medical Education and Research	Danville
G0475	Hahnemann Medical College and Hospital	Philadelphia
G0936	H. K. Cooper Institute for Oral–Facial Anomalies and Communicative Disorders (Lancaster Cleft Palate Clinic)	Lancaster
G0273XB	Lankenau Hospital	Philadelphia
G0189XB	Magee-Womens Hospital	Pittsburgh
G1378	Marriage Council of Philadelphia, Inc. (Reviewed by G0012)	Philadelphia
G2121XM	Marywood College	Scranton
G0057XB	Montefiore Hospital	Pittsburgh
G0338	Norristown State Hospital	Norristown
G0333	Pennsylvania Hospital	Philadelphia
G0222	Pennsylvania, Medical College of	Philadelphia
G0573	Pennsylvania State University & Commonwealth Campus	University Park
G1668	Pennsylvania State University, Milton S. Hershey Medical Center	Hershey
G0012	Pennsylvania, University of	Philadelphia
G1531	Philadelphia Department of Public Health	Philadelphia
G0324	Philadelphia Geriatric Center	Philadelphia
G0344	Philadelphia Psychiatric Center	Philadelphia
G0881	Philadelphia State Hospital	Philadelphia
G1532	Pittsburgh Child Guidance Center	Pittsburgh
G2929	Pittsburgh, University Health Center of	Pittsburgh
G0418	Pittsburgh, University of	Pittsburgh
G2200	Presbyterian-University Hospital	Pittsburgh
G0465	Presbyterian-University of Pennsylvania Medical Center	Philadelphia
G2264XM	Research for Better Schools, Inc.	Philadelphia
G0795	St Joseph's Hospital	Lancaster
G2880	Temple University, excluding the Health Sciences Center	Philadelphia
G2879	Temple University Health Sciences Center	Philadelphia
G0049	Thomas Jefferson University, all components	Philadelphia

General Assurance Number	Institution	City
G0362XB	Wills Eye Hospital and Research Institute	Philadelphia
G0054XB	Wistar Institute	Philadelphia
	PUERTO RICO	
G1348XM	Catholic University of Puerto Rico	Ponce
G0256XB	Puerto Rico, University of, Medical Sciences Campus	San Juan Rio Piedras
	RHODE ISLAND	
G0467	Brown University	Providence
*G0318	Rhode Island Hospital	Providence
G1252	Rhode Island, University of	Kingston
G2212XB	Roger Williams General Hospital	Providence
	SOUTH CAROLINA	
G2621	Clemson University	Clemson
G0533	South Carolina Department of Mental Health	Columbia
G0240	South Carolina, Medical University of	Charleston
G0629XM	South Carolina, University of	Columbia
	TENNESSEE	
G2869	East Tennessee State University	Johnson City
G0490XM	George Peabody College for Teachers	Nashville
G0393	Meharry Medical College	Nashville
G1716	Oak Ridge Associated Universities	Oak Ridge
G1716	Oak Ridge National Laboratory	Oak Ridge
G0648	St Jude Children's Research Hospital	Memphis
G0170	Tennessee, University of, Center for the Health Sciences	Memphis
G0124	Tennessee, University of—Knoxville	Knoxville
G0401	Vanderbilt University	Nashville
	TEXAS	
G2255	Audie L. Murphy Veterans Administration Hospital	San Antonio
G0307	Baylor College of Medicine	Houston
G2255	Bexar County Hospital District	San Antonio
G0398XM	Houston, University of	Houston

(*continued*)

General Assurance Number	Institution	City
*G2344XO	Southwest Educational Development Corporation (AKA Southwest Educational Development Lab)	Austin
G1045	Texas A & M University and Research Foundation, College Station Campus only	College Station
G3027	Texas Department of Public Welfare	Austin
G0564	Texas Tech University	Lubbock
G3154	Texas Tech University School of Medicine at Lubbock (including Academic Health Centers at El Paso and Amarillo)	Lubbock
*G1121XO	Texas, University of, at Arlington	Arlington
G0123	Texas, University of, at Austin	Austin
G1815	Texas, University of, at Dallas	Richardson
G0113	Texas, University of, Cancer Center (and its M.D. Anderson Hospital and Tumor Institute)	Houston
G2615	Texas, University of, Health Science Center at Dallas	Dallas
G2760	Texas, University of, Health Science Center at Houston	Houston
G2255	Texas, University of, Health Science Center at San Antonio, including Medical School, Dental School, and Graduate School of Biomedical Sciences, School of Nursing, School of Allied Health Sciences, and Graduate School of Pharmacy	San Antonio
G0107	Texas, University of, Medical Branch at Galveston	Galveston
G0010	USPHS Hospital, Galveston	Galveston
	UTAH	
G2426	LDS Hospital	Salt Lake City
G2426	LDS Hospital—Deseret Foundation	Salt Lake City
G0462	Utah State University	Logan
G0235	Utah, University of	Salt Lake City
G0235	Utah, University of, Research Institute	Salt Lake City
G0235	Veterans Administration Hospital	Salt Lake City

General Assurance Number	Institution	City
	VERMONT	
G0194	Vermont, University of, and State Agricultural College	Burlington
G2335XB	Veterans Administration Center—White River Junction	White River Junction
	VIRGINIA	
G2684	Eastern Virginia Medical Authority and Eastern Virginia Medical School	Norfolk
G2275XM	Human Resources Research Organization	Alexandria
G0010	USPHS Hospital, Norfolk	Norfolk
G0239	Virginia Commonwealth University (all schools and divisions)	Richmond
G0306	Virginia, University of	Charlottesville
	WASHINGTON	
G0522	Children's Orthopedic Hospital and Medical Center	Seattle
G2627XB	Fred Hutchinson Cancer Research Center	Seattle
G0029XB	Pacific Northwest Research Foundation	Seattle
G0913XB	Swedish Hospital Medical Center	Seattle
G0010	USPHS Hospital, Seattle	Seattle
*G0900	Virginia Mason Research Center, including Mason Clinic and Virginia Mason Hospital	Seattle
G1717	Washington Department of Social and Health Services, all components	Olympia
G0105	Washington State University	Pullman
G0299	Washington, University of	Seattle
	WEST VIRGINIA	
G0431	West Virginia University	Morgantown
	WISCONSIN	
G0315	Central Wisconsin Colony and Training School	Madison
G2891XB	Columbia Hospital	Milwaukee

(*continued*)

General Assurance Number	Institution	City
G0387XM	Marquette University	Milwaukee
G0763	Mendota Mental Health Institute	Madison
G3010XB	Mount Sinai Medical Center of Milwaukee	Milwaukee
G0503	Southern Wisconsin Center for the Developmentally Disabled	Union Grove
G3051	St Mary's Hospital	Milwaukee
G0077	Wisconsin, Medical College of, Inc	Milwaukee
G0319	Wisconsin, University of, Madison Campus	Madison
G2106XM	Wisconsin, University of, Parkside	Kenosha
	BANGLADESH	
G3061	International Centre for Diarrheal Disease Research, Bangladesh	Dacca
	SWITZERLAND	
*G1137	World Health Organization	Geneva

Total number of institutions and components with acceptable general assurances—564

APPENDIX 4.2

Regulations by the Department of Health and Human Services on General Assurance

[Note: These regulations are from the *Federal Register*—Vol. 46, No. 16, Monday, January 26, 1981, pp. 8387–8388, and contain the current rules on the approved general assurance mechanism (IRBs) now used by the federal government. Although these regulations appeared prior to this chapter as part of other materials, they appear separately here to aid in reading Chapter 4 of this book.]

DEPARTMENT OF HEALTH AND HUMAN SERVICES

Office of the Secretary
45 CFR Part 46

Final Regulations Amending Basic HHS Policy for the Protection of Human Research Subjects

AGENCY: Department of Health and Human Services
ACTION: Final rule

[Note: Deleted here are extensive introductory and explanatory sections, and sections about other aspects of subject protection.]

Sec. 46.103 Assurances.

(a) Each institution engaged in research covered by these regulations shall provide written assurance satisfactory to the Secretary that it will comply with the requirements set forth in these regulations.

(b) The Department will conduct or fund research covered by these regulations only if the institution has an assurance approved as provided in this section, and only if the institution has certified to the Secretary that the research has been reviewed and approved by an IRB provided for in the assurance, and will be subject to continuing review by the IRB. This assurance shall at a minimum include:

(1) A statement of principles governing the institution in the discharge of its responsibilities for protecting the rights and welfare of human subjects of research conducted at or sponsored by the institution, regardless of source of funding. This may include an appropriate existing code, declaration, or statement of ethical principles, or a statement formulated by the institution itself. This requirement does not preempt provisions of these regulations applicable to Department-funded research and is not applicable to any research in an exempt category listed in Sec. 46.101.

(2) Designation of one or more IRBs established in accordance with the requirements of this subpart, and for which provisions are made for meeting space and sufficient staff to support the IRB's review and recordkeeping duties.

(3) A list of the IRB members identified by name; earned degrees; representative capacity; indications of experience such as board certifications, licenses, etc., sufficient to describe each member's chief anticipated contributions to IRB deliberations; and any employment or other relationship between each member and the institution; for example: full-time employee, part-time employee, member of governing panel or board, stockholder, paid or unpaid consultant. Changes in IRB membership shall be reported to the Secretary.

(4) Written procedures which the IRB will follow (i) for conducting its initial and continuing review of research and for reporting its findings and actions to the investigator and the institution; (ii) for determining which projects require review more often than annually and which projects need verification from sources other than the investigators that no material changes have occurred since previous IRB review; (iii) for insuring prompt reporting to the IRB of proposed changes in a research activity, and for insuring that changes in approved research, during the period for which IRB approval has already been given, may not be initiated without IRB review and approval except where necessary to eliminate apparent immediate hazards to the subject; and (iv) for insuring prompt reporting to the IRB and to the Secretary of unanticipated problems involving risks to subjects or others.

(c) The assurance shall be executed by an individual authorized to act for the institution and to assume on behalf of the institution the obligations imposed by these regulations, and shall be filed in such form and manner as the Secretary may prescribe.

(d) The Secretary will evaluate all assurances submitted in accordance with these regulations through such officers and employees of the Department and such experts or consultants engaged for this purpose as the Secretary determines to be appropriate. The Secretary's evaluation will take into consideration the adequacy of the proposed IRB in light of the anticipated scope of the institution's research activities and the types of subject populations likely to be involved, the appropriateness of the proposed initial and continuing review procedures in light of the probable risks, and the size and complexity of the institution.

(e) On the basis of this evaluation, the Secretary may approve or disapprove the assurance, or enter into negotiations to develop an approvable one. The Secretary may limit the period during which any particular approved assurance or class of approved assurances shall remain effective or otherwise condition or restrict approval.

(f) Within 60 days after the date of submission to HHS of an application or proposal, an institution with an approved assurance covering the proposed research shall certify that the application or proposal has been reviewed and approved by the IRB. Other institutions shall certify that the application or proposal has been approved by the IRB within 30 days after receipt of a request for such a certification from the Department. If the certification is not submitted within these time limits, the application or proposal may be returned to the institution.

CHAPTER 5

Freedom of Information and Privacy

Again, we have a topic that could fill an entire volume by itself. In keeping with the spirit of this book, however, this chapter will contain brief overviews of areas related to information access and privacy, and the interested reader may consult the references for more detail. As we will see later in this chapter, the area of privacy contains rules affecting who may obtain data gathered by a researcher and rules for how scientists may obtain data gathered by the government.

The privacy area in general is quite complex, with as many levels to it as the general concept of *consent* addressed earlier. For example, there are federal rules as to who may have access to various types of information stored in government files, including the much-publicized Freedom of Information Act and the Privacy Act. There are state rules that also apply, professional guidelines (such as those for physicians governing who may view their patients' records), and case law where suits have been filed when a party felt that someone should or should not have access to some kind of information. For our purposes, as before, we are going to emphasize federal rules and their effects on the professional and occasionally on the lay citizen.

There are two primary pieces of legislation that we will consider: the Freedom of Information Act (often referred to as FOI or FOIA) and the Privacy Act. Although federal agencies subsequently published myriad rules on how each agency complies with the requirements that were in the acts, we will deal with the acts themselves.

Persons wishing to know how any one particular agency deals with the acts' requirements should contact that agency. One should also be prepared for some confusion over what the most current version might be of the agency's rules. This is because of the increasing number of exemptions from the acts (particularly from the FOIA) sought by agencies and the resulting changes in agency rules. Thus, someone trying to acquire federal records (through the FOIA) or attempting to control what information is shared between agencies (Privacy Act) should contact the relevant agencies directly. The regulatory and legislative processes being what they are, individual agencies are likely to modify their own rules far more often, and in greater detail, than is the Congress to review the enabling legislation (the acts) that led to agency rules in the first place.

In addition to the two primary acts, a secondary act (the Sunshine Act) will be discussed in its relation to the general concept of privacy. Finally, we will examine confidentiality of information that is achieved through a confidentiality certificate that can be granted by the federal government in certain circumstances to protect certain information gathered by researchers.

The reader should keep in mind as we proceed that many of the terms used actually relate to the same phenomena, but the viewpoints of the parties are different. For example, if I want to make sure that you cannot view confidential records on subjects who participate in my research project, I may worry about the confidentiality of my own data. The subjects may worry about their general privacy or, specifically, about the confidentiality of my private files. If you, for whatever reason, wish to review these files, you may speak of your freedom to have access to my information.

Because of the interrelatedness of these issues, despite the use of different words, it is no accident that the two primary acts (FOIA and the Privacy Act) both are contained in the same title of the U.S. Code.

Government Files Only

One final note before we begin. Although I have heard colleagues quote the Privacy Act or FOIA as legal justifications for protecting files, they are incorrect. These acts govern only access to, and sharing of, information in *government* files. They have *nothing whatever* to do with private records, that is, files maintained by individuals or agencies not under direct federal supervision. Nevertheless, in keeping with the spirit of the legislation, professionals may do well to adopt at least some of the acts' provisions, although professional associations' code of ethics usually handle these concerns adequately.

THE FREEDOM OF INFORMATION ACT (FOIA)

The Freedom of Information Act essentially states that federal agencies must publish or otherwise make available (e.g., at the request of the public) indexes of information it maintains, steps for the public to take in obtaining agency information, the role of the courts in such processes, exemptions of some agency records from the FOIA, and how agencies report their compliance to FOIA to the Congress.

Although the FOIA as a specific piece of legislation is relatively recent, the federal government has recognized the need for access to agency information by the public for decades.

History

On July 4, 1966, Public Law 89–487 was passed by the Congress. This was one of the initial pieces of legislation to be passed in the era of increasing concern over government intrusion into private lives and access to government information. This untitled act was passed to modify part of the considerably older Administrative Procedure Act (June 11, 1946) concerning the rights of the public to information. PL 89–487 was quite brief (about 1,000 words) but did lay the ground work for current rules. The intervention of the court was authorized for instances in which the public complained about denial of access to agency records, and a list of exemptions was presented for which public access was not approved (e.g., trade secrets, investigatory law enforcement files).

The next year, on June 5, 1967, Public Law 90–23 was passed, amending section 552 of Title 5 of the United States Code. It is this section 552 of Title 5, U.S.C., that has been the focus of the recent legislation (i.e., FOIA, Privacy Act, Sunshine Act). Essentially, PL 90–23 took the text from the 1966 Act (PL 89–487), reworded it slightly, and inserted it in Section 552 of Title 5, U.S.C., repealing PL 89–487 at the same time. From this time on, it has been Section 552 of Title 5 that has been variously amended to guarantee access to information yet still preserve privacy of individuals with regard to government information.

It should be noted that specific exemptions to public access to information were thus in effect at least as far back as 1966 and 1967. However, it appears that the public did not generally use these provisions substantially until the 1970s, when the FOIA and the Privacy Act were enacted. Thus the exceptions perhaps were not well known as well. In contrast, one can often read today of an individual or an organization attempting to block an agency's effort to exempt additional information from the FOIA in particular.

FOIA Today

On November 21, 1974, Public Law 93–502 was passed. It was at this point that Section 552 of Title 5, U.S.C., became better known as the Freedom of Information Act following the substantial changes and additions made by PL 93–502. The reader should note that, currently, FOIA usually refers to the *current* set of rules contained in Section 552 of Title 5, U.S.C., rather than any one particular act passed by Congress on one particular day. Amendments to the FOIA can be expected to continue, but FOIA will usually refer to current rules, not to an historical event.

In considerable detail, PL 93–502 described how individuals may gain access to federal records, including how the cost of duplicating such files is to be handled and who shall pay in various circumstances. Specific deadlines were instituted covering how long an agency can take to respond in various steps to the public's request for information. The role of the court in assisting the public lawfully to obtain information was expanded, even to the point of permitting the court to assess attorney fees and other litigation costs against the gevernment in cases where the public "substantially prevailed" in suing the government to obtain information!

Basic Elements of the FOIA. The actual application of the FOIA has been quite broad, although the act itself is relatively brief. The complete text of the FOIA is contained in the appendix of this chapter. We will address only the basic elements of the FOIA in this chapter, and the reader is referred to the following two excellent sources for additional information. For a general description of the FOIA, how it relates to the Privacy Act, and considerable detail on how to use both the FOIA and the Privacy Act, see Bouchard and Franklin's *Guidebook to The Freedom of Information and Privacy Acts* (1980).[1] For a much briefer (24 pages) guide aimed mainly at journalists who want to use the FOIA to obtain federally controlled information, see Lovenheim, Landau, Neville, and Rosenthal's *How to Use the Federal FOI Act* (1980).[2]

In brief, the following elements of the FOIA are especially relevant:

1. *Anyone* can request information by using the FOIA. This includes United States citizens, citizens of other countries, corporations, and any other entity.
2. The FOIA covers information kept by almost every agency, department, and any other unit of the executive branch of the federal government. Except for some types of information, the FOIA does not cover the legislative branch (Congress) or the judicial branch (courts) of the government.
3. The information covered by the FOIA is very loose in definition (in contrast to the more specifically defined Privacy Act) and includes reports, audiovisual recordings, correspondence, and almost every conceivable form of storing information.
4. Acquiring access to information is straightforward and may take the form of a phone call or a formal, written request. Generally, the agency must

[1]Bouchard, R. F., and Franklin, J. D. *Guidebook to the Freedom of Information and Privacy Acts.* New York: Clark Boardman, 1980.

[2]Lovenheim, P. C., Landau, J. C., Neville, S., and Rosenthal, K. *How to use the Federal FOI Act.* Prepared by the Society of Professional Journalists (Sigma Delta Chi) and the Reporters Committee for Freedom of the Press. Available from the FOI Service Center, c/o Reporters Committee, 1125 15th St., N.W., Washington, D.C., 20005.

respond to the request within ten days, and appeals by the requestor are guaranteed if the agency does not respond satisfactorily, including legal suits against the government. The requestor can expect, usually, to pay some copying fees to obtain the desired information.

Exemptions to the FOIA. There are, of course, exemptions to the act that would otherwise open almost all government files to public view, including perusal by persons from other countries. There are nine general exemptions, one of which is broad enough to allow agencies to point to various statutes that apparently exempt part or all of certain types of records. Taken from the current FOIA (see appendix for this chapter), these nine major exemptions to the FOIA are:

> (b) This section does not apply to matters that are—
> (1) (A) specifically authorized under criteria established by an Executive order to be kept secret in the interest of national defense or foreign policy and (B) are in fact properly classified pursuant to such Executive order;
> (2) related solely to the internal personnel rules and practices of an agency;
> (3) specifically exempted from disclosure by statute (other than section 552b of this title), provided that such statute (A) requires that the matters be withheld from the public in such a manner as to leave no discretion on the issue, or (B) establishes particular types of matters to be withheld;
> (4) trade secrets and commercial or financial information obtained from a person and privileged or confidential;
> (5) inter-agency or intra-agency memorandums or letters which would not be available by law to a party other than an agency in litigation with the agency;
> (6) personnel and medical files and similar files the disclosure of which would constitute a clearly unwarranted invasion of personal privacy;
> (7) investigatory records compiled for law enforcement purposes, but only to the extent that the production of such records would (A) interfere with enforcement proceedings, (B) deprive a person of a right to a fair trial or an impartial adjudication, (C) constitute an unwarranted invasion of personal privacy, (D) disclose the identity of a confidential source and, in the case of a record compiled by a criminal law enforcement authority in the course of a criminal investigation, or by an agency conducting a lawful national security intelligence investigation, confidential information furnished only by the confidential source, (E) disclose investigative techniques and procedures, or (F) endanger the life or physical safety of law enforcement personnel;
> (8) contained in or related to examination, operating, or condition reports prepared by, on behalf of, or for the use of an agency responsible for the regulation or supervision of financial institutions; or
> (9) geological and geophysical information and data, including maps concerning wells.

It should be pointed out here that if any of the exemptions appear particularly specific or narrow, it is because construction and eventual passing by the Congress of the FOIA went through the normal lobbying process. That is, any interest group (within or outside the government) that wished to have a specific exemption made part of the act could have affected the wording of the eventual

FOIA. Readers may find the final chapter of this book informative as to how bills become law and how they can be influenced by those who know how to do so.

Before leaving these exemptions to the FOIA, a final note is warranted. Exemption 3 is very general and opened the door for agencies to claim exemption of various types of information because of requirements imposed by other statutes. In fact, Lovenheim *et al.*[3] noted that federal agencies have cited almost 100 statutes to justify withholding of various types of documents (e.g., the Central Intelligence Agency—18 U.S.C. 798; trade secrets—18 U.S.C. 1905). Essentially, agencies are saying to the person who requests certain types of information under the FOIA that "We're sorry, but other federal statutes prohibit us from releasing that information, as authorized by Exemption 3 of the FOIA."

How fair, honest, or moral one considers this protective stance by federal agencies undoubtedly depends, to some extent, on the vested interest of the viewer. That is, would the viewer (or reader) be helped or hurt by the release of the information in question?

The FOIA and The Privacy Act

How much a person may be hurt, in any way, by release of information is one consideration that influenced Congress during the formulation of the Privacy Act. Both the FOIA and the Privacy Act deal with information and protection of individual rights, but they do so in slightly different fashions.

The FOIA was passed as a measure to guarantee the public's right to know, with exemptions that we have already examined. The Privacy Act, on the other hand, was formed to give individuals more control over information specifically *about them* that an agency may share with other agencies and restrict others' access to this data. Thus the Privacy Act gives private persons the right to obtain records about themselves and modify or challenge the accuracy or completeness of their files. Further, the Privacy Act then restricts agencies in various ways from sharing the information on a person with other agencies or individuals without the consent of the person whose file may be shared. As with the FOIA, there are exceptions to the Privacy Act, one of which is the FOIA! Thus the FOIA is an exemption to the Privacy Act and permits—under certain conditions—an individual's file to be seen by someone else.

Another contrast between the two Acts lies in the definitions of what types of information are covered. The FOIA covers all agency records, but the Privacy Act is limited to records that are considered part of a system of records, with a system defined as a group of records containing information that can be retrieved by using the individual's name, social security number, or other specific identification item for each individual.

[3]Ibid., pp. 8–10.

In essence, both acts deal with the same general topic, protection of information. But they address different types of information, require different procedural steps for an individual to use the acts, charge different types of fees for agency efforts to comply with individual requests, and so on. Someone interested in obtaining information from agencies under these acts would be wise in citing both acts for authorization.

The Privacy Act

On December 31, 1974, Congress passed Public Law 93–579 to address several privacy related issues. PL 93–579 established a new Privacy Protection Study Commission to help develop and monitor privacy issues, especially issues on data banks and retention of information on individuals, whether by public or private agencies, organizations, or individuals. But PL 93–579 is perhaps best known for the portion that was then known as the "Privacy Act of 1974" and later, after common usage, as simply the Privacy Act.

The reader will note that the Privacy Act was passed only one month after PL 93–502 had been passed. PL 93–502 modified and established the FOIA as a dominant piece of legislation regarding freedom of individuals to acquire access to federal information. Such freedom, of course, has as its corollary the possibility of permitting such easy access that the privacy of individuals' data may be compromised. Thus, in this sense, the Privacy Act can be viewed as a control on the FOIA, and vice-versa. Which act governs in any one case depends on the type of information sought from the government, who is seeking the information, and other criteria mentioned in both acts.

The Privacy Act Today

The Privacy Act is considerably longer than the FOIA, containing a significant amount of procedural detail on when and how information will or will not be released by federal agencies. As we saw before, the Privacy Act does not refer to all information possessed by an agency, but rather to systems of records.

> (a) (4) the term "record" means any item, collection, or grouping of information about an individual that is maintained by an agency, including, but not limited to, his education, financial transactions, medical history, and criminal or employment history and that contains his name, or the identifying number, symbol, or other identifying particular assigned to the individual, such as a finger or voice print or a photograph.[4]

The full text of the Privacy Act is contained in the appendix to this chapter for readers who wish to become more familiar with its provisions. Essentially, the

[4]Title 5, *United States Code*, Section 552a. Washington, D.C.: U.S. Government Printing Office, 1977.

Privacy Act forbids agencies to disclose any record within a system of records to anyone unless the agency has written consent from the individual to whom the record pertains.

Basic Elements of the Privacy Act

Briefly, the Privacy Act contains several basic provisions and procedural steps:

1. Eleven provisions whereby written consent of the individual is *not* required (e.g., where the disclosure would be to the Bureau of the Census for census purposes);
2. Steps for agencies to follow in accounting for their responses to disclosure requests;
3. Detailed procedures for permitting individuals to view and/or obtain copies of and/or modify agency records that involve that individual;
4. Requirements for informing individuals at the time of collecting information about the intended uses of the information, and for publishing a list or index of the agency records annually in the *Federal Register* so that individuals may know more easily whom to contact to search various agency files; and
5. Provisions for the courts to handle disputes so that individuals can resort to judicial proceedings in seeking to access relevant agency files.

Of course, there are far more details on these and related areas in the Privacy Act itself, as can be seen from the text in the appendix. Anyone actually seeking to check his own files, as maintained by federal agencies, should read the entire act carefully. Searching the indexes published by federal agencies in the *Federal Register* will provide clues about the different types of files maintained by agencies, although agencies can be expected to be helpful in organizing such a search. The individual conducting such a search should be prepared to have to furnish any agency with enough descriptive information so that it can conclude what files the data sought for might involve. How much detail a person wishes to reveal to an agency as part of this process is a matter of individual decision. As an extreme example, a person may or may not wish to reveal sufficient information to the Federal Bureau of Investigation to enable the FBI to search its files to tell the requestor just what information the FBI has on file about that requestor. The obvious ironic risk to privacy in such a case is revealing information to the agency, as part of a search, that the agency never had before.

The impact of this act on research, as well as the effects of the FOIA, will be summarized later in the chapter, following discussion of two more federal actions regarding privacy: the Sunshine Act and confidentiality certificates.

The Sunshine Act

In the same spirit that the FOIA and the Privacy Act were passed by Congress, the "Government in the Sunshine Act" was passed as Public Law 94–409 on September 13, 1976. The primary purpose of the Sunshine Act was to establish various rules regarding federal agency meetings so that the public would be aware of, and possibly participate in, such government meetings. This was another measure adopted by Congress to ensure that the public could influence agency deliberations, when appropriate, and generally make sure that government bodies could not secretly establish policies or procedures that might adversely affect the public.

The Sunshine Act thus protects individuals' rights by requiring agencies, in various ways, to inform the public of its deliberations. Presumably, such knowledge would enable individuals who feel that they will be affected by agency meetings to provide some kind of input to the decision-making that might occur in the meetings.

Basic Elements of the Sunshine Act

The full text of the Sunshine Act is contained in the appendix to this chapter. However, to provide a brief overview of the main parts of the act, the following points can be made:

1. As for other regulations governing control of information, there are several exemptions of the act, such as exempting from the act meetings that would disclose trade secrets, expose confidential personal information, reveal investigatory law enforcement data, and so on.
2. Provisions are made for public notices to be made on the time, place, and subject matter of relevant agency meetings.
3. Transcripts are kept of all meetings and portions of meetings that are not exempt from the act and are available to the public at cost (e.g., a duplication fee if the recordings of the meeting are in the form of typed minutes).
4. Remedy through the courts is available to persons who have legitimate claim against any agency that does not conform to the requirements of the act.
5. Each agency covered by the act must report annually to the Congress its compliance with the act, including a record of the total number of agency meetings open to the public, the total number of closed meetings, the reasons for closing meetings, and a description of any litigation brought against the agency under this act.

Sunshine Act and the Researcher

The Sunshine Act is less immediately relevant to a researcher than the FOIA, the Privacy Act, and confidentiality certificates. The placement of the Sunshine Act[5] in the United States Code of course relates it to the other controls on information and access; therefore, at least in that respect, the researcher should be aware of his or her rights to participate in agency decision-making. The less politically active researcher can obtain post-hoc transcripts of agency meetings to monitor the activities of agencies of interest.

However, the researcher can be expected to have somewhat more of an interest in the Sunshine Act than the average citizen. Most researchers who depend on contracts or grants from federal agencies would be well advised to monitor closely such meetings of potential funding agencies that determine goals or priorities. Not all such meetings are open to the public, of course, but those that are can be instructive for the researcher hoping for future funding.

As for other contacts with the federal government, a distinct advantage accrues to persons who are in or near Washington, D.C., since that is where most meetings are held. In this respect the Sunshine Act aids those who are not so close, since the notices (required by the act) that are published in the *Federal Register* also contain the address to use in order to obtain subsequent transcripts of the meetings. This author has taken advantage of this aspect of the Sunshine Act in obtaining minutes of meetings for a more thorough understanding of how the Department of Health, Education, and Welfare (now Health and Human Services) has developed the current informed consent and IRB regulations.

Confidentiality Certificates

One of the problems inherent in the new freedoms of giving access to government information is the likelihood of releasing material that is of such a private nature that it should be better protected. For researchers, the concern is over who might acquire data about individuals that are collected as part of a research project. Obviously, researchers who cannot offer some type of guarantee of confidentiality are less likely to obtain consent and participation of potential research subjects. Further, most researchers feel an ethical obligation to protect such information, both to allay subject fears and to ensure proprietary control over the research data.

Research on Sensitive Issues

Such concern over protecting research data can be expected to be most intense, for both researcher and subject, for research topics of a sensitive nature.

[5]Title 5, *United States Code*, Section 552b. Washington, D.C.: U.S. Government Printing Office.

As we have already seen, research on topics that pose less than minimal risk (e.g., demographic survey questionnaires) usually generate data that are so innocuous that informed consent may not even be necessary. In contrast, research on drug use, alcohol use, and law offenses typically pose far more risk to the willing subject if such data is ever subsequently linked to their identity.

There are several mechanisms within the federal government for granting researchers the right to withhold from various persons or agencies such sensitive data. Our discussion will focus on the confidentiality certificates available from the Department of Health and Human Services.

Confidentiality Certificates in Health and Human Services

On April 4, 1979, the former Department of Health, Education, and Welfare published a new set of rules regarding the protection of the identity of research subjects.[6] As usual for this book, the full text of the rules is presented in the appendix at the end of this chapter, but a brief summary is presented here.

Applicability of Protection. The rules apply to a wide group of researchers, with the added bonus that the researchers do not even have to be receiving federal funds to obtain such protection for the data. The protection is in the form of a confidentiality certificate issued by the Secretary of HEW (now HHS) that permits researchers to refuse access to the data to almost anyone, including law enforcement officials. In general, the researcher usually must grant access to the data only to HHS staff as part of their normal and necessary duties (see rules in the appendix for more detail).

The intent and scope of applicability of the confidentiality rules are quite specific[7]:

> 2a.1 Applicability
>
> (a) Section 303(a) of the Public Health Service Act (42 U.S.C. 242a(a)) provides that "(t)he Secretary (of Health, Education, and Welfare) may authorize persons engaged in research on mental health, including research on the use and effect of alcohol and other psychoactive drugs, to protect the privacy of individuals who are the subject of such research by withholding from all persons not connected with the conduct of such research the names or other identifying characteristics of such individuals. Persons so authorized to protect the privacy of such individuals may not be compelled in any Federal, State, or local civil, criminal, administrative, legislative, or other proceedings to identify such individuals." The regulations in this part establish procedures under which any person engaged in research on mental health including research on the use and effect of alcohol and other psychoactive drugs (whether or not the research is Federally funded) may, subject to the exceptions set forth in paragraph (b) of this section, apply for such an authorization of confidentiality.

Exceptions. There are two exceptions to this type of confidentiality protection. First, the rules do not apply to a different type of confidentiality that is

[6]*Federal Register*, Vol. 44, No. 66, Wednesday, April 4, 1979, pp. 20382–20387.
[7]Ibid., p. 20384.

authorized for research that requires an investigational new drug exemption or for work with approved new drugs such as methadone. Such research already has separate confidentiality provisions governing patient records.[8,9]

Second, these rules by HHS do not apply to research that is related to law enforcement activities or other such research that comes under the separate confidentiality rules of the Controlled Substances Act (21 U.S.C. 872(c)).

Time Limit. A confidentiality certificate, as issued by HHS, is in effect only for the period so stated on the certificate. That is, during the period when the certificate is in force, the researcher may refuse to release data obtained about any subject who participated in the research between the certificate's start and termination dates. Further, the researcher is subsequently authorized to refuse access to data on subjects who participated in research during any time the authorization was in effect, even after expiration of the certificate itself.

Finally, it should be noted that, in common with previously discussed consent procedures, these rules at no time supersede the right of a subject to release voluntarily such research data if he or she (or the legal guardian) so chooses.

Impact of Privacy Rules on Research

For this discussion, we shall subsume all of the topics of this chapter (the Privacy Act, the Freedom of Information Act, the Sunshine Act, and confidentiality certificates) under the general heading of "privacy rules." Indeed, the knowledgeable reader, at this point, may have already realized that human service professionals occasionally do not specify a particular set of rules when referring to privacy legislation in the literature. As we have seen from this chapter, however, there are a variety of privacy-related rules that each address specific aspects of privacy. Researchers are advised to keep in mind the distinct nature of the different rules, particularly if they wish to cite one in seeking information from a government agency.

From a researcher's viewpoint, the FOIA is perhaps the most relevant of the rules since it regulates what kinds of government-controlled data the researcher may wish to obtain. The Privacy Act also can come into play, as we have already seen, and the Sunshine Act has made it easier for researchers and others to learn of upcoming federal agency meetings that may affect the funding of research by grants or contracts. Confidentiality certificates obviously can benefit researchers dealing with sensitive research topics.

[8]Federal Food, Drug, and Cosmetic Act, Section 505(i).
[9]Title 21, *Code of Federal Regulations*, Section 291.505(g). Washington, D.C.: U.S. Government Printing Office.

But it is also clear that the research community and others (including government agencies) do not unequivocally favor some aspect of privacy rules. Some of the disadvantages and related advantages of privacy rules and their impact on research are contained in the following summary sections. Readers interested in more detail on these topics may wish to consult the references thoughout the chapter.

Impact on Federal Agencies

From the government's point of view, and perhaps that of the taxpayer as well, implementation of privacy rules has been costly. A 1978 issue of *ACCESS Reports*[10] noted that the Government Accounting Office (GAO) conducted an investigation of the effect of the Privacy Act and the FOIA on 13 law enforcement agencies. The study was done at the request of Senator James Eastland (Democrat, Mississippi) on behalf of the Subcommittee on Criminal Laws and Procedures. The subcommittee felt that perhaps the privacy rules were unduly hampering law enforcement agencies.

The GAO report contained the following findings:

1. The 13 agencies spent a total of $35.9 million from 1975–1977 to comply with the rules.
2. Initial implementation of the Privacy Act of 1974 cost $594,000.
3. Operating costs after that ranged from $159,000 at the Postal Service's Inspection Service to $13.8 million for the Federal Bureau of Investigation.
4. The 13 agencies reported a total of 147,000 information requests from 1975–1977, with the Immigration and Naturalization Service at the top of the list with 51,000 requests.
5. "The most dominant category of requestors identified by many of the agencies was individuals who have been or are subjects of investigation by the agencies. . . . Some of these requestors were also identified as being criminals."[11]
6. In compiling future projections, the agencies expected an 85% increase in the number of requests from 1977–1982.

Clearly, the privacy rules are costing us all a considerable amount of money.

In addition to private researchers who might wish to obtain agency data for purposes of research, government agencies themselves conduct a significant amount of statistical analysis of a wide variety of data for numerous purposes.

[10]*ACCESS Reports*, July 11, 1978, pp. 11–12. Washington, D.C.: Plus Publications.
[11]Ibid., p. 12.

Duncan (1975), as Deputy Associate Director for Statistical Policy in OMB in Washington, D.C., pointed out that—despite its usefulness in protecting the public—the Privacy Act greatly restricts the sharing of data between agencies.[12] Such restrictions may lead to the creation of duplicate data files, and most researchers are aware of the complications that can result from duplicate files. Among other suggestions, Duncan recommended that "sound scientific information" collected for scientific or statistical purposes should be protected from mandatory legal process that requires the disclosure of individual records (i.e., through the FOIA). He further suggested that there should be a controlled, effective process for exchanging data among statistical agencies to insure the accuracy of sampling frames and conducted so that disclosure of any individual records would remain confidential within the system.

One outcome of concerns such as Duncan's has been continual efforts by various agencies to exempt specific record systems from the FOIA and the Privacy Act. These efforts can be followed in the daily publication of the *Federal Register*. The individual researcher, wondering whether or not desired information can be obtained, should contact directly the agency in question.

Impact on Researchers

In the immediate years following passage of privacy rules, research concern appeared to focus primarily on how the rules would hinder researchers' access to government information (e.g., by survey statisticians) and a heightened awareness of confidentiality in general. For example, Hofferbert cautioned researchers about publication of research results and suggested among other procedures, that researchers consider grouping data or deleting certain data to protect the privacy of research subjects.[13]

Subsequently, Morris, Sales, and Berman wrote of broader concerns for the researcher and cited court cases that have been tried in which privacy rules and research were the focal issues.[14] Morris *et al.* referred to four types of information *that researchers provide to the government* and that might be acquired by others through the FOIA: a grant or contract proposal, raw data, results of preliminary analyses, or a final report. The highlights of this article are presented here and the reader is urged to examine the original article for more detail.

Research Proposal. In the case of the Washington Research Project vs. Department of Health, Education, and Welfare (504 F.2d 238, D.C. Cir. 1974)

[12]Duncan, J. W. The impact of privacy legislation on the federal statistical system. *Public Data Use*, 1975 (January), 3 (1), 51–53.

[13]Hofferbert, R. I. Social science archives and confidentiality. *American Behavioral Scientist*, 1976, *19*, 467–488.

[14]Morris, R. A., Sales, B. D., & Berman, J. J. Research and the Freedom of Information Act. *American Psychologist*, 1981, 36 (8), 819–826.

the court found that a grant proposal could not be exempted from the FOIA, thus opening up such proposals to the scrutiny of anyone else, including competing funding-seekers. The court did conclude that the review comments concerning the proposal and other related materials were exempt from the FOIA, but the proposal itself, with its hypotheses, design, sampling techniques, and so forth, is open to public access. Obviously, researchers now worry about others' "stealing" new theories or methods, unless there is some way to construct a proposal with sufficient information to appeal to funding agencies but not each detail to divulge their efforts to colleagues or to be misunderstood by a curious press.

Morris *et al.* correctly point out an advantage here as well for researchers who use the FOIA to obtain copies of funded proposals to learn how to write better proposals themselves. Still, this procedure was subsequently challenged in the recommendations of the National Commission for the Protection of Human Subjects of Biomedical and Behavioral Research, which recommended that only *funded* proposals be accessible under the FOIA.[15] It is true that agencies that publish lists of funded proposals (thereby providing enough information for others to use the FOIA to obtain the proposals) do not publish such information for nonfunded proposals. As a practical matter, this does make it more difficult for someone to obtain a copy of an unfunded proposal using the FOIA. In any event, a researcher planning to submit a grant or contract proposal to a federal agency would be well advised to inquire about the extent to which any material he or she might submit would be accessible to others through the FOIA.

In a related matter, this author was involved in an effort to encourage selected colleagues to forward views about the ownership of grant materials by researchers to the National Commission when it was originally seeking input on the issue. The 1977 report by the Commission noted, among other such respondents, the views held by the author and his colleagues, and the final recommendations appear to have been at least moderately influenced by their suggestions. Fully 10% of all the comments received from the public by the Commission were from the author's colleagues. What is important to note in this case is not that the particular views of the author may have played a role, but rather that a researcher can influence the rules that affect research. Methods for providing such input have been discussed previously and are presented in greater detail in the later chapter in this book on how to involve oneself in the process of federal rule-making.

Raw Data. The raw data from a study can be expected to constitute an area in which a researcher might be especially concerned lest others gain access to the data. Although it can be beneficial to allow others to reanalyze data, either as a

[15] *Disclosure of Research Information under the Freedom of Information Act.* Report of the National Commission for the Protection of Human Subjects of Biomedical and Behavioral Research, 1977. Washington, D.C.: DHEW Pub. No. (OS)77–0003.

reliability measure or to generate new ideas not conceived by the original researchers, it is also true that misinterpretation of the data is more likely when one was not involved in the original conception, design, or conduct of the study.

In general, without exploring the intricate details of who truly "owns" data when public funds are involved, courts have maintained in at least two cases that the public does *not* have access to raw data through the FOIA when private groups obtain data with financial support from a federal agency (*Ciba-Geigy* v. *Mathews*[16], and *Forsham* v. *Califano*[17]—later becoming *Forsham* v. *Harris* in the Supreme Court[18]).

One of the central issues in these cases was whether the data could be considered agency records for purposes of the FOIA, since the federal agency itself did not collect the data, but instead had funded private individuals (as is customary for most grants) to conduct the research. It can be assumed that the data *may* have had to be released through FOIA if the funding agency had ever taken possession of the raw data. This was implied in a notice to recipients of National Institutes of Health funds[19] where it was emphasized that, in the *Forsham* v. *Harris* case, Justice Rehnquist wrote, "But in this context FOIA applies to records which have been *in fact* obtained and not to records which merely *could have been* obtained."[20]

Of course, then, the primary reason that raw data is not available to the public through the FOIA is that seldom are raw data provided to the agency by a researcher. If the raw data are provided to the agency, it seems quite likely they would then be readily obtained by the public.

Preliminary Analyses and Final Reports. In concluding their article, Morris *et al.* cautioned researchers that, unlike raw data, analyses, reports, summaries, and even letters sent by researchers to agencies are available to others through the FOIA. Knowing this, the cautious researcher may need to explain adequately even preliminary findings in progress reports lest they be misunderstood by others. Researchers working with any topic deemed controversial, or of particular interest to a curious press (where understanding of scientific data may not be paramount) should be especially conservative in any such correspondence with any agency.

Until a new specific exemption for research proposals and data is added to the FOIA, or a special bill passed by the Congress, researchers should be aware that, when they write anything to send to an agency, they should consider that they are writing to the world.

[16]428 F. Supp. 523 (S.D.N.Y. 1977).
[17]587 F.2d 1128 (D.C. Cir. 1978).
[18]100 S. Ct. 978 (1980).
[19]*NIH Guide for Grants and Contracts*, April 25, 1980, 9 (6), p. 3. Bethesda, Maryland: National Institutes of Health.
[20]Ibid.

Summary

Because of the scope of the privacy area, this chapter was limited to the impact of some specific federal acts, rather than addressing the entire issue of privacy. For example, there was no discussion of the area of computers and privacy, although it is readily apparent that most of the American public is aware of the dangers of computers in the fields of finance and law enforcement[21] to say nothing of the particular concerns of researchers. Instead, we examined the acts and related rules that, in the author's opinion, most affect what information researchers would like to obtain from the federal government and the information researchers might turn over to the government.

Although often discussed in the literature, we have seen that the Privacy Act is primarily concerned with how information may be shared between agencies and what restrictions are placed on such sharing processes to protect the confidentiality of individuals and information about them in some (not all) agency systems of records. If a researcher is concerned with what information is passed to which agencies after the researcher has forwarded information to a funding agency, then the Privacy Act spells out how the researcher can trace, and potentially correct or modify, such shared information.

However, for most researchers, it is the Freedom of Information Act (FOIA) that is of primary importance in the privacy field. The FOIA permits public access to a wide variety of government files, and the files need not be a clearly defined system of records, in contrast to the Privacy Act. Like the Privacy Act, the FOIA contains a variety of exemptions for certain types of information, but the FOIA is still wide in its scope.

On one hand, the FOIA represents an advantage for those researchers who might wish to study information gathered by the government but, prior to the FOIA, may have found it difficult or impossible to obtain the data from agencies. An analysis of FOIA requests conducted by the Office of Management and Budget, however, revealed that many of the requestors were persons who had been or were being investigated by those agencies. Presumably, the requestors wanted to see what the government knew about them. Since this analysis involved law enforcement agencies, the analysis may not be representative of typical requestors but it appears likely that most FOIA requestors are either persons wishing to know what the government knows about them or members of the press seeking news items.

Thus it is likely that researchers may well face the disadvantages of the FOIA, that is, the chance that information sent by researchers to agencies will be obtained by others who use the FOIA to obtain the information. We have seen that this potential danger applies to grant proposals, progress reports, final reports, and even correspondence. The very existence of the FOIA, although it

[21]Crooks, J. Civil liberties, libraries, and computers. *Library Journal*, 1976 (February), 482–487.

may certainly be for the public good, enjoins researchers from informing agencies of anything they don't want everybody to know.

One protection researchers can use that is related to privacy is a confidentiality certificate. Such a certificate provides confidentiality to researchers such that almost no one (excepting the authorizing agency for certain reasons) may obtain individually identifiable data from a project, even if the project is privately financed. These certificates are especially useful for researchers working on alcohol or drug abuse programs as a way of assuring subjects of confidentiality.

Finally, the Sunshine Act bears some relevance to the privacy field, since it requires, in most circumstances, that federal agencies give public notice to agency meetings. Such notice can enable the alert researcher to attend meetings that might involve topics germane to his or her research or even on funding, especially if the researcher can travel to the Washington area, where most of the meetings are held. For researchers who do not have such access, the Sunshine Act provides that a contact person be designated in each meeting notice and minutes or transcripts of the meeting be made available to anyone desiring them.

The subject of privacy in general continues to be in a state of flux, since it is influenced by broader social concerns and fears wherein research is simply one small aspect. Despite the emphasis in earlier chapters of this book on the importance of consent and of assuring subjects of confidentiality, portions of this chapter clearly reveal that there are limitations on the confidentiality researchers can promise—especially if they hand individually identifiable data over to a federal agency. With that in mind, researchers should be aware of the provisions of both the FOIA and the Privacy Act (see complete texts in this chapter's appendix).

APPENDIX 5.1

The Freedom of Information Act

[Note: Text of FOIA is taken from the United States Code, Title 5—Government Organization and Employees—Section 552, Washington, D.C., U.S. Government Printing Office, 1977, as amended.]

Sec. 552. Public information; agency rules, opinions, records, and proceedings

(a) Each agency shall make available to the public information as follows:

(1) Each agency shall separately state and currently publish in the Federal Register for the guidance of the public—

(A) descriptions of its central and field organization and the established places at which, the employees (and in the case of a uniformed service, the members) from whom, and the methods whereby, the public may obtain information, make submittals or requests, or obtain decisions;

(B) statements of the general course and method by which its functions are channeled and determined, including the nature and requirements of all formal and informal procedures available:

(C) rules of procedure, descriptions of forms available or the places at which forms may be obtained, and instructions as to the scope and contents of all papers, reports, or examinations;

(D) substantive rules of general applicability adopted as authorized by law, and statements of general policy or interpretations of general applicability formulated and adopted by the agency; and

(E) each amendment, revision, or repeal of the foregoing.

Except to the extent that a person has actual and timely notice of the terms thereof, a person may not in any manner be required to resort to, or be adversely affected by, a matter required to be published in the Federal Register and not so published. For the purpose of this paragraph, matter reasonably available to the class of persons affected thereby is deemed published in the Federal Register when incorporated by reference therein with the approval of the Director of the Federal Register.

(2) Each agency, in accordance with published rules, shall make available for public inspection and copying—

(A) final opinions, including concurring and dissenting opinions, as well as orders, made in the adjudication of cases;
(B) those statements of policy and interpretations which have been adopted by the agency and are not published in the Federal Register; and
(C) administrative staff manuals and instructions to staff that affect a member of the public; unless the materials are promptly published and copies offered for sale.

To the extent required to prevent a clearly unwarranted invasion of personal privacy, an agency may delete identifying details when it makes available or publishes an opinion, statement of policy, interpretation, or staff manual or instruction. However, in each case the justification for the deletion shall be explained fully in writing. Each agency shall also maintain and make available for public inspection and copying current indexes providing identifying information for the public as to any matter issued, adopted, or promulgated after July 4, 1967, and required by this paragraph to be made available or published. Each agency shall promptly publish, quarterly or more frequently, and distribute (by sale or otherwise) copies of each index or supplements thereto unless it determines by order published in the Federal Register that the publication would be unnecessary and impracticable, in which case the agency shall nonetheless provide copies of such index on request at a cost not to exceed the direct cost of duplication. A final order, opinion, statement of policy, interpretation, or staff manual or instruction that affects a member of the public may be relied on, used, or cited as precedent by an agency against a party other than an agency only if—

(i) it has been indexed and either made available or published as provided by this paragraph; or
(ii) the party has actual and timely notice of the terms thereof.

(3) Except with respect to the records made available under paragraphs (1) and (2) of this subsection, each agency, upon any request for records which (A) reasonably describes such records and (B) is made in accordance with published rules stating the time, place, fees (if any), and procedures to be followed, shall make the records promptly available to any person.

(4) (A) In order to carry out the provisions of this section, each agency shall promulgate regulations, pursuant to notice and receipt of public comment, specifying a uniform schedule of fees applicable to all constituent units of each agency. Such fees shall be limited to reasonable standard charges for document search and duplication and provide for recovery of only the direct costs of such search and duplication. Documents shall be furnished without charge or at a reduced charge where the agency determines that waiver or reduction of the fee is in the public interest because furnishing the information can be considered as primarily benefiting the general public.

(B) On complaint, the district court of the United States in the district in which the complainant resides, or has his principal place of business, or in which the agency records are situated, or in the District of Columbia, has jurisdiction to enjoin the agency from withholding agency records and to order the production of any agency records improperly withheld from the complainant. In such a case the court shall determine the matter de novo, and may examine the contents of such agency records in camera to determine whether such records or any part thereof shall be withheld under any of the exemptions set forth in subsection (b) of this section, and the burden is on the agency to sustain its action.

(C) Notwithstanding any other provision of law, the defendant shall serve an answer or otherwise plead to any complaint made under this subsection within thirty days after service upon the defendant of the pleading in which such complaint is made, unless the court otherwise directs for good cause shown.

(D) Except as to cases the court considers of greater importance, proceedings before the district court, as authorized by this subsection, and appeals therefrom, take precedence on the docket over all cases and shall be assigned for hearing and trial or for argument at the earliest practicable date and expedited in every way.

(E) The court may assess against the United States reasonable attorney fees and other litigation costs reasonably incurred in any case under this section in which the complainant has substantially prevailed.

(F) Whenever the court orders the production of any agency records improperly withheld from the complainant and assesses against the United States reasonable attorney fees and other litigation costs, and the court additionally issues a written finding that the circumstances surrounding the withholding raise questions whether agency personnel acted arbitrarily or capriciously with respect to the withholding, the Special Counsel, after investigation and consideration of the evidence submitted, shall submit its findings and recommendations to the administrative authority of the agency concerned and shall send copies of the findings and recommendations to the officer or employee or his representative. The administrative authority shall take the corrective action that the Special Counsel recommends.

(G) In the event of noncompliance with the order of the court, the district court may punish for contempt the responsible employee, and in the case of a uniformed service, the responsible member.

(5) Each agency having more than one member shall maintain and make available for public inspection a record of the final votes of each member in every agency proceeding.

(6) (A) Each agency, upon any request for records made under paragraph (1), (2), or (3) of this subsection, shall—

(i) determine within ten days (excepting Saturdays, Sundays, and legal public holidays) after the receipt of any such request whether to comply with such request and shall immediately notify the person making such request of such determination and the reasons therefor, and of the right of such person to appeal to the head of the agency any adverse determination; and

(ii) make a determination with respect to any appeal within twenty days (excepting Saturdays, Sundays, and legal public holidays) after the receipt of such appeal. If on appeal the denial of the request for records is in whole or in part upheld, the agency shall notify the person making such request of the provisions for judicial review of that determination under paragraph (4) of this subsection.

(B) In unusual circumstances as specified in this subparagraph, the time limits prescribed in either clause (i) or clause (ii) of subparagraph (A) may be extended by written notice to the person making such request setting forth the reasons for such extension and the date on which a determination is expected to be dispatched. No such notice shall specify a date that would result in an extension for more than ten working days. As used in this subparapraph, "unusual circumstances" means, but only to the extent reasonably necessary to the proper processing of the particular request—

(i) the need to search for and collect the requested records from field facilities or other establishments that are separate from the office processing the request;
(ii) the need to search for, collect, and appropriately examine a voluminous amount of separate and distinct records which are demanded in a single request; or
(iii) the need for consultation, which shall be conducted with all practicable speed, with another agency having a substantial interest in the determination of the request or among two or more components of the agency having substantial subject-matter interest therein.

(C) Any person making a request to any agency for records under paragraph (1), (2), or (3) of this subsection shall be deemed to have exhausted his administrative remedies with respect to such request if the agency fails to comply with the applicable time limit provisions of this paragraph. If the Government can show exceptional circumstances exist and that the agency is exercising due diligence in responding to the request, the court may retain jurisdiction and allow the agency additional time to complete its review of the records. Upon any determination by an agency to comply with a request for records, the records shall be made promptly available to such person making such request. Any notification of denial of any request for records under this subsection shall set forth the names and titles or positions of each person responsible for the denial of such request.

(b) This section does not apply to matters that are—

(1) (A) specifically authorized under criteria established by an Executive order to be kept secret in the interest of national defense or foreign policy and (B) are in fact properly classified pursuant to such Executive order;

(2) related solely to the internal personnel rules and practices of an agency;

(3) specifically exempted from disclosure by statute (other than section 552b of this title), provided that such statute (A) requires that the matters be withheld from the public in such a manner as to leave no discretion on the issue, or (B) establishes particular criteria for withholding or refers to particular types of matters to be withheld;

(4) trade secrets and commercial or financial information obtained from a person and privileged or confidential;

(5) inter-agency or intra-agency memorandums or letters which would not be available by law to a party other than an agency in litigation with the agency;

(6) personnel and medical files and similar files the disclosure of which would constitute a clearly unwarranted invasion of personal privacy;

(7) investigatory records compiled for law enforcement purposes, but only to the extent that the production of such records would (A) interfere with enforcement proceedings, (B) deprive a person of a right to a fair trial or an impartial adjudication, (C) constitute an unwarranted invasion of personal privacy, (D) disclose the identity of a confidential source and, in the case of a record compiled by a criminal law enforcement authority in the course of a criminal investigation, or by an agency conducting a lawful national security intelligence investigation, confidential information furnished only by the confidential source, (E) disclose investigative techniques and procedures, or (F) endanger the life or physical safety of law enforcement personnel;

(8) contained in or related to examination, operating, or condition reports prepared by, on behalf of, or for the use of an agency responsible for the regulation or supervision of financial institutions; or

(9) geological and geophysical information and data, including maps, concerning wells.

Any reasonably segregable portion of a record shall be provided to any person requesting such record after deletion of the portions which are exempt under this subsection.

(c) This section does not authorize withholding of information or limit the availability of records to the public, except as specifically stated in this section. This section is not authority to withhold information from Congress.

(d) On or before March 1 of each calendar year, each agency shall submit a report covering the preceding calendar year to the Speaker of the House of Representatives and President of the Senate for referral to the appropriate committees of the Congress. The report shall include—

(1) the number of determinations made by such agency not to comply with requests for records made to such agency under subsection (a) and the reasons for each such determination;

(2) the number of appeals made by persons under subsection (a) (6), the result of such appeals, and the reason for the action upon each appeal that results in a denial of information;

(3) the names and titles or positions of each person responsible for the denial of records requested under this section, and the number of instances of participation for each;

(4) the results of each proceeding conducted pursuant to subsection (a)(4)(F), including a report of the disciplinary action taken against the officer or employee who was primarily responsible for improperly withholding records or an explanation of why disciplinary action was not taken;

(5) a copy of every rule made by such agency regarding this section;

(6) a copy of the fee schedule and the total amount of fees collected by the agency for making records available under this section; and

(7) such other information as indicates efforts to administer fully this section.

The Attorney General shall submit an annual report on or before March 1 of each calendar year which shall include for the prior calendar year a listing of the number of cases arising under this section, the exemption involved in each case, the disposition of such case, and the cost, fees, and penalties assessed under subsections (a)(4)(E), (F), and (G). Such reports shall also include a description of the efforts undertaken by the Department of Justice to encourage agency compliance with this section.

(e) For purposes of this section, the term "agency" as defined in section 551(1) of this title includes any executive department, military department, Government corporation, Government controlled corporation, or other establishment in the executive branch of the Government (including the Executive Office of the President), or any independent regulatory agency.

(Pub. L. 89–554, Sept. 6, 1966, 80 Stat. 383; Pub. L. 90–23, Sec. 1, June 5, 1967, 81 Stat. 54; Pub. L. 93–502, Secs. 1–3, Nov. 21, 1974, 88 Stat. 1561–1564; Pub. L. 94–409, Sec. 5(b), Sept. 13, 1976, 90 Stat. 1247.)

APPENDIX 5.2

The Privacy Act of 1974

[Note: The following text is taken from Public Law 93–579 which was passed on December 31, 1974. Although some sources do not include the establishment rules for the Privacy Protection Study Commission at the end of this act, it was in fact part of the "Privacy Act of 1974" and is included here.]

AN ACT

To amend title 5, United States Code, by adding a section 552a to safeguard individual privacy from the misuse of Federal records, to provide that individuals be granted access to records concerning them which are maintained by Federal agencies, to establish a Privacy Protection Study Commission, and for other purposes.

Be it enacted by the Senate and House of Representatives of the United States of America in Congress assembled, That this Act may be cited as the "Privacy Act of 1974."

Sec. 2. (a) The Congress finds that—

(1) the privacy of an individual is directly affected by the collection, maintenance, use, and dissemination of personal information by Federal agencies;

(2) the increasing use of computers and sophisticated information technology, while essential to the efficient operations of the Government, has greatly magnified the harm to individual privacy that can occur from any collection, maintenance, use or dissemination of personal information;

(3) the opportunities for an individual to secure employment, insurance, and credit, and his right to due process, and other legal protections are endangered by the misuse of certain information systems;

(4) the right to privacy is a personal and fundamental right protected by the Constitution of the United States; and

(5) in order to protect the privacy of individuals identified in information systems maintained by Federal agencies, it is necessary and proper for the Congress to regulate the collection, maintenance, use, and dissemination of information by such agencies.

(b) The purpose of this Act is to provide certain safeguards for an individual against an invasion of personal privacy by requiring Federal agencies, except as otherwise provided by law, to—

(1) permit an individual to determine what records pertaining to him are collected, maintained, used, or disseminated by such agencies;

(2) permit an individual to prevent records pertaining to him obtained by such agencies for a particular purpose from being used or made available for another purpose without his consent;

(3) permit an individual to gain access to information pertaining to him in Federal agency records, to have a copy made of all or any portion thereof, and to correct or amend such records;

(4) collect, maintain, use, or disseminate any record of identifiable personal information in a manner that assures that such action is for a necessary and lawful purpose, that the information is current and accurate for its intended use, and that adequate safeguards are provided to prevent misuse of such information;

(5) permit exemptions from the requirements with respect to records provided in this Act only in those cases where there is an important public policy need for such exemption as has been determined by specific statutory authority; and

(6) be subject to civil suit for any damages which occur as a result of willful or intentional action which violates any individual's rights under this Act.

Sec. 3. Title 5. United States Code, is amended by adding after section 552 the following new section:

Sec. 552a. Records maintained on individuals

(a) Definitions. For purposes of this section—

(1) the term "agency" means agency as defined in section 552(e) of this title;

(2) the term "individual" means a citizen of the United States or an alien lawfully admitted for permanent residence;

(3) the term "maintain" includes maintain, collect, use, or disseminate;

(4) the term "record" means any item, collection, or grouping of information about an individual that is maintained by an agency, including, but not limited to, his education, financial transactions, medical history, and criminal or employment history and that contains his name, or the identifying number, symbol, or other identifying particular assigned to the individual, such as a finger or voice print or a photograph;

(5) the term "system of records" means a group of any records under the control of any agency from which information is retrieved by the name of the individual or by some identifying number, symbol, or other identifying particular assigned to the individual;

(6) the term "statistical record" means a record in a system of records maintained for statistical research or reporting purposes only and not used in whole or in part in making any determination about an identifiable individual, except as provided by section 8 of title 13; and

(7) the term "routine use" means, with respect to the disclosure of a record, the use of such record for a purpose which is compatible with the purpose for which it was collected.

(b) Condition of Disclosure. No agency shall disclose any record which is contained in a system of records by any means of communication to any person, or to another agency, except pursuant to a written request by, or with the prior written consent of, the individual to whom the record pertains, unless disclosure of the record would be—

(1) to those officers and employees of the agency which maintains the record who have a need for the record in the performance of their duties;

(2) required under section 552 of this title;

(3) for a routine use as defined in subsection (a)(7) of this section and described under subsection (e)(4)(D) of this section;

(4) to the Bureau of the Census for purposes of planning or carrying out a census or survey or related activity pursuant to the provisions of title 13;

(5) to a recipient who has provided the agency with advance adequate written assurance that the record will be used solely as a statistical research or reporting record, and the record is to be transferred in a form that is not individually identifiable;

(6) to the National Archives of the United States as a record which has sufficient historical or other value to warrant its continued preservation by the United States Government, or for evaluation by the Administrator of General Services or his designee to determine whether the record has such value;

(7) to another agency or to an instrumentality of any governmental jurisdiction within or under the control of the United States for a civil or criminal law enforcement activity if the activity is authorized by law, and if the head of the agency or instrumentality has made a written request to the agency which maintains the record specifying the particular portion desired and the law enforcement activity for which the record is sought;

(8) to a person pursuant to a showing of compelling circumstances affecting the health or safety of an individual if upon such disclosure notification is transmitted to the last known address of such individual;

(9) to either House of Congress, or, to the extent of matter within its jurisdiction, any committee or subcommitte thereof, any joint committee of Congress or subcommittee of any such joint committee;

(10) to the Comptroller General or any of his authorized representatives, in the course of the performance of the duties of the General Accounting Office; or

(11) pursuant to the order of a court of competent jurisdiction.

(c) Accounting of Certain Disclosures. Each agency, with respect to each system of records under its control, shall—

(1) except for disclosures made under subsections (b)(1) or (b)(2) of this section, keep an accurate accounting of—

(A) the date, nature, and purpose of each disclosure of a record to any person or to another agency made under subsection (b) of this section; and

(B) the name and address of the person or agency to whom the disclosure is made;

(2) retain the accounting made under paragraph (1) of this subsection for at least five years or the life of the record, whichever is longer, after the disclosure for which the accounting is made;

(3) except for disclosures made under subsection (b)(7) of this section, make the accounting made under paragraph (1) of this subsection available to the individual named in the record at his request; and

(4) inform any person or other agency about any correction or notation of dispute

made by the agency in accordance with subsection (d) of this section of any record that has been disclosed to the person or agency if an accounting of the disclosure was made.

(d) Access to Records. Each agency that maintains a system of records shall—

(1) upon request by an individual to gain access to his record or to any information pertaining to him which is contained in the system, permit him and upon his request, a person of his own choosing to accompany him, to review the record and have a copy made of all or any portion thereof in a form comprehensible to him, except that the agency may require the individual to furnish a written statement authorizing discussion of that individual's record in the accompanying person's presence;

(2) permit the individual to request amendment of a record pertaining to him and—

(A) not later than 10 days (excluding Saturdays, Sundays, and legal public holidays) after the date of receipt of such request, acknowledge in writing such receipt; and

(B) promptly, either—

(i) make any correction of any portion thereof which the individual believes is not accurate, relevant, timely, or complete; or

(ii) inform the individual of its refusal to amend the record in accordance with his request, the reason for the refusal, the procedures established by the agency for the individual to request a review of that refusal by the head of the agency or an officer designated by the head of the agency, and the name and business address of that official;

(3) permit the individual who disagrees with the refusal of the agency to amend his record to request a review of such refusal, and not later than 30 days (excluding Saturdays, Sundays, and legal public holidays) from the date on which the individual requests such review, complete such review and make a final determination unless, for good cause shown, the head of the agency extends such 30-day period; and if, after his review, the reviewing official also refuses to amend the record in accordance with the request, permit the individual to file with the agency a concise statement setting forth the reasons for his disagreement with the refusal of the agency, and notify the individual of the provisions for judicial review of the reviewing official's determination under subsection (g)(1)(A) of this section;

(4) in any disclosure, containing information about which the individual has filed a statement of disagreement, occurring after the filing of the statement under paragraph (3) of this subsection, clearly note any portion of the record which is disputed and provide copies of the statement and, if the agency deems it appropriate, copies of a concise statement of the reasons of the agency for not making the amendments requested, to persons or other agencies to whom the disputed record has been disclosed; and

(5) nothing in this section shall allow an individual access to any information compiled in reasonable anticipation of a civil action or proceeding.

(e) Agency Requirements. Each agency that maintains a system of records shall—

(1) maintain in its records only such information about an individual as is relevant and necessary to accomplish a purpose of the agency required to be accomplished by statute or by executive order of the President;

(2) collect information to the greatest extent practicable directly from the subject individual when the information may result in adverse determinations about an individual's rights, benefits, and privileges under Federal programs;

(3) inform each individual whom it asks to supply information, on the form which it uses to collect the information or on a separate form that can be retained by the individual—

(A) the authority (whether granted by statute, or by executive order of the President) which authorizes the solicitation of the information and whether disclosure of such information is mandatory or voluntary;

(B) the principal purpose or purposes for which the information is intended to be used;

(C) the routine uses which may be made of the information, as published pursuant to paragraph (4)(D) of this subsection; and

(D) the effects on him, if any, of not providing all or any part of the requested information;

(4) subject to the provisions of paragraph (11) of this subsection, publish in the Federal Register at least annually a notice of the existence and character of the system of records, which notice shall include—

(A) the name and location of the system;

(B) the categories of individuals on whom records are maintained in the system;

(C) the categories of records maintained in the system;

(D) each routine use of the records contained in the system, including the categories of users and the purpose of such use;

(E) the policies and practices of the agency regarding storage, retrievability, access controls, retention, and disposal of the records;

(F) the title and business address of the agency official who is responsible for the system of records;

(G) the agency procedures whereby an individual can be notified at his request if the system of records contains a record pertaining to him;

(H) the agency procedures whereby an individual can be notified at his request how he can gain access to any record pertaining to him contained in the system of records, and how he can contest its content; and

(I) the categories of sources of records in the system;

(5) maintain all records which are used by the agency in making any determination about any individual with such accuracy, relevance, timeliness, and completeness as is reasonably necessary to assure fairness to the individual in the determination;

(6) prior to disseminating any record about an individual to any person other than an agency, unless the dissemination is made pursuant to subsection (b)(2) of this section, make reasonable efforts to assure that such records are accurate, complete, timely, and relevant for agency purposes;

(7) maintain no record describing how any individual exercises rights guaranteed by the First Amendment unless expressly authorized by statute or by the individual about whom the record is maintained or unless pertinent to and within the scope of an authorized law enforcement activity;

(8) make reasonable efforts to serve notice on an individual when any record on such individual is made available to any person under compulsory legal process when such process becomes a matter of public record;

(9) establish rules of conduct for persons involved in the design, development, operation, or maintenance of any system of records, or in maintaining any record, and

instruct each such person with respect to such rules and the requirements of this section, including any other rules and procedures adopted pursuant to this section and the penalties for noncompliance;

(10) establish appropriate administrative, technical, and physical safeguards to insure the security and confidentiality of records and to protect against any anticipated threats or hazards to their security or integrity which could result in substantial harm, embarrassment, inconvenience, or unfairness to any individual on whom information is maintained; and

(11) at least 30 days prior to publication of information under paragraph (4)(D) of this subsection, publish in the Federal Register notice of any new use or intended use of the information in the system, and provide an opportunity for interested persons to submit written data, views, or arguments to the agency.

(f) Agency Rules. In order to carry out the provisions of this section, each agency that maintains a system of records shall promulgate rules, in accordance with the requirements (including general notice) of section 553 of this title, which shall—

(1) establish procedures whereby an individual can be notified in response to his request if any system of records named by the individual contains a record pertaining to him;

(2) define reasonable times, places, and requirements for identifying an individual who requests his record or information pertaining to him before the agency shall make the record or information available to the individual;

(3) establish procedures for the disclosure to an individual upon his request of his record or information pertaining to him, including special procedure, if deemed necessary, for the disclosure to an individual of medical records, including psychological records, pertaining to him;

(4) establish procedures for reviewing a request from an individual concerning the amendment of any record or information pertaining to the individual, for making a determination on the request, for an appeal within the agency of an initial adverse agency determination, and for whatever additional means may be necessary for each individual to be able to exercise fully his rights under this section; and

(5) establish fees to be charged, if any, to any individual for making copies of his record, excluding the cost of any search for and review of the record.

The Office of the Federal Register shall annually compile and publish the rules promulgated under this subsection and agency notices published under subsection (e)(4) of this section in a form available to the public at low cost.

(g) (1) Civil Remedies. Whenever any agency

(A) makes a determination under subsection (d)(3) of this section not to amend an individual's record in accordance with his request, or fails to make such review in conformity with that subsection;

(B) refuses to comply with an individual request under subsection (d)(1) of this section;

(C) fails to maintain any record concerning any individual with such accuracy, relevance, timeliness, and completeness as is necessary to assure fairness in any determination relating to the qualifications, character, rights, or opportunities of, or benefits to the individual that may be made on the basis of such record, and consequently a determination is made which is adverse to the individual; or

(D) fails to comply with any other provision of this section, or any rule promulgated thereunder, in such a way as to have an adverse effect on an individual,

the individual may bring a civil action against the agency, and the district courts of the United States shall have jursidiction in the matters under the provisions of this subsection.

(2) (A) In any suit brought under the provisions of subsection (g)(1)(A) of this section, the court may order the agency to amend the individual's record in accordance with his request or in such other way as the court may direct. In such a case the court shall determine the matter de novo.

(B) The court may assess against the United States reasonable attorney fees and other litigation costs reasonably incurred in any case under this paragraph in which the complainant has substantially prevailed.

(3) (A) In any suit brought under the provisions of subsection (g)(1)(B) of this section, the court may enjoin the agency from withholding the records improperly withheld from him. In such a case the court shall determine the matter de novo, and may examine the contents of any agency records in camera to determine whether the records or any portion thereof may be withheld under any of the exemptions set forth in subsection (k) of this section, and the burden is on the agency to sustain its action.

(B) The court may assess against the United States reasonable attorney fees and other litigation costs reasonably incurred in any case under this paragraph in which the complainant has substantially prevailed.

(4) In any suit brought under the provisions of subsection (g)(1)(C) or (D) of this section in which the court determines that the agency acted in a manner which was intentional or willful, the United States shall be liable to the individual in an amount equal to the sum of—

(A) actual damages sustained by the individual as a result of the refusal or failure, but in no case shall a person entitled to recovery receive less than the sum of $1,000; and

(B) the costs of the action together with reasonable attorney fees as determined by the court.

(5) An action to enforce any liability created under this section may be brought in the district court of the United States in the district in which the complainant resides, or has his principal place of business, or in which the agency records are situated, or in the District of Columbia, without regard to the amount in controversy, within two years from the date on which the cause of action arises, except that where an agency has materially and willfully misrepresented any information required under this section to be disclosed to an individual and the information so misrepresented is material to establishment of the liability of the agency to the individual under this section, the action may be brought at any time within two years after discovery by the individual of the misrepresentation. Nothing in this section shall be construed to authorize any civil action by reason of any injury sustained as the result of a disclosure of a record prior to the effective date of this section.

(h) Rights of Legal Guardians. For the purpose of this section, the parent of any minor, or the legal guardian of any individual who has been declared to be incompetent due to physical or mental incapacity or age by a court of competent jurisdiction, may act on behalf of the individual.

(i)(1) Criminal Penalties. Any officer or employee of an agency, who by virtue of his employment or official position, has possession of, or access to, agency records which contain individually identifiable information the disclosure of which is prohibited by this section or by rules or regulations established thereunder, and who knowing that disclosure of the specific material is so prohibited, willfully discloses the materials in any manner to any person or agency not entitled to receive it, shall be guilty of a misdemeanor and fined not more than $5,000.

(2) Any officer or employee of any agency who willfully maintains a system of records without meeting the notice requirements of subsection (e)(4) of this section shall be guilty of a misdemeanor and fined no more than $5,000.

(3) Any person who knowingly and willfully requests or obtains any record concerning an individual from an agency under false pretenses shall be guilty of a misdemeanor and fined not more than $5,000.

(j) General Exemptions. The head of any agency may promulgate rules, in accordance with the requirements (including general notice) of sections 553(b)(1), (2), and (3), (c), and (e) of this title, to exempt any system of records within the agency from any part of this section except subsections (b), (c)(1) and (2), (e)(4)(A) through (F), (e)(6), (7), (9), (10), and (11), and (i) of the system of records is—

(1) maintained by the Central Intelligence Agency; or

(2) maintained by an agency or component thereof which performs as its principal function any activity pertaining to the enforcement of criminal laws, including police efforts to prevent, control, or reduce crime or to apprehend criminals, and the activities of prosecutors, courts, correctional, probation, pardon, or parole authorities, and which consists of (A) information compiled for the purpose of identifying individual criminal offenders and alleged offenders and consisting only of identifying data and notations of arrests, the nature and disposition of criminal charges, sentencing, confinement, release, and parole and probation status; (B) information compiled for the purpose of a criminal investigation, including reports of informants and investigators, and associated with an identifiable individual; or (C) reports identifiable to an individual compiled at any stage of the process of enforcement of the criminal laws from arrest or indictment through release from supervision.

At the time rules are adopted under this subsection, the agency shall include in the statement required under section 553(c) of this title, the reasons why the system of records is to be exempted from a provision of this section.

(k) Specific Exemptions. The head of any agency may promulgate rules in accordance with the requirements (including general notice) of sections 553(b)(1), (2), and (3), (c), and (e) of this title, to exempt any system of records within the agency from subsections (c)(3), (d), (e)(1), (e)(4)(G), (H), and (I) and (f) of this section if the system of records is—

(1) subject to the provisions of section 552(b)(1) of this title;

(2) investigatory material compiled for law enforcement purposes, other than material within the scope of subsection (j)(2) of this section: Provided, however, That if any individual is denied any right, privilege, or benefit that he would otherwise be entitled to by Federal law, or for which he would otherwise be eligible, as a result of the maintenance of such material, such material shall be provided to such individual, except to the extent that the disclosure of such material would reveal the identity of a source who

furnished information to the Government under an express promise that the identity of the source would be held in confidence, or, prior to the effective date of this section, under an implied promise that the identity of the source would be held in confidence;

(3) maintain in connection with providing protective services to the President of the United States or other individuals pursuant to section 3056 of title 18;

(4) required by statute to be maintained aand used solely as statistical records;

(5) investigatory material compiled solely for the purpose of determining suitability, eligibility, or qualifications for Federal civilian employment, military service, Federal contracts, or access to classified information, but only to the extent that the disclosure of such material would reveal the identity of a source who furnished information to the Government under an express promise that the identity of the source would be held in confidence, or, prior to the effective date of this section, under an implied promise that the identify of the source would be held in confidence;

(6) testing or examination material used solely to determine individual qualifications for appointment or promotion in the Federal service the disclosure of which would compromise the objectivity or fairness of the testing or examination process; or

(7) evaluation materials used to determine potential for promotion in the armed services, but only to the extent that the disclosure of such material would reveal the identity of a source who furnished information to the Government under an express promise that the identity of the source would be held in confidence, or, prior to the effective date of this section, under an implied promise that the identity of the source would be held in confidence.

At the time rules are adopted under this subsection, the agency shall include in the statement required under section 553(c) of this title, the reasons why the system of records is to be exempted from a provision of this section.

(l)(1) Archival Records. Each agency record which is accepted by the Administrator of General Services for storage, processing, and servicing in accordance with section 3103 of title 44 shall, for the purposes of this section, be considered to be maintained by the agency which deposited the record and shall be subject to the provisions of this section. The Administrator of General Services shall not disclose the record except to the agency which maintains the record, or under rules established by that agency which are not inconsistent with the provisions of this section.

(2) Each agency record pertaining to an identifiable individual which was transferred to the National Archives of the United States as a record which has sufficient historical or other value to warrant its continued preservation by the United States Government, prior to the effective date of this section, shall, for the purposes of this section, be considered to be maintained by the National Archives and shall not be subject to the provisions of this section, except that a statement generally describing such records (modeled after the requirements relating to records subject to subsections (e)(4)(A) through (G) of this section) shall be published in the Federal Register.

(3) Each agency record pertaining to an identifiable individual which is transferred to the National Archives of the United States as a record which has sufficient historical or other value to warrant its continued preservation by the United States Government, on or after the effective date of this section, shall, for the purposes of this section, be considered to be maintained by the National Archives and shall be exempt from the requirements of this section except subsections (e)(4)(A) through (G) and (e)(9) of this section.

(m) Government Contractors. When an agency provides by a contract for the operation by or on behalf of the agency of a system of records to accomplish an agency function, the agency shall, consistent with its authority, cause the requirements of this section to be applied to such system. For purposes of subsection (i) of this section any such contractor and any employee of such contractor, if such contract is agreed to on or after the effective date of this section, shall be considered to be an employee of an agency.

(n) Mailing Lists. An individual's name and address may not be sold or rented by an agency unless such action is specifically authorized by law. This provision shall not be construed to require the withholding of names and addresses otherwise permitted to be made public.

(o) Report on New Systems. Each agency shall provide adequate advance notice to Congress and the Office of Management and Budget of any proposal to establish or alter any system of records in order to permit an evaluation of the probable or potential effect of such proposal on the privacy and other personal or property rights of individuals or the disclosure of information relating to such individuals, and its effect on the preservation of the constitutional principles of federalism and separation of powers.

(p) Annual Report. The President shall submit to the Speaker of the House and the President of the Senate, by June 30 of each calendar year, a consolidated report, separately listing for each Federal agency the number of records contained in any system of records which were exempted from the application of this section under the provisions of subsections (j) and (k) of this section during the preceding calendar year, and the reasons for the exemptions, and such other information as indicates efforts to administer fully this section.

Sec. 4 The chapter analysis of chapter 5 of title 5, United States Code, is amended by inserting:

552a. Records about individuals. *Immediately below*:

552. Public information; agency rules, opinions, orders, and proceedings.

Sec. 5 (a)(1) There is established a Privacy Protection Study Commission (hereafter referred to as the "Commission") which shall be composed of seven members as follows:

(A) three appointed by the President of the United States.

(B) two appointed by the President of the Senate, and

(C) two appointed by the Speaker of the House of Representatives.

Members of the Commission shall be chosen from among persons who, by reason of their knowledge and expertise in any of the following areas—civil rights and liberties, law, social sciences, computer technology, business, records, management, and state and local government—are well qualified for service on the Commission.

(2) The members of the Commission shall elect a Chairman from among themselves.

(3) Any vacancy in the membership of the Commission, as long as there are four members in office, shall not impair the power of the Commission but shall be filled in the same manner in which the original appointment was made.

(4) A quorum of the Commission shall consist of a majority of the members, except that the Commission may establish a lower number as a quorum for the purpose of taking testimony. The Commission is authorized to establish such committees and delegate such

authority to them as may be necessary to carry out its functions. Each member of the Commission, including the Chairman, shall have equal responsibility and authority in all decisions and actions of the Commission, shall have full access to all information necessary to the performance of their functions, and shall have one vote. Action of the Commission shall be determined by a majority vote of the members present. The Chairman (or a member designated by the Chairman to be acting Chairman) shall be the official spokesman of the Commission in its relations with the Congress, Government agencies, other persons, and the public, and, on behalf of the Commission, shall see to the faithful execution of the administrative policies and decisions of the Commission, and shall report thereon to the Commission from time to time or as the Commission may direct.

(5) (A) Whenever the Commission submits any budget estimate or request to the President or the Office of Management and Budget, it shall concurrently transmit a copy of that request to Congress.

(B) Whenever the Commission submits any legislative recommendations, or testimony, or comments on legislation to the President or Office of Management and Budget, it shall concurrently transmit a copy thereof to the Congress. No officer or agency of the United States shall have any authority to require the Commission to submit its legislative recommendations, or testimony, or comments on legislation, to any officer or agency of the United States for approval, comments, or review, prior to the submission of such recommendations, testimony, or comments to the Congress.

(b) The Commission shall—

(1) make a study of the data banks, automated data processing programs, and information systems of governmental, regional, and private organizations, in order to determine the standards and procedures in force for the protection of personal information; and

(2) recommend to the President and the Congress the extent, if any, to which the requirements and principles of section 552a of title 5, United States Code, should be applied to the information practices of those organizations by legislation, administrative action, or voluntary adoption of such requirements and principles, and report on such other legislative recommendations as it may determine to be necessary to protect the privacy of individuals while meeting the legitimate needs of government and society for information.

(c) (1) In the course of conducting the study required under subsection (b) (1) of this section, and in its reports thereon, the Commission may research, examine, and analyze—

(A) interstate transfer of information about individuals that is undertaken through manual files or by computer or other electronic or telecommunications means;

(B) data banks and information programs and systems the operations of which significantly or substantially affect the enjoyment of the privacy and other personal and property rights of individuals;

(C) the use of social security numbers, license plate numbers, universal identifiers, and other symbols to identify individuals in data banks and to gain access to, integrate, or centralize information systems and files; and

(D) the matching and analysis of statistical data, such as Federal census data, with other sources of personal data, such as automobile registries and telephone directories, in order to reconstruct individual responses to statistical questionnaires for commercial or other purposes, in a way which results in a violation of the implied or explicitly recognized confidentiality of such information.

(2) (A) The Commission may include in its examination personal information activities in the following areas: medical; insurance; education; employment and personnel; credit, banking and financial institutions; credit bureaus; the commercial reporting industry; cable television and other telecommunications media; travel, hotel and entertainment reservations; and electronic check processing.

(B) The Commission shall include in its examinations a study of—

(i) whether a person engaged in interstate commerce who maintains a mailing list should be required to remove an individual's name and address from such list upon request of that individual;

(ii) whether the Internal Revenue Service should be prohibited from transferring individually identifiable data to other agencies and to agencies of State governments;

(iii) whether the Federal Government should be liable for general damages incurred by an individual as the result of a willful or intentional violation of the provisions of sections 552a (g) (1) (C) or (D) of title 5, United States Code; and

(iv) whether and how the standards for security and confidentiality of records required under section 552a (e) (10) of such title should be applied when a record is disclosed to a person other than an agency.

(C) The Commission may study such other personal information activities necessary to carry out the congressional policy embodied in this Act, except that the Commission shall not investigate information systems maintained by religious organizations.

(3) In conducting such study, the Commission shall—

(A) determine what laws, Executive orders, regulations, directives, and judicial decisions govern the activities under study and the extent to which they are consistent with the rights of privacy, due process of law, and other guarantees in the Constitution;

(B) determine to what extent governmental and private information systems affect Federal-State relations or the principle of separation of powers;

(C) examine the standards and criteria governing programs, policies, and practices relating to the collection, soliciting, processing, use, access, integration, dissemination, and transmission of personal information; and

(D) to the maximum extent practicable, collect and utilize findings, reports, studies, hearing transcripts, and recommendations of governmental, legislative and private bodies, institutions, organizations, and individuals which pertain to the problems under study by the Commission.

(d) In addition to its other functions the Commission may—

(1) request assistance of the heads of appropriate departments, agencies, and instrumentalities of the Federal Government, of State and local governments, and other persons in carrying out its functions under this Act;

(2) upon request, assist Federal agencies in complying with the requirements of section 552a of title 5, United States Code;

(3) determine what specific categories of information, the collection of which would violate an individual's right of privacy, should be prohibited by statute from collection by Federal agencies; and

(4) upon request, prepare model legislation for use by State and local governments in establishing procedures for handling, maintaining, and disseminating personal information at the State and local level and provide such technical assistance to State and local governments as they may require in the preparation and implementation of such legislation.

(e) (1) The Commission may, in carrying out its functions under this section, conduct such inspections, sit and act at such times and places, hold such hearings, take such testimony, require by subpoena the attendance of such witnesses and the production of such books, records, papers, correspondence, and documents, administer such oaths, have such printing and binding done, and make such expenditures as the Commission deems advisable. A subpoena shall be issued only upon an affirmative vote of a majority of all members of the Commission. Subpoenas shall be issued under the signature of the Chairman or any member of the Commission designated by the Chairman and shall be served by any person designated by the Chairman or any such member. Any member of the Commission may administer oaths or affirmations to witnesses appearing before the Commission.

(2) (A) Each department, agency, and instrumentality of the executive branch of the Government is authorized to furnish to the Commission, upon request made by the Chairman, such information, data, reports and such other assistance as the Commission deems necessary to carry out its functions under this section. Whenever the head of any such department, agency, or instrumentality submits a report pursuant to section 552a (o) of title 5, United States Code, a copy of such report shall be transmitted to the Commission.

(B) In carrying out its functions and exercising its powers under this section, the Commission may accept from any such department, agency, independent instrumentality, or other person any individually identifiable data if such data is necessary to carry out such powers and functions. In any case in which the Commission accepts any such information, it shall assure that the information is used only for the purpose for which it is provided, and upon completion of that purpose such information shall be destroyed or returned to such department, agency, independent instrumentality, or person from which it is obtained, as appropriate.

(3) The Commission shall have the power to—

(A) appoint and fix the compensation of an executive director, and such additional staff personnel as may be necessary, without regard to the provisions of title 5, United States Code, governing appointments in the competitive service, and without regard to chapter 51 and subchapter III of chapter 53 of such title relating to classification and General Schedule pay rates, but at rates not in excess of the maximum rate for GS-18 of the General Schedule under section 5332 of such title; and

(B) procure temporary and intermittent services to the same extent as is authorized by section 3109 of title 5, United States Code.

The Commission may delegate any of its functions to such personnel of the Commission as the Commission may designate and may authorize such successive redelegations of such functions as it may deem desirable.

(4) The Commission is authorized—

(A) to adopt, amend, and repeal rules and regulations governing the manner of its operations, organization, and personnel;

(B) to enter into contracts or other arrangements or modifications thereof, with any government, any department, agency, or independent instrumentality of the United States, or with any person, firm, association, or corporation, and such contracts or other arrangements, or modifications thereof may be entered into without legal consideration, without performance or other bonds, and without regard to section 3709 of the Revised Statutes, as amended (41 U.S.C. 5);

(C) to make advance, progress, and other payments which the Commission deems necessary under this Act without regard to the provisions of section 3648 of the Revised Statutes, as amended (31 U.S.C. 529); and

(D) to take such other action as may be necessary to carry out its functions under this section.

(f) (1) Each (the) member of the Commission who is an officer or employee of the United States shall serve without additional compensation, but shall continue to receive the salary of his regular position when engaged in the performance of the duties vested in the Commission.

(2) A member of the Commission other than one to whom paragraph (1) applies shall receive per diem at the maximum daily rate for GS–18 of the General Schedule when engaged in the actual performance of the duties vested in the Commission.

(3) All members of the Commission shall be reimbursed for travel, subsistence, and other necessary expenses incurred by them in the performance of the duties vested in the Commission.

(g) The Commission shall, from time to time, and in an annual report, report to the President and the Congress on its activities in carrying out the provisions of this section. The Commission shall make a final report to the President and to the Congress on its findings pursuant to the study required to be made under subsection (b) (1) of this section not later than two years from the date on which all of the members of the Commission are appointed. The Commission shall cease to exist thirty days after the date on which its final report is submitted to the President and the Congress.

(h) (1) Any member, officer, or employee of the Commission, who by virtue of his employment or official position, has possession of, or access to, agency records which contain individually identifiable information the disclosure of which is prohibited by this section, and who knowing that disclosure of the specific material is so prohibited, willfully discloses the material in any manner to any person or agency not entitled to receive it, shall be guilty of a misdemeanor and fined not more than $5,000.

(2) Any person who knowingly and willfully requests or obtains any record concerning an individual from the Commission under false pretenses shall be guilty of a misdemeanor and fined not more than $5,000.

Sec. 6 The Office of Management and Budget shall—

(1) develop guidelines and regulations for the use of agencies in implementing the provisions of section 552a of title 5, United States Code, as added by section 3 of this Act; and

(2) provide continuing assistance to and oversight of the implementation of the provisions of such section by agencies.

Sec. 7 (a) (1) It shall be unlawful for any Federal, State or local government agency to deny to any individual any right, benefit, or privilege provided by law because of such individual's refusal to disclose his social security account number.

(2) the provisions of paragraph (1) of this subsection shall not apply with respect to—

(A) any disclosure which is required by Federal statute, or

(B) the disclosure of a social security number to any Federal, State, or local agency maintaining a system of records in existence and operating before January 1, 1975, if such disclosure was required under statute or regulation adopted prior to such date to verify the identity of an individual.

(b) Any Federal, State, or local government agency which requests an individual to disclose his social secrity account number shall inform that individual whether that disclosure is mandatory or voluntary, by what statutory or other authority such number is solicited, and what uses will be made of it.

Sec. 8 The provisions of this Act shall be effective on and after the date of enactment, except that the amendments made by sections 3 and 4 shall become effective 270 days following the day on which this Act is enacted.

Sec. 9 There is authorized to be appropriated to carry out the provisions of section 5 of this Act for fiscal years 1975, 1976, and 1977 the sum of $1,500,000, except that not more than $750,000 may be expended during any such fiscal year.

Approved December 31, 1974.

APPENDIX 5.3

Government in the Sunshine Act

[Note: Popularly known as the "Sunshine Act," this was passed as Public Law 94–409 on September 13, 1976.]

AN ACT

To provide that meetings of Government agencies shall be open to the public, and for other purpose.

Be it enacted by the Senate and House of Representatives of the United States of America in Congress assembled, That this Act may be cited as the "Government in the Sunshine Act."

DECLARATION OF POLICY

Sec. 2 It is hereby declared to be the policy of the United States that the public is entitled to the fullest practicable information regarding the decisionmaking processes of the Federal Government. It is the purpose of this Act to provide the public with such information while protecting the rights of individuals and the ability of the Government to carry out its responsibilities.

OPEN MEETINGS

Sec. 3. (a) Title 5, United States Code, is amended by adding after section 552a the following new section: Section 552b. Open meetings

(a) For purposes of this section—

(1) the term "agency" means any agency, as defined in section 552(e) of this title, headed by a collegial body composed of two or more individual members, a majority of whom are appointed to such position by the President with the advice and consent of the Senate, and any subdivision thereof authorized to act on behalf of the agency;

(2) the term "meeting" means the deliberations of at least the number of individual

agency members required to take action on behalf of the agency where such deliberations determine or result in the joint conduct or disposition of official agency business, but does not include deliberations required or permitted by subsection (d) or (e); and

(3) the term "member" means an individual who belongs to a collegial body heading an agency.

(b) Members shall not jointly conduct or dispose of agency business other than in accordance with this section. Except as provided in subsection (c), every portion of every meeting of an agency shall be open to public observation.

(c) Except in a case where the agency finds that the public interest requires otherwise, the second sentence of subsection (b) shall not apply to any portion of an agency meeting, and the requirements of subsections (d) and (e) shall not apply to any information pertaining to such meeting otherwise required by this section to be disclosed to the public, where the agency properly determines that such portion or portions of its meeting or the disclosure of such information is likely to—

(1) disclose matters that are (A) specifically authorized under criteria established by an Executive order to be kept secret in the interests of national defense or foreign policy and (B) in fact properly classified pursuant to such Executive order;

(2) relate solely to the internal personnel rules and practices of an agency;

(3) disclose matters specifically exempted from disclosure by statute (other than section 552 of this title), provided that such statute (A) requires that the matters be withheld from the public in such a manner as to leave no discretion on the issue, or (B) establishes particular criteria for withholding or refers to particular types of matters to be withheld;

(4) disclose trade secrets and commercial or financial information obtained from a person and privileged or confidential;

(5) involve accusing any person of a crime, or formally censuring any person;

(6) disclose information of a personal nature where disclosure would constitute a clearly unwarranted invasion of personal privacy;

(7) disclose investigatory records compiled for law enforcement purposes, or information which if written would be contained in such records, but only to the extent that the production of such records or information would (A) interfere with enforcement proceedings, (B) deprive a person of a right to a fair trial or an impartial adjudication, (c) constitute an unwarranted invasion of personal privacy, (D) disclose the identity of a confidential source and, in the case of a record compiled by a criminal law enforcement authority in the course of a criminal investigation, or by an agency conducting a lawful national security intelligence investigation, confidential information furnished only by the confidential source, (E) disclose investigative techniques and procedures, or (F) endanger the life or physical safety of law enforcement personnel;

(8) disclose information contained in or related to examination, operating, or condition reports prepared by, on behalf of, or for the use of an agency responsible for the regulation or supervision of financial institutions;

(9) disclose information the premature disclosure of which would—

(A) in the case of an agency which regulates currencies, security, commodities, or financial institutions, be likely to (i) lead to significant financial speculation in currencies, securities, or commodities, or (ii) significantly endanger the stability of any financial institution; or

(B) in the case of any agency, be likely to significantly frustrate implementation of a proposed agency action, except that subparagraph (B) shall not apply in any instance where the agency has already disclosed to the public the content or nature of its proposed action, or where the agency is required by law to make such disclosure on its own initiative prior to taking final agency action on such proposal; or

(10) specifically concern the agency's issuance of a subpoena, or the agency's participation in a civil action or proceeding, an action in a foreign court or international tribunal, or an arbitration, or the initiation, conduct, or disposition by the agency of a particular case of formal agency adjudication pursuant to the procedures in section 554 of this title or otherwise involving a determination on the record after opportunity for a hearing.

(d) (1) Action under subsection (c) shall be taken only when a majority of the entire membership of the agency (as defined in subsection (a) (1)) votes to take such action. A separate vote of the agency members shall be taken with respect to each agency meeting a portion or portions of which are proposed to be closed to the public pursuant to subsection (c), or with respect to any information which is proposed to be withheld under subsection (c). A single vote may be taken with respect to a series of meetings, a portion or portions of which are proposed to be closed to the public, or with respect to any information concerning such series of meetings, so long as each meeting in such series involves the same particular matters and is scheduled to be held no more than thirty days after the initial meeting in such series. The vote of each agency member participating in such vote shall be recorded and no proxies shall be allowed.

(2) Whenever any person whose interests may be directly affected by a portion of a meeting requests that the agency close such portion to the public for any of the reasons referred to in paragraph (5), (6), or (7) of subsection (c), the agency, upon request of any one of its members shall vote by recorded vote whether to close such meeting.

(3) Within one day of any vote taken pursuant to paragraph (1) or (2), the agency shall make publicly available a written copy of such vote reflecting the vote of each member on the question. If a portion of a meeting is to be closed to the public, the agency shall, within one day of the vote taken pursuant to paragraph (1) or (2) of this subsection, make publicly available a full written explanation of its action closing the portion together with a list of all persons expected to attend the meeting and their affiliation.

(4) Any agency, a majority of whose meetings may properly be closed to the public pursuant to paragraph (4), (8), (9) (A), or (10) of subsection (c), or any combination thereof, may provide by regulation for the closing of such meetings or portions thereof in the event that a majority of the members of the agency votes by recorded vote at the beginning of such meeting, or portion thereof, to close the exempt portion or portions of the meeting, and a copy of such vote, reflecting the vote of each member on the question, is made available to the public. The provisions of paragraphs (1), (2), and (3) of this subsection and subsection (e) shall not apply to any portion of a meeting to which such regulations apply: Provided, That the agency shall, except to the extent that such information is exempt from disclosure under the provisions of subsection (c), provide the public with public announcement of the time, place, and subject matter of the meeting and of each portion thereof at the earliest practicable time.

(e) (1) In the case of each meeting, the agency shall make public announcement, at least one week before the meeting, of the time, place, and subject matter of the meeting, whether it is to be open or closed to the public, and the name and phone number of the official designated by the agency to respond to requests for information about the meeting. Such announcement shall be made unless a majority of the members of the agency determines by a recorded vote that agency business requires that such meeting be called at an earlier date, in which case the agency shall make public announcement of the time, place, and subject matter of such meeting, and whether open or closed to the public, at the earliest practicable time.

(2) The time or place of a meeting may be changed following the public announcement required by paragraph (1) only if the agency publicly announces such change at the earliest practicable time. The subject matter of a meeting, or the determination of the agency to open or close a meeting, or portion of a meeting, to the public, may be changed following the public announcement required by this subsection only if (A) a majority of the entire membership of the agency determines by a recorded vote that agency business so requires and that no earlier announcement of the change was possible, and (B) the agency publicly announces such change and the vote of each member upon such change at the earliest practicable time.

(3) Immediately following such public announcement required by this subsection, notice of the time, place, and subject matter of a meeting, whether the meeting is open or closed, any change in one of the preceding, and the name and phone number of the official designated by the agency to respond to requests for information about the meeting, shall also be submitted for publication in the Federal Register.

(f) (1) For every meeting closed pursuant to paragraphs (1) through (10) of subsection (c), the General Counsel or chief legal officer of the agency shall publicly certify that, in his or her opinion, the meeting may be closed to the public and shall state each relevant exemptive provision. A copy of such certification, together with a statement from the presiding officer of the meeting setting forth the time and place of the meeting, and the persons present, shall be retained by the agency. The agency shall maintain a complete transcript or electronic recording adequate to record fully the proceedings of each meeting, or portion of a meeting, closed to the public, except that in the case of a meeting, or portion of a meeting, closed to the public pursuant to paragraph (8), (9) (A), or (10) of subsection (c), the agency shall maintain either such a transcript or recording, or a set of minutes. Such minutes shall fully and clearly describe all matters discussed and shall provide a full and accurate summary of any actions taken, and the reasons therefore, including a description of each of the views expressed on any item and the record of any roll call vote (reflecting the vote of each member on the question). All documents considered in connection with any action shall be identified in such minutes.

(2) The agency shall make promptly available to the public, in a place easily accessible to the public, the transcript, electronic recording, or minutes (as required by paragraph (1)) of the discussion of any item on the agenda, or of any item of the testimony of any witness received at the meeting, except for such item or items of such discussion or testimony as the agency determines to contain information which may be withheld under subsection (c). Copies of such transcript, or minutes, or a transcription of such recording disclosing the identity of each speaker, shall be furnished to any person at the actual cost of duplication or transcription. The agency shall maintain a complete verbatim copy of the transcript, a complete copy of the minutes, or a complete electronic recording of each

meeting, or portion of a meeting, closed to the public, for a period of at least two years after such meeting, or until one year after the conclusion of any agency proceeding with respect to which the meeting or portion was held, whichever occurs later.

(g) Each agency subject to the requirements of this section shall, within 180 days after the date of enactment of this section, following consultation with the Office of the Chairman of the Administrative Conference of the United States and published notice in the Federal Register of at least thirty days and opportunity for written comment by any person, promulgate regulations to implement the requirements of subsections (b) through (f) of this section. Any person may bring a proceeding in the United States District Court for the District of Columbia to require an agency to promulgate such regulations if such agency has not promulgated such regulations within the time period specified herein. Subject to any limitations of time provided by law, any person may bring a proceeding in the United States Court of Appeals for the District of Columbia to set aside agency regulations issued pursuant to this subsection that are not in accord with the requirements of subsections (b) through (f) of this section and to require the promulgation of regulations that are in accord with such subsections.

(h) (1) The district courts of the United States shall have jurisdiction to enforce the requirements of subsections (b) through (f) of this section by declaratory judgment, injunctive relief, or other relief as may be appropriate. Such actions may be brought by any person against an agency prior to, or within sixty days after, the meeting out of which the violation of this section arises, except that if public announcement of such meeting is not initially provided by the agency in accordance with the requirements of this section, such action may be instituted pursuant to this section at any time prior to sixty days after any public announcement of such meeting. Such actions may be brought in the district court of the United States for the district in which the agency meeting is held or in which the agency in question has its headquarters, or in the District Court for the District of Columbia. In such actions a defendant shall serve his answer within thirty days after the service of the complaint. The burden is on the defendant to sustain his action. In deciding such cases the court may examine in camera any portion of the transcript, electronic recording, or minutes of a meeting closed to the public, and may take such additional evidence as it deems necessary. The court, having due regard for orderly administration and the public interest, as well as the interests of the parties, may grant such equitable relief as it deems appropriate, including granting an injunction against future violations of this section or ordering the agency to make available to the public such portion of the transcript, recording, or minutes of a meeting as is not authorized to be withheld under subsection (c) of this section.

(2) Any Federal court otherwise authorized by law to review agency action may, at the application of any person properly participating in the proceeding pursuant to other applicable law, inquire into violations by the agency of the requirements of this section and afford such relief as it deems appropriate. Nothing in this section authorizes any Federal court having jurisdiction solely on the basis of paragraph (1) to set aside, enjoin, or invalidate any agency action (other than an action to close a meeting or to withhold information under this section) taken or discussed at any agency meeting out of which the violation of this section arose.

(i) The court may assess against any party reasonable attorney fees and other litigation costs reasonably incurred by any other party who substantially prevails in any action brought in accordance with the provisions of subsection (g) or (h) of this section, except that costs may be assessed against the plaintiff only where the court finds that the suit was initiated by the plaintiff primarily for frivolous or dilatory purposes. In the case of assessment of costs against an agency, the costs may be assessed by the court against the United States.

(j) Each agency subject to the requirements of this section shall annually report to Congress regarding its compliance with such requirements, including a tabulation of the total number of agency meetings open to the public, the total number of meetings closed to the public, the reasons for closing such meetings, and a description of any litigation brought against the agency under this section, including any costs assessed against the agency in such litigation (whether or not paid by the agency).

(k) Nothing herein expands or limits the present rights of any person under section 552 of this title, except that the exemptions set forth in subsection (c) of this section shall govern in the case of any request made pursuant to section 552 to copy or inspect the transcripts, recordings, or minutes described in subsection (f) of this section. The requirements of chapter 33 of title 44, United States Code, shall not apply to the transcripts, recordings, and minutes described in subsection (f) of this section.

(1) This section does not constitute authority to withhold any information from Congress, and does not authorize the closing of any agency meeting or portion thereof required by any other provisions of law to be open.

(m) Nothing in this section authorizes any agency to withhold from any individual any record, including transcripts, recordings, or minutes required by this section, which is otherwise accessible to such individual under section 552a of this title.

(b) The chapter analysis of chapter 5 of title 5, United States Code, is amended by inserting:

552b. Open meetings. *Immediately below*:

552a. Records about individuals.

EX PARTE COMMUNICATIONS

Sec. 4. (a) Section 557 of title 5, United States Code, is amended by adding at the end thereof the following new subsection:

(d) (1) In any agency proceeding which is subject to subsection (a) of this section, except to the extent required for the disposition of ex parte matters as authorized by law—

(A) no interested person outside the agency shall make or knowingly cause to be made to any member of the body comprising the agency, administrative law judge, or other employee who is or may reasonably be expected to be involved in the decisional process of the proceeding, an ex parte communication relevant to the merits of the proceeding;

(B) no member of the body comprising the agency, administrative law judge, or other employee who is or may reasonably be expected to be involved in the decisional process of the proceeding, shall make or knowingly cause to be made to any interested person outside the agency an ex parte communication relevant to the merits of the proceeding;

(C) a member of the body comprising the agency, administrative law judge, or other employee who is or may reasonably be expected to be involved in the decisional process of such proceeding who receives, or who makes or knowingly causes to be made, a communication prohibited by this subsection shall place on the public record of the proceeding:

(i) all such written communications;
(ii) memoranda stating the substance of all such oral communications; and
(iii) all written responses, and memoranda stating the substance of all oral responses, to the materials described in clauses (i) and (ii) of this subparagraph;

(D) upon receipt of a communication knowingly made or knowingly caused to be made by a party in violation of this subsection, the agency, administrative law judge, or other employee presiding at the hearing may, to the extent consistent with the interests of justice and the policy of the underlying statutes, require the party to show cause why his claim or interest in the proceeding should not be dismissed, denied, disregarded, or otherwise adversely affected on account of such violation; and

(E) the prohibitions of this subsection shall apply beginning at such time as the agency may designate, but in no case shall they begin to apply later than the time at which a proceeding is noticed for hearing unless the person responsible for the communication has knowledge that it will be noticed, in which case the prohibitions shall apply beginning at the time of his acquisition of such knowledge.

(2) This subsection does not constitute authority to withhold information from Congress.

(b) Section 551 of title 5, United States Code, is amended—

(1) by striking out "and" at the end of paragraph (12);

(2) by striking out the "act," at the end of paragraph (13) and inserting in lieu thereof "act; and"; and

(3) by adding at the end thereof the following new paragraph: (14) "ex parte communication" means an oral or written communication not on the public record with respect to which reasonable prior notice to all parties is not given, but it shall not include requests for status reports on any matter or proceeding covered by this subchapter.

(c) Section 556(d) of title 5, United States Code, is amended by inserting between the third and fourth sentences thereof the following new sentence: The agency may, to the extent consistent with the interests of justice and the policy of the underlying statutes administered by the agency, consider a violation of section 557(d) of this title sufficient grounds for a decision adverse to a party who has knowingly committed such violation or knowingly caused such violation to occur.

CONFORMING AMENDMENTS

Sec. 5 (a) Section 410(b)(1) of title 39, United States Code, is amended by inserting after "Section 552 (public information)," the words "section 552a (records about individuals), section 552b (open meetings)."

(b) Section 552(b)(3) of title 5, United States Code, is amended to read as follows:

(3) specifically exempted from disclosure by statute (other than section 552b of this title), provided that such statute (A) requires that the matters be withheld from the public in such a manner as to leave no discretion on the issue, or (B) establishes particular criteria for withholding or refers to particular types of matters to be withheld;

(c) Subsection (d) of section 10 of the Federal Advisory Committee Act is amended by striking out the first sentence and inserting in lieu thereof the following: "Subsections (a)(1) and (a)(3) of this section shall not apply to any portion of an advisory committee meeting where the President, or the head of the agency to which the advisory committee reports, determines that such portion of such meeting may be closed to the public in accordance with subsection (c) of section 552b of title 5, United States Code."

EFFECTIVE DATE

Sec. 6 (a) Except as provided in subsection (b) of this section, the provisions of this Act shall take effect 180 days after the date of its enactment.

(b) Subsection (g) of section 552b of title 5, United States Code, as added by section 3(a) of this Act, shall take effect upon enactment.

Approved September 13, 1976.

APPENDIX 5.4

Protection of Identity of Research Subjects

[Note: The following text is taken from the *Federal Register*, Vol. 44, No. 66, Wednesday, April 4, 1979, pp. 20382–20387. Introductory material that appeared with the regulations have been deleted.]

DEPARTMENT OF HEALTH, EDUCATION, AND WELFARE

Public Health Service
42 CFR Part 2a
Protection of Identity—Research Subjects

AGENCY: Public Health Service, HEW
ACTION: Final regulations
SUMMARY: These final regulations set forth procedures under which persons engaged in research on mental health, including research on the use and effect of alcohol and other psychoactive drugs, may apply for an authorization under section 303(a) of the Public Health Service Act, as amended by Public Law 93–282 (42 U.S.C. 242a(a)). Such an authorization affords the person to whom it is given a privilege to protect the privacy of research subjects by withholding the names or other identifying characteristics of such research subjects from all persons not connected with the conduct of the research. Persons so authorized may not be compelled in any Federal, State, or local civil, criminal, administrative, legislative, or other proceedings to identify such individuals.

[Note: Introductory material deleted here.]

Subchapter A of Chapter I, Title 42 CFR, is amended by adding a new Part 2a to read as follows:

PART 2a—PROTECTION OF IDENTITY: RESEARCH SUBJECTS

Sec.
2a.1 Applicability.
2a.2 Definitions.

Authority: Sec. 3(a), Pub. L. 91–513 as amended by sec. 122(b), Pub. L. 93–282; 84 Stat. 1241 (42 U.S.C. 242a(a)), as amended by 88 Stat. 132.

Sec. 2a.1 Applicability.

(a) Section 303(a) of the Public Health Service Act (42 U.S.C. 242a(a)) provides that "(t)he Secretary (of Health, Education, and Welfare) may authorize persons engaged in research on mental health, including research on the use and effect of alcohol and other psychoactive drugs, to protect the privacy of individuals who are the subject of such research by withholding from all persons not connected with the conduct of such research the names or other identifying characteristics of such individuals. Persons so authorized to protect the privacy of such individuals may not be compelled in any Federal, State, or local civil, criminal, administrative, legislative, or other proceedings to identify such individuals." The regulations in this part establish procedures under which any person engaged in research on mental health including research on the use and effect of alcohol and other psychoactive drugs (whether or not the research is federally funded) may, subject to the exceptions set forth in paragraph (b) of this section, apply for such an authorization of confidentiality.

(b) These regulations do not apply to:

(1) Authorizations of confidentiality for research requiring an Investigational New Drug exemption under section 505(i) of the Federal Food, Drug, and Cosmetic Act (21 U.S.C. 355(i)) or to approved new drugs, such as methadone, requiring continuation of long-term studies, records, and reports. Attention is called to 21 CFR 291.505(g) relating to authorization of confidentiality for patient records maintained by methadone treatment programs.

(2) Authorizations of confidentiality for research which are related to law enforcement activities or otherwise within the purview of the Attorney General's authority to issue authorizations of confidentiality pursuant to section 502(c) of the Controlled Substances Act (21 U.S.C. 872(c)) and 21 CFR 1316.21.

(c) The Secretary's regulations on confidentiality of alcohol and drug abuse patient records (42 CFR Part 2) and the regulations of this part may, in some instances, concurrently cover the same transaction. As explained in 42 CFR 2.24 and 2.24–1, 42 CFR Part 2 restricts voluntary disclosures of information from applicable patient records while a Confidentiality Certificate issued pursuant to the regulations of this part protects a person engaged in applicable research from being compelled to disclose identifying characteristics of individuals who are the subject of such research.

Sec. 2a.2 Definitions.

(a) "Secretary" means the Secretary of Health, Education, and Welfare and any other officer or employee of the Department of Health, Education, and Welfare to whom the authority involved has been delegated.

(b) "Person" means any individual, corporation, government, or governmental subdivision or agency, business trust, partnership, association, or other legal entity.

(c) "Research" means systematic study directed toward new or fuller knowledge and understanding of the subject studied. The term includes, but is not limited to, behavioral science studies, surveys, evaluations, and clinical investigations.

(d) "Drug" has the meaning given that term by section 201(g)(1) of the Federal Food, Drug, and Cosmetic Act (21 U.S.C. 321(g)(1)).

(e) "Controlled drug" means a drug which is included in schedule I, II, III, IV, or V of Part B of the Controlled Substance Act (21 U.S.C. 811–812).

(f) "Administer" refers to the direct application of a drug to the body of a human research subject, whether such application be by injection, inhalation, ingestion, or any other means, by (1) a qualified person engaged in research (or, in his or her presence, by his or her authorized agent), or (2) a research subject in accordance with instructions of a qualified person engaged in research, whether or not in the presence of a qualified person engaged in research.

(g) "Identifying characteristics" refers to the name, address, any identifying number, fingerprints, voiceprints, photographs or any other item or combination of data about a research subject which could reasonably lead directly or indirectly by reference to other information to identification of that research subject.

(h) "Psychoactive drug" means, in addition to alcohol, any drug which has as its principal action an effect on thought, mood, or behavior.

Sec. 2a.3 Application; coordination.

(a) Any person engaged in (or who intends to engage in) the research to which this part applies, who desires authorization to withhold the names and other identifying characteristics of individuals who are the subject of such research from any person or authority not connected with the conduct of such research may apply to the Office of the Director, National Institute on Drug Abuse, National Institute of Mental Health, or the Office of the Director, National Institute on Alcohol Abuse and Alcoholism, 5600 Fishers Lane, Rockville, Maryland 20857 for an authorization of confidentiality.

(b) If there is uncertainty with regard to which Institute is appropriate or if the research project falls within the purview of more than one Institute, an application need be submitted only to one Institute. Persons who are uncertain with regard to the applicability of these regulations to a particular type of research may apply for an authorization of confidentiality under the regulations of this part to one of the Institutes. Requests which are within the scope of the authorities described in Sec. 2a.1(b) will be forwarded to the appropriate agency for consideration and the person will be advised accordingly.

(c) An application may accompany, precede, or follow the submission of a request for DHEW grant or contract assistance, though it is not necessary to request DHEW grant or contract assistance in order to apply for a Confidentiality Certificate. If a person has previously submitted any information required in this part in connection with a DHEW grant or contract, he or she may substitute a copy of information thus submitted, if the information is current and accurate. If a person requests a Confidentiality Certificate at the same time he or she submits an application for DHEW grant or contract assistance, the application for a Confidentiality Certificate may refer to the pertinent section(s) of the DHEW grant or contract application which provide(s) the information required to be submitted under this part. (See Sections 2a.4 and 2a.5.)

(d) A separate application is required for each research project for which an authorization of confidentiality is requested.

Sec. 2a.4 Contents of application; in general.

In addition to any other pertinent information which the Secretary may require, each application for an authorization of confidentiality for a research project shall contain:

(a) The name and address of the individual primarily responsible for the conduct of the research and the sponsor or institution with which he or she is affiliated, if any. Any application from a person affiliated with an institution will be considered only if it contains or is accompanied by documentation of institutional approval. This documentation may consist of a written statement signed by a responsible official of the institution or of a copy of or reference to a valid certification submitted in accordance with 45 CFR Part 46.

(b) The location of the research project and a description of the facilities available for conducting the research, including the name and address of any hospital, institution, or clinical laboratory facility to be utilized in connection with the research.

(c) The names, addresses, and summaries of the scientific or other appropriate training and experience of all personnel having major responsibilities in the research project and the training and experience requirements for major positions not yet filled.

(d) An outline of the research protocol for the project including a clear and concise statement of the purpose and rationale of the research project and the general research methods to be used.

(e) The date on which research will begin or has begun and the estimated date for completion of the project.

(f) A specific request, signed by the individual primarily responsible for the conduct of the research, for authority to withhold the names and other identifying characteristics of the research subjects and the reasons supporting such request.

(g) An assurance (1) From persons making application for a Confidentiality Certificate for a research project for which DHEW grant or contract support is received or sought that they will comply with all the requirements of 45 CFR Part 46, "Protection of Human Subjects," or (2) From all other persons making application that they will comply with the informed consent requirements of 45 CFR 46.103(c) and document legally effective informed consent in a manner consistent with the principles stated in 45 CFR 46.110, if it is determined by the Secretary, on the basis of information submitted by the person making application, that subjects will be placed at risk. If a modification of paragraphs (a) or (b) of 45 CFR 46.110 is to be used, as permitted under paragraph (c) of that section, the applicant will describe the proposed modification and submit it for approval by the Secretary.

(h) An assurance that if an authorization of confidentiality is given it will not be represented as an endorsement of the research project by the Secretary or used to coerce individuals to participate in the research project.

(i) An assurance that any person who is authorized by the Secretary to protect the privacy of research subjects will use that authority to refuse to disclose identifying characteristics of research subjects in any Federal, State, or local civil, criminal, administrative, legislative, or other proceedings to compel disclosure of the identifying characteristics of research subjects.

(j) An assurance that all research subjects who participate in the project during the period the Confidentiality Certificate is in effect will be informed that:

(1) A Confidentiality Certificate has been issued;

(2) The persons authorized by the Confidentiality Certificate to protect the identity of research subjects may not be compelled to identify research subjects in any civil, criminal, administrative, legislative, or other proceedings whether Federal, State, or local;

(3) If any of the following conditions exist the Confidentiality Certificate does not authorize any person to which it applies to refuse to reveal identifying information concerning research subjects: (i) The subject consents in writing to disclosure of identifying information, (ii) Release is required by the Federal Food, Drug, and Cosmetic Act (21 U.S.C. 301) or regulations promulgated thereunder (Title 21, Code of Federal Regulations), or (iii) Authorized personnel of DHEW request identifying information for audit or program evaluation of a research project funded by DHEW or for investigation of DHEW grantees or contractors and their employees or agents carrying out such a project. (See Sec. 2a.7(b));

(4) The Confidentiality Certificate does not govern the voluntary disclosure of identifying characteristics of research subjects;

(5) The Confidentiality Certificate does not represent an endorsement of the research project by the Secretary.

(k) An assurance that all research subjects who enter the project after the termination of the Confidentiality Certificate will be informed that the authorization of confidentiality has ended and that the persons authorized to protect the identity of research subjects by the Confidentiality Certificate may not rely on the Certificate to refuse to disclose identifying characteristics of research subjects who were not participants in the project during the period the Certificate was in effect. (See Sec. 2a.8(c)).

Sec. 2a.5 Contents of application; research projects in which drugs will be administered.

(a) In addition to the information required by Sec. 2a.4 and any other pertinent information which the Secretary may require, each application for an authorization of confidentiality for a research project which involves the administering of a drug shall contain:

(1) Identification of the drugs to be administered in the research project and a description of the methods for such administration, which shall include a statement of the dosages to be administered to the research subjects;

(2) Evidence that individuals who administer drugs are authorized to do so under applicable Federal and State law; and

(3) In the case of a controlled drug, a copy of the Drug Enforcement Administration Certificate of Registration (BND Form 223) under which the research project will be conducted.

(b) An application for an authorization of confidentiality with respect to a research project which involves the administering of a controlled drug may include a request for exemption of persons engaged in the research from State or Federal prosecution for possession, distribution, and dispensing of controlled drugs as authorized under section 502(d) of the Controlled Substances Act (21 U.S.C. 872(d)) and 21 CFR 1316.22. If the request is in such form, and is supported by such information, as required by 21 CFR

1316.22, the Secretary will forward it, together with his or her recommendation that such request be approved or disapproved, for the consideration of the Administrator of the Drug Enforcement Administration.

Sec. 2a.6 Issuance of Confidentiality Certificates; single project limitation.

(a) In reviewing the information provided in the application for a Confidentiality Certificate, the Secretary will take into account:

(1) The scientific or other appropriate training and experience of all personnel having major responsibilities in the research project;

(2) Whether the project constitutes bona fide "research" which is within the scope of the regulations of this part; and

(3) Such other factors as he or she may consider necessary and appropriate. All applications for Confidentiality Certificates shall be evaluated by the Secretary through such officers and employees of the Department and such experts or consultants engaged for this purpose as he or she determines to be appropriate.

(b) After consideration and evaluation of an application for an authorization of confidentiality, the Secretary will either issue a Confidentiality Certificate or a letter denying a Confidentiality Certificate, which will set forth the reasons for such denial, or will request additional information from the person making application. The Confidentiality Certificate will include:

(1) The name and address of the person making application;

(2) The name and address of the individual primarily responsible for conducting the research, if such individual is not the person making application;

(3) The location of the research project;

(4) A brief description of the research project;

(5) A statement that the Certificate does not represent an endorsement of the research project by the Secretary;

(6) The Drug Enforcement Administration registration number of the project, if any; and

(7) The date or event upon which the Confidentiality Certificate becomes effective, which shall not be before the later of either the commencement of the research project or the date of issuance of the Certificate, and the date or event upon which the Certificate will expire.

(c) A Confidentiality Certificate is not transferable and is effective only with respect to the names and other identifying characteristics of those individuals who are the subjects of the single research project specified in the Confidentiality Certificate. The recipient of a Confidentiality Certificate shall, within 15 days of any completion or discontinuance of the research project which occurs prior to the expiration date set forth in the Certificate, provide written notification to the Director of the Institute to which application was made. If the recipient determines that the research project will not be completed by the expiration date set forth in the Confidentiality Certificate he or she may submit a written request for an extension of the expiration date which shall include a justification for such extenstion and a revised estimate of the date for completion of the project. Upon approval of such a request, the Secretary will issue an amended Confidentiality Certificate.

(d) The protection afforded by a Confidentiality Certificate does not extend to significant changes in the research project as it is described in the application for such Certificate (e.g., changes in the personnel having major responsibilities in the research project, major changes in the scope or direction of the research protocol, or changes in the drugs to be administered and the persons who will administer them). The recipient of a Confidentiality Certificate shall notify the Director of the Institute to which application was made of any proposal for such a significant change by submitting an amended application for a Confidentiality Certificate in the same form and manner as an original application. On the basis of such application and other pertinent information the Secretary will either:

(1) Approve the amended application and issue an amended Confidentiality Certificate together with a Notice of Cancellation terminating the original Confidentiality Certificate in accordance with Sec. 2a.8; or

(2) Disapprove the amended application and notify the applicant in writing that adoption of the proposed significant changes will result in the issuance of a Notice of Cancellation terminating the original Confidentiality Certificate in accordance with Sec. 2a.8.

Sec. 2a.7 Effect of Confidentiality Certificate.

(a) A Confidentiality Certificate authorizes the withholding of the names and other identifying characteristics of individuals who participate as subjects in the research project specified in the Certificate while the Certificate is in effect. The authorization applies to all persons who, in the performance of their duties in connection with the research project, have access to information which would identify the subjects of the research. Persons so authorized may not, at any time, be compelled in any Federal, State, or local civil, criminal, administrative, legislative, or other proceedings to identify the research subjects encompassed by the Certificate, except in those circumstances specified in paragraph (b) of this section.

(b) A Confidentiality Certificate granted under this part does not authorize any person to refuse to reveal the name or other identifying characteristics of any research subject in the following circumstances: (1) The subject (or, if he or she is legally incompetent, his or her guardian) consents, in writing, to the disclosure of such information, (2) Authorized personnel of DHEW request such information for audit or program evaluation of a research project funded by DHEW or for investigation of DHEW grantees or contractors and their employees or agents carrying out such a project. (See 45 CFR 5.71 for confidentiality standards imposed on such DHEW personnel), or (3) Research of such information is required by the Federal Food, Drug, and Cosmetic Act (21 U.S.C. 301) or the regulations promulgated thereunder (Title 21, Code of Federal Regulations).

(c) Neither a Confidentiality Certificate nor the regulations of this part govern the voluntary disclosure of identifying characteristics of research subjects.

Sec. 2a.8 Termination.

(a) A Confidentiality Certificate is in effect from the date of its issuance until the effective date of its termination. The effective date of termination shall be the earlier of:

(1) The expiration date set forth in the Confidentiality Certificate; or

(2) Ten days from the date of mailing a Notice of Cancellation to the applicant, pursuant to a determination by the Secretary that the research project has been completed or discontinued or that retention of the Confidentiality Certificate is otherwise no longer necessary or desirable.

(b) A Notice of Cancellation shall include: an identification of the Confidentiality Certificate to which it applies; the effective date of its termination; and the grounds for cancellation. Upon receipt of a Notice of Cancellation the applicant shall return the Confidentiality Certificate to the Secretary.

(c) Any termination of a Confidentiality Certificate pursuant to this section is operative only with respect to the names and other identifying characteristics of individuals who begin their participation as research subjects after the effective date of such termination. (See Sec. 2a.4(k) requiring researchers to notify subjects who enter the project after the termination of the Confidentiality Certificate of termination of the Certificate). The protection afforded by a Confidentiality Certificate is permanent with respect to subjects who participated in research during any time the authorization was in effect.

PART II

Research with Special Populations

Part II contains Chapters 6–10, each chapter dealing with protection of subjects who come from specific types of backgrounds or share some characteristics that constitute a special class of potential research subjects. Specifially, these special populations are children, students, prisoners, fetuses and pregnant women, and the institutionalized mentally disabled. The latter group consists largely, but not entirely, of persons more commonly referred to as "mentally retarded."

In general, these specific populations are mutually distinct, with the rather obvious exception of students and children, since many students are also children. What is shared in common by these groups is that they have been deemed, for various reasons, to be at risk as research subjects simply because of certain characteristics of the group rather than to any injurious procedure proposed by any researcher. For example, children may be considered to be at risk simply because, as children, they may be unduly influenced by any adult figure (e.g., the prestigious researcher) to consent to something that may not be in their best interest. We have already seen in the chapter on informed consent that at least one provision for this risk was included in the general HHS rules by requiring the signature of a legal guardian on a consent form whenever the subject was a minor. Presumably, the guardian, as another adult, can protect the child and prevent any undue influence from being exerted by the researcher.

However, even adults may be at risk because of special circumstances. An example of this is represented by prisoners. Because of the nature of being a prisoner (i.e., incarceration in a highly controlled setting), researchers may be able to obtain consent by taking unfair advantage of persons who may be in no position to refuse to cooperate.

Although these features paint an extreme picture, they are among the reasons why special rules have been formulated to guarantee the maximum individual liberty that is possible for such populations. The impact on research goes beyond the general rules for research discussed in the previous chapters of this book. Since the general research issues for protection have been presented already, however, the subsequent chapters of Part II will be significantly briefer than the chapters of Part I. Essentially, chapters in Part II consist of a brief explanation of each population as it pertains to research, followed by an appendix containing the full text of the relevant rules for research with that population.

Also, as noted before, it should be stressed that practitioners who work with these special populations will find these chapters useful although the emphasis in this book is on research aspects in protecting individual rights. This is true in two respects. First, provisions for insuring client rights can be written into *treatment plans* as well as informed consent forms. Thus a practitioner may wish to describe the treatment procedures, what the client can expect to gain from the treatment, and so on, just as a researcher might propose a research project to a potential research subject. Second, confidentiality of information applies to both research and treatment, and the careful practitioner may be avoiding embarrassment or even litigation by setting up procedural safeguards governing access to client files. Such safeguards could include the types of restrictions on access and/or distribution that have been placed on many government files (see the chapter on freedom of information and privacy) or on files of specific populations (e.g., as for students).

Part II commences with a look at a subject population that has been the focus of considerable attention from the public and private sectors with regards to protection of individual rights: children.

CHAPTER 6

Children

In addition to all provisions discussed previously for research with any subject population, special protections have been instituted for children who serve as research subjects. These protections involve various consent factors and extra responsibilities of institutional review boards (IRBs) when children are involved. Before we examine these two areas, let us look at a brief overview of the history of these rules. It should be noted that the rules discussed in this chapter are the *final* HHS rules for research with children. As we will see, these rules were delayed for several years, due primarily to a freeze placed by President Reagan on regulations in general. However, they were made final just before this book was printed. Given the amount of time and deliberation devoted to these rules before they were made final in 1983, it is likely that the rules (as appended to this chapter) will remain in effect for the forseeable future.

History

On January 13, 1978, the Department of Health, Education, and Welfare published a discussion on various aspects of research involving children.[1] This discussion was a report by the National Commission for the Protection of Human Research Subjects which had commissioned various smaller reports from various professionals. The Commission also held hearings to compile its own recommendations. Thus the publication of the report in the *Federal Register* did not contain any proposed rules but rather consisted of extensive comments on a multitude of issues surrounding children as research subjects. DHEW sought public opinion about the report to aid in constructing subsequent proposed rules, later published in July of 1978.

[1]*Federal Register*, Vol. 43, No. 9, Friday, January 13, 1978, pp. 2084–2114.

As is usual for such a process, some of the recommendations by the National Commission were added to the proposed rules, while others were not. For example, the National Commission's detailed recommendations on how IRBs should differentially handle minimal risk and more-than-minimal risk research with children[2] were not fully implemented in the proposed rules, although a scaled-down version was adopted. On the other hand, the National Commission's views on the need for parental *permission* and children's *assent* were generally adopted. Even in this latter case there was an exception, however. Whereas the National Commission recommended seeking assent from children seven years of age or older,[3] DHEW was less certain about the age at which a child can be judged capable of understanding research well enough to grant assent. As we shall see, DHEW later proposed five different options in handling this issue.

1978 Rules

In July, 1978, DHEW published proposed rules to govern the use of children as research subjects.[4] The proposed rules were based on the Commission's recommendations. In proposing these rules, DHEW noted that they had received comments on the commission's report from 132 persons and organizations. This indicates that various professional groups were better informed about government rule-making procedures than was true for the early 1970s, but more professionals could benefit—as could resultant rules—from involvement by providing input to Washington.

The July, 1978, proposed rules formed the basis of the only available guidelines for the next five years. One of President Reagan's mandates to federal agencies was the reduction of federal regulations in general, as a way of lessening the costs of government and reducing obstacles to business enterprises in the United States. One of the results of this mandate has been the indefinite suspension of many federal regulations, which included for several years the special protections for children in research. As we will see later, this freeze on regulations has also placed in limbo the special regulations for the institutionalized mentally disabled.[5]

1983 Rules

The current final regulations were published on March 8, 1983, and are appended to this chapter.[6] Since the 1978 proposed rules were the only guide-

[2]Ibid., p. 2086.
[3]Ibid., p. 2087.
[4]*Federal Register*, Vol. 43, No. 141, Friday, July 21, 1978, pp. 31786–31794.
[5]*APA Monitor*, March, 1981, Section 2, p. 12. Washington, D.C.: American Psychological Association.
[6]*Federal Register*, Vol. 48, No. 46, Tuesday, March 8, 1983, pp. 9814–9820.

lines in use for several years, it is likely that many researchers still use those now obsolete proposed rules as their guidelines. To dispell any incorrect notions, two aspects of the current final rules should be noted. First, there is no longer a requirement for an advocate to be involved in the consent process in addition to the parent(s), the child, and the relevant IRB. The only exception to this is the use of an advocate for those children who are legally wards of the state. Second, the earlier discriminations between minimal risk and slightly more risk are no longer relevant, as the current rules are much simpler and easier to understand. In fact, one of the notable differences between the 1978 and the 1983 rules is that the 1983 rules are considerably shorter. In this respect at least, President Reagan's insistence on more manageable regulations has definitely affected rules for protection of human research subjects. Of course, the rules were also modified by HHS in response to public comments aimed at reducing the complexity of the government rules. Readers interested in more details about the changes that led to the current final rules should consult the full text of the *Federal Register* (Vol. 48, March 8, 1983, p. 9814) to read HHS' reaction to various public comments on the previous proposed rules. For the sake of brevity, the final rules appended to the end of this chapter do not contain these discussions.

Consent Issues for Children

As noted previously, there are essentially two areas of main concern within the rules for research with children: consent issues and extra responsibilities for IRBs when children are involved. The first area, that of consent, can be further divided into two issues: (a) the concept of *assent*, as opposed to consent or permission and (b) what to do when the protector of the child (i.e., the parent or legal guardian) is incompetent or dangerous to the child.

Assent versus Consent versus Permission

Persons interested in the details of this topic are urged to read the introductory materials in the *Federal Register* that preceded the proposed rules of 1978.[7] To put it briefly, *permission* can be granted for an action by a responsible person (e.g., a parent) for someone else (e.g., a child) to undertake the action. A child who is old enough may be of sufficient capacity to understand research and therefore *consent* to participation but as a minor might still legally require the parent's *permission* to participate. A young child (e.g., five years of age) might understand some aspects of research and *assent* to participate, but by no stretch of the imagination could the child be expected to comprehend the research

[7] *Federal Register*, July 21, 1978, pp. 31786–31792.

sufficiently to *consent* adequately. Thus parent permission would be especially crucial in this latter case.

Obviously, the most prudent course for the researcher to take when involving children as subjects would be to obtain written parental permission as well as consent from the child (requiring all elements of informed consent) or, simply, assent from the child.

That the age at which a child is judged to be capable of actual consent was no easy matter for HEW to determine in 1978 was clear from the request by HEW for public comments on five different options. These options were presented in the proposed 1978 rules rather than in the rules themselves, and HHS later (1983) chose not to try to set any age criterion on capability for assent, but rather to let the IRB make such a decision on a case-by-case basis. The 1978 options[8] were:

1. Requiring assent from all children who are 12 years of age or older, if the research is not expected to be of direct benefit to the health or well-being of the particular child.
2. Same as above, but setting the age at 7.
3. In the regulation itself, leaving it to the discretion of the board, but in the preamble to the regulation and in implementing policy statements recommending that assent normally be secured if the children are above a certain age (e.g., 12 or 7). Depending on the type of research and the types of candidates involved as subjects, the board may wish to take a flexible approach in selecting ages at which assent may be required.
4. Leaving it to the discretion of the board, with no guidance either in the preamble or in implementing policy statements.
5. Other alternatives.

From an historical standpoint, it is interesting to note that HHS was never able to establish a magical age at which a child is capable of assent, nor did public commentators agree on what that age should be. In fact, public views gave a range of age 3–16 as possible! Undoubtedly, HHS's decision to rely on the judgment of local IRBs has handily solved this problem, since circumstances can be expected to vary widely between research projects, to say nothing of the variation in mental capacity between individual children.

Actual experience also indicates that attempting to establish a set age for assent is not feasible. The author has found that children differ widely in the ability to understand an enterprise as complicated as scientific research. It is not possible to set the same age criterion for everyone. Others agree. O'Sullivan (1982)[9] has related the experiences at Boston Children's Hospital, where their work is entirely in the pediatrics field. There, too, it has been the conclusion of

[8]Ibid., p. 31787.
[9]O'Sullivan, G. Studies involving children. In R. A. Greenwald, M. K., Ryan, and J. E. Mulvill (Eds.), *Human subjects research.* New York: Plenum Press, 1982.

the IRB that no such criterion on age can be set. Instead, they involve both the parents and children as much as possible in the consent process. If there is conflict between parents and children regarding participation in the research, hospital staff simply do not involve the child in research unless there is some obvious benefit to the child.

Age of the child is not the only variable that can influence research responsibility. Factors governing the protection of children in research can also vary from setting to setting. For example, it is not unusual for children to be the least informed members in family counseling, even when it is the child who has the most at stake as the primary therapy client (Christian, Clark, and Luke, 1981).[10] Such situations, in which the focus on other family members can eclipse the need to inform children adequately, call for more attention to be given to providing all necessary information to the child. Although such a situation involves treatment rather than research, we saw previously that the line between treatment and research—especially treatment that is relatively new—is a very thin line.

It should be kept in mind that all children are not the same, and that different environments and children with special problems may call for special researcher and IRB efforts. For example, in a discussion of protecting the rights of autistic children in residential treatment settings, McClannahan and Krantz (1981)[11] describe how language characteristics make even normal communication difficult for such children. In such instances, researchers may need to spend significantly more time in drafting any written consent documents. As we have seen previously, formulas do exist for computing the reading level of such documents, but the formulas were based on normal children and are measured according to standard educational grade levels. Researchers working with children such as the autistic must rely on their own professional training and clinical expertise to construct consent or assent documents for children who do not fit standard educational achievement criteria. The paramount goal in such documents must be that the children will fully understand the information presented by the researcher.

Settings in which treatment and research both occur call for special vigilance by the researcher. An example of this was presented by Timbers, Jones, and Davis (1981)[12] for a regional group home network for disturbed and delin-

[10]Christian, W. P., Clark, H. B., and Luke, D. E. Client rights in clinical counseling services for children. In G. T. Hannah, W. P. Christian, and H. B. Clark (Eds.), *Preservation of client rights*. New York: The Free Press, 1981.

[11]McClannahan, L. E., and Krantz, P. J. Accountability systems for protection of the rights of autistic children and youth. In G. T. Hannah, W. P. Christian, and H. B. Clark (Eds.), *Preservation of client rights*. New York: The Free Press, 1981.

[12]Timbers, G. D., Jones, R. J., and Davis, J. L. Safeguarding the rights of children and youth in group-home treatment settings. In G. T. Hannah, W. P. Christian, and H. B. Clark (Eds.), *Preservation of client rights*. New York: The Free Press, 1981.

quent adolescents. Although research is a natural companion to the counseling and behavioral treatment, they have set up two distinct consent mechanisms. Children's consent for research is sought but is not required for acceptance into the program for treatment. The types of information presented to the children for the research and treatment aspects of the program are different, as is the level of child consent. For example, on the treatment aspect, child consent may not be relevant if placement for residential treatment is made through a court order. However, for voluntary placements, assent of the child and consent of the parent or guardian is considered crucial to protect the rights of all involved.

Children as Wards of the State

For various reasons, children often end up as wards of the state. Thus the government, in essence, acts as the parent for such children. For these cases, the National Commission recommended that research with children specify the use of an advocate, so recognized by the IRB involved, who otherwise acts as a parent by granting (or refusing) necessary permission on behalf of the child. The responsibilities of such an advocate for wards of the state are presented in Section 46.409 of the rules appended to this chapter.[13] Essentially, the advocate must be appointed by the IRB as someone who will look out for the best interests of the child. The advocate can speak for many children, thus avoiding the potential problem with earlier rules where the IRB might have had to appoint a separate advocate for every such ward involved in a research project. The advocate may also be a member of an IRB. It should be reemphasized that the final rules (1983) require the use of an advocate *only* in those instances in which the child is a ward of the state. Public comments and HHS deliberations resulted in the removal of previously proposed rules wherein advocates would have been required in many other instances as well.

Extra Responsibilities for an IRB

As can be clearly seen from the final rules appended for this chapter, one primary mechanism used to protect the rights of children as research subjects is to require extra responsibilities of an IRB. In the proposed 1978 rules a variety of strictures were placed on IRBs to take special care in research with children. For example, it was proposed at that time that the IRB must determine that the research in question was always conducted first on animals, then on adults, and so on before the research under review was attempted with children. However, this issue was not handled by incorporating a separate action in the final rules. Instead, in the preamble to the rules appears the opinion of HHS (concurring

[13]*Federal Register*, Vol. 48, March 8, 1983, p. 9820.

with that of the National Commission) that researchers approach research with children cautiously. In this newer approach, investigators were *encouraged* first to conduct their research on animals, adult humans, and older children before involving younger children as research subjects.

Research with Children and Minimal Risk

The issue of risk in excess of minimal risk for children in research was mentioned earlier in this chapter. Although dealt with specificially in Sections 46.405–46.407, it can be seen that these final rules[14] do not forbid such research. Rather, the rules provide for some extra caution by an IRB, on specific dimensions, when risk greater than minimal risk is involved. For example, in such situations IRBs must pay close attention to the relationship between any benefit to the child and the amount of risk. However, it is not necessary that the direct benefit (if any) outweigh the amount of risk, so long as safeguards such as parental, guardian, or advocate monitoring of the research are provided.

Assessment of just how much risk is presented has always been a difficult task. The IRB responsibilities in research with children (as with any special population) are more serious in assessing risk level because of the vulnerability of the subject class. One approach to ascertain risk level is to rely on the expertise of the individual IRB members to determine whether, and by how much, a research proposal will exceed minimal risk. Another factor to aid in the assessment is the documentation provided by the researcher to substantiate his or her statements on the level of risk that children will experience in the research.

Still another factor to consider is the combined experiences of researchers and IRB members *at that particular institution*. For example, let us suppose that hundreds of projects, all using procedures similar to the ones in the proposed research under review, have been conducted at the institution in the past. If these projects have proceeded with very few or no difficulties, then the IRB has an actual record upon which to base their assessment of risk level. Such a consideration has been noted by O'Sullivan (1982)[15] who further suggested that researchers could even add such statistical information to the consent form. Such a notation could go a long way toward reassuring children and parents regarding the risk level of the proposed research.

SUMMARY

This chapter represents our first set of rules written to protect a special population of potential research subjects. In this case, where the population is

[14]Ibid., p. 9819.
[15]O'Sullivan, op. cit.

children, the government's primary concern is the probable diminished capacity for the population to *consent* to research. The cause of this diminished capacity is age, whereas for other populations (e.g., institutionalized mentally disabled) such diminished capacity may arise from other causes (e.g., impaired mental development) that are not necessarily linked to age. That age alone can reduce capacity to consent is quite likely since it has been shown (Bower and Gasparis, 1978) that it is sometimes hard even for adults to understand that they participated in research.[16]

Thus a distinction was drawn between *consent*, *assent*, and *permission*, relegating each function according to the capacity of the person and their relationship to a child. Further, since there are instances when the typical grantor of permission for research participation should not be a parent, the role of an advocate for the child was discussed.

Finally, we saw that IRBs have extra responsibilities in cases where the research subjects are children. These responsibilities include appointing advocates for children, when appropriate, and permitting research involving greater-than-minimal risk for children only if various contingencies are met. In the opinion of this author, these contingencies are not prohibitive since they permit such higher-risk research with children even when there is no direct benefit expected to accrue to the subjects. Still, the steps to be followed by an IRB when the research subjects are children definitely go beyond what is required of an IRB for most other populations who are capable of giving their own consent to participate in research.

[16]Bower, R. T., and Gasparis, P. *Ethics of social research: Protecting the interests of human subjects.* New York: Praeger, 1978.

APPENDIX 6.1

Protection of Human Subjects in Research Involving Children

[Note: The following text is from the *Federal Register*, Vol. 48, No. 46, Tuesday, March 8, 1983, pp. 9814–9820.]

DEPARTMENT OF HEALTH AND HUMAN SERVICES

Office of the Secretary
45 CFR Part 46

Additional Protections for Children
Involved as Subjects in Research

AGENCY: Department of Health and Human Services
ACTION: Final rule
SUMMARY: The Department of Health and Human Services (Department or HHS) is prescribing additional requirements for protection of children involved as subjects in research. These regulations adopt, with some changes, the recommendations of the National Commission for the Protection of Human Subjects of Biomedical and Behavioral Research (National Commission) as were presented in the notice of proposed rule making (NPRM) 43 FR 31786 which preceded this final rule.

Specifically, these regulations impose certain added responsibilities on Instutitional Review Boards (IRBs) depending on the degree of risk involved in the research and the extent that the research is likely to be a benefit to the subject or relate to a subject's illness. The regulations also set forth requirements for obtaining permission by parents and guardians and, except under certain circumstances, assent by the children themselves. When the child is a ward of the state, the appointment of an advocate is required under some circumstances. The regulations exempt from coverage most social, economic, and educational research in which the only involvement of children as subjects will be in one or more of the following categories: (a) Research conducted in established or commonly accepted educational settings, involving normal educational practices; (b) Research involving the observation of public behavior; (c) Research involving the use of educational tests; (d) Research involving the collection or study of existing data, documents, records or specimens.

EFFECTIVE DATE: These regulations are applicable to all research reviewed after June 6, 1983 by institutional review boards established under an HHS approved Assurance of Compliance.

[Note: Introductory material is deleted here.]

PART 46—(Amended)

Accordingly, Part 46 of 45 CFR is amended by adding a new Subpart D to read as follows:

Subpart D—Additional Protections for Children Involved as Subjects in Research

Sec.
46.401 To what do these regulations apply?
46.402 Definitions.
46.403 IRB duties.
46.404 Research not involving greater than minimal risk.
46.405 Research involving greater than minimal risk but presenting the prospect of direct benefit to the individual subjects.
46.406 Research involving greater than minimal risk and no prospect of direct benefit to individual subjects, but likely to yield generalizable knowledge about the subject's disorder or condition.
46.407 Research not otherwise approvable which presents an opportunity to understand, prevent, or alleviate a serious problem affecting the health or welfare of children.
46.408 Requirements for permission by parents or guardians and for assent by children.
46.409 Wards.

Subpart D—Additional Protections for Children Involved as Subjects in Research

Sec. 46.401 To what do these regulations apply?

(a) This subpart applies to all research involving children as subjects, conducted or supported by the Department of Health and Human Services.

(1) This includes research conducted by Department employees, except that each head of an Operating Division of the Department may adopt such nonsubstantive, procedural modifications as may be appropriate from an administrative standpoint.

(2) It also includes research conducted or supported by the Department of Health and Human Services outside the United States, but in appropriate circumstances, the Secretary may, under paragraph (e) of Sec. 46.101 of Subpart A, waive the applicability of some or all of the requirements of these regulations for research of this type.

(b) Exemptions (1), (2), (5), and (6) as listed in Subpart A at Sec. 46.101 (b) are applicable to this subpart. Exemption (4), research involving the observation of public behavior, listed at Sec. 46.101 (b), is applicable to this subpart where the investigator(s) does not participate in the activities being observed. Exemption (3), research involving survey or interview procedures, listed at Sec. 46.101 (b) does not apply to research covered by this subpart.

(c) The exceptions, additions, and provisions for waiver as they appear in paragraphs (c) through (i) of Sec. 46.101 of Subpart A are applicable to this subpart.

Sec. 46.402 Definitions.

The definitions in Sec. 46.102 of Subpart A shall be applicable to this subpart as well. In addition, as used in this Subpart:

(a) "Children" are persons who have not attained the legal age of consent to treatments or procedures involved in the research, under the applicable law of the jurisdiction in which the research will be conducted.

(b) "Assent" means a child's affirmative agreement to participate in research. Mere failure to object should not, absent affirmative agreement, be construed as assent.

(c) "Permission" means the agreement of parent(s) or guardian to the participation of their child or ward in research.

(d) "Parent" means a child's biological or adoptive parent.

(e) "Guardian" means an individual who is authorized under applicable State or local law to consent on behalf of a child to general medical care.

Sec. 46.403 IRB duties.

In addition to all other responsibilities assigned to IRBs under this part, each IRB shall review research covered by this subpart and approve only research which satisfies the conditions of all applicable sections of this subpart.

Sec. 46.404 Research not involving greater than minimal risk.

HHS will conduct or fund research in which the IRB finds no greater than minimal risk to children is presented, only if the IRB finds that adequate provisions are made for soliciting the assent of the children and the permission of their parent or guardians, as set forth in Sec. 46.408.

Sec. 46.405 Research involving greater than minimal risk but presenting the prospect of direct benefit to the individual subjects.

HHS will conduct or fund research in which the IRB finds that more than minimal risk to children is presented by an intervention or procedure that holds out the prospect of direct benefit for the individual subject, or by a monitoring procedure that is likely to contribute to the subject's well-being, only if the IRB finds that:

(a) The risk is justified by the anticipated benefit to the subjects;

(b) The relation of the anticipated benefit to the risk is at least as favorable to the subjects as that presented by available alternative approaches; and

(c) Adequate provisions are made for soliciting the assent of the children and permission of their parents or guardians, as set forth in Sec. 46.408.

Sec. 46.406 Research involving greater than minimal risk and no prospect of direct benefit to individual subjects, but likely to yield generalizable knowledge about the subject's disorder or condition.

HHS will conduct or fund research in which the IRB finds that more than minimal risk to children is presented by an intervention or procedure that does not hold out the prospect of direct benefit for the individual subject, or by a monitoring procedure which is not likely to contribute to the well-being of the subject, if the IRB finds that:

(a) The risk represents a minor increase over minimal risk;

(b) The intervention or procedure presents experiences to subjects that are reasonably commensurate with those inherent in their actual or expected medical, dental, psychological, social, or educational situations;

(c) The intervention or procedure is likely to yield generalizable knowledge about the subjects' disorder or condition which is of vital importance for the understanding or amelioration of the subjects' disorder or condition; and

(d) Adequate provisions are made for soliciting assent of the children and permission of their parents or guardians, as set forth in section 46.408.

Sec. 46.407 Research not otherwise approvable which presents an opportunity to understand, prevent, or alleviate a serious problem affecting the health or welfare of children.

HHS will conduct or fund research that the IRB does not believe meets the requirements of Sections 46.404, 46.405, or 46.406 only if:

(a) The IRB finds that the research presents a reasonable opportunity to further the understanding, prevention, or alleviation of a serious problem affecting the health or welfare of children; and

(b) The Secretary, after consultation with a panel of experts in pertinent disciplines (for example: science, medicine, education, ethics, law) and following opportunity for public review and comment, has determined either: (1) That the research in fact satisfies the conditions of Sections 46.404, 46.405, or 46.406, as applicable, or (2) the following:

> (i) The research presents a reasonable opportunity to further the understanding, prevention, or alleviation of a serious problem affecting the health or welfare of children;
>
> (ii) The research will be conducted in accordance with sound ethical principles;
>
> (iii) Adequate provisions are made for soliciting the assent of children and the permission of their parents or guardians, as set forth in Section 46.408.

Sec. 46.408 Requirements for permission by parents or guardians and for assent by children.

(a) In addition to the determinations required under other applicable sections of this subpart, the IRB shall determine that adequate provisions are made for soliciting the assent of the children, when in the judgment of the IRB the children are capable of providing assent. In determining whether children are capable of assenting, the IRB shall take into account the ages, maturity, and psychological state of the children involved. This judgment may be made for all children to be involved under a particular protocol, or for each child, as the IRB deems appropriate. If the IRB determines that the capability of some or all of the children is so limited that they cannot reasonably be consulted or that the intervention or procedure involved in the research holds out a prospect of direct benefit that is important to the health or well-being of the children and is available only in the context of the research, the assent of the children is not a necessary condition for proceeding with the research. Even where the IRB determines that the subjects are capable of assenting, the IRB may still waive the assent requirement under circumstances in which consent may be waived in accord with Sec. 46.116 of Subpart A.

(b) In addition to the determinations required under other applicable sections of this subpart, the IRB shall determine, in accordance with and to the extent that consent is required by Sec. 46.116 of Subpart A, that adequate provisions are made for soliciting the permission of each child's parents or guardian. Where parental permission is to be obtained, the IRB may find that the permission of one parent is sufficient for research to be conducted under Sec. 46.404 or 46.405. Where research is covered by Sections 46.406 and 46.407 and permission is to be obtained from parents, both parents must give their permission unless one parent is deceased, unknown, incompetent, or not reasonably available, or when only one parent has legal responsibility for the care and custody of the child.

(c) In addition to the provisions for waiver contained in Sec. 46.116 of Subpart A, if the IRB determines that a research protocol is designed for conditions or for a subject population for which parental or guardian permission is not a reasonable requirement to protect the subjects (for example, neglected or abused children) it may waive the consent requirements in Subpart A of this part and paragraph (b) of this section, provided an appropriate mechanism for protecting the children who will participate as subjects in the research is substituted, and provided further that the waiver is not inconsistent with Federal, State or local law. The choice of an appropriate mechanism would depend upon the nature and purpose of the activities described in the protocol, the risk and anticipated benefit to the research subjects, and their age, maturity, status, and condition.

(d) Permission by parents or guardians shall be documented in accordance with and to the extent required by Sec. 46.117 of Subpart A.

(e) When the IRB determines that assent is required, it shall also determine whether and how assent must be documented.

Sec. 46.409 Wards.

(a) Children who are wards of the state or any other agency, institution, or entity can be included in research approved under Sec. 46.406 or 46.407 only if such research is:

(1) Related to their status as wards; or

(2) Conducted in schools, camps, hospitals, institutions, or similar settings in which the majority of children involved as subjects are not wards.

(b) If the research is approved under paragraph (a) of this section, the IRB shall require appointment of an advocate for each child who is a ward, in addition to any other individual acting on behalf of the child as guardian or in loco parentis. One individual may serve as advocate for more than one child. The advocate shall be an individual who has the background and experience to act in, and agrees to act in, the best interests of the child for the duration of the child's participation in the research and who is not associated in any way (except in the role as advocate or member of the IRB) with the research, the investigator(s), or the guardian organization.

CHAPTER 7

Students

If there is any one aspect about research with students that stands out, it is that there appears to be more confusion about what is required by law than is really warranted. In reality, the federal regulations governing students and educational records in general actually address research in a tangential sense only. Specifically, there is only one very brief section in the rules we shall examine that deals with research. This information is placed in this part of the volume because it can be considered to deal with a particular population—and the author has heard colleagues mistakenly refer to rules for research with students as a separate topic in its own right. Thus interested readers are likely still to consider students as a special population to be included in Part II of this book.

At the same time, the rules governing educational records, who may examine these records, and what limitations have been placed on access do affect the researcher. But it must be noted that the effects such rules have on researchers are the same as exist for anyone who wishes to make use of educational records (e.g., intelligence test scores, attendance data) for any purpose.

In general, parental consent must be obtained for anyone to view such records pertaining to students under the age of eighteen. Indeed, one of the objections raised in the legal field is that parents appear to have more control over their children's educational records than do the children themselves! This point and other aspects of the education field as it relates to access to records is discussed later in this chapter.

In keeping with the intent of Part II of this book, this chapter will be brief. Researchers should note that, as a general rule for educational research, research goals would be most expeditiously served by securing parental consent (see the chapter on children) and using complete informed consent agreements as discussed in numerous sections of Part I chapters of this book.

History

Although we will look at only the highlights of the legislative history of rules governing educational records and educational research, we will touch upon somewhat more detail than is typical for other chapters of Part II. This has been done because of the rather unusual way in which some education rules—especially the Buckley Amendment—were passed, with almost no participation by educators or the professional research community. In turn, such inattention, as we shall see later, was marked by and due in some part to the apparent general lack of knowledge on the part of educators and researchers about federal rules in this area.

1965–1970

In the late 1960s and in 1970, Congress passed various acts related to support of education in the United States. These acts governed the amounts of federal aid that were approved to be spent on various educational programs. Of course, along with such financial aid has come a variety of "strings," including reporting requirements and strictures on the types of students and families to benefit from various educational programs. It was this series of related acts that eventually was amended in the 1970s to include requirements regarding privacy and access to student records.

Arbitrarily beginning with 1965 to keep our discussion of rules within recent history, we find Public Law 89–10, also known as the Elementary and Secondary Education Act of 1965.[1] This act contained a variety of titles, each of which addressed a different area of financial assistance to education or some other aspect of federal rules about education. For example, Title IV was entitled "Educational Research and Training" and established rules for the types of educational research that could be funded by the Office of Education, among other things. This serves as a good example of alerting researchers and others of the professional community that, should it be needful to find the exact rules to support litigation or to obtain funds, there is no such thing as "Title IV monies." One must know what particular *act* is being referred to in such discussions, or the less-than-knowledgeable researcher or funds-seeker may be at a disadvantage when it comes time to discuss a proposal in detail with a government bureaucrat.

Public Law 90–247 contained a lengthy list of amendments to PL 89–10. This new law (PL 90–247) was referred to as the Elementary and Secondary Education Amendments of 1967.[2] A variety of issues were addressed in this act,

[1]Public Law 89–10; April 11, 1965.
[2]Public Law 90–247; January 2, 1968.

including some new definitions of the word *handicapped* to govern rules for educating the handicapped.

Next on our list of highlighted legislation is Public Law 91–230.[3] Once again, this act addressed a wide variety of education issues. Of special historical interest was Title IV of this act. Title IV had a separate name, known as the General Education Provisions Act, or GEPA. GEPA, in turn, modified Title IV of the earlier PL 90-247. Essentially, GEPA addressed a variety of administrative issues regarding federal control of education. Part C of GEPA, entitled "Advisory Councils," described the responsibilities of various boards, committees, and agencies and the authoritative relationship between the government and educational institutions. Of particular interest for our purposes is that the subsequent rules on privacy and student records were added to this section of GEPA.

Privacy and Student Records

This first significant body of federal legislation affecting protection of student records and involving rules on access was passed in 1974. The overall act that was passed was the Education Amendments of 1974,[4] with the legal title of Public Law 93–380. As usual, there were several titles within this act, each dealing with a different aspect of federal involvement with education. In particular, Sections 513–515 of Title V ("Education Administration") dealt with the rights and privacy of parents and students. These sections, as noted before, modified Part C of GEPA (passed in 1970).

Because of the newness of rules governing student records and parental consent for others to access such records, Sections 513–515 were given their own separate heading as an act. This act is referred to as the Family Educational Rights and Privacy Act of 1974, abbreviated as FERPA. FERPA actually laid the groundwork for almost all of the basic protection and consent procedures now required for anyone using school records. This author has heard professionals mistakenly refer to some subsequent "Buckley Amendments" as the main rules governing protection of students' privacy. In reality, FERPA was also termed the Buckley Amendment. What has been confusing is that some subsequent amendments to FERPA (or the Buckley Amendment) were also cosponsored by the same Senator James Buckley (Republican, New York) who sponsored FERPA.

1974: An Active Year

FERPA was passed on August 21, 1974. Barely four months later, some further amendments were passed as part of Public Law 93–568.[5] As we will see,

[3]Public Law 91–230; April 13, 1970.
[4]Public Law 93–380; August 21, 1974.
[5]Public Law 93–568; December 31, 1974.

the way in which the original Buckley Amendment was introduced to Congress was an excellent example of how practitioners and researchers—in this case, of the education field—learned too late of how their lives were being radically modified by legislative fiat.

To keep our terminology consistent, we will, from this point on, refer to the Buckley Amendment rather than to FERPA. The Buckley Amendment was then quickly modified a few months later by Section 2 of a joint resolution of the Congress. Even the most astute Washington observer or politically savvy researcher would not have been likely to look for education amendments on student rights in this resolution that was passed as a:

> Joint Resolution to Authorize and Request the President to call a White House Conference on Library and Information Services not later than 1978, and for other purposes.[6]

1979

Finally, in 1979, another brief change was made in the Buckley Amendment by a portion of PL 96–46 on education matters.[7] This new section essentially made explicit that state and local officials do have access to student records when auditing or evaluating various educational programs, providing certain steps are taken. Readers can view the precise requirements in the text of the rules appended to this chapter.

The Buckley Amendment

The Buckley Amendment can be briefly summarized as originally having had the following major points regarding protection of educational records and consent requirements:

1. In general, no school or educational agency could receive federal funds if the school blocked parental access to a wide variety of educational records such as intelligence test scores, standardized test results, attendance data, teacher evaluations of the student, and so forth.
2. Parents had the right to challenge, in a formal hearing if desired, the accuracy or completeness of any of the educational records of their child.
3. No federal funds would be made available to any educational institution that permitted others to examine student records without the written consent of the parent (with some exceptions).

[6]Ibid.
[7]Public Law 96–46; August 6, 1979.

4. Outside agencies could not gain access to personally identifiable student records unless specifically authorized to do so by federal law.
5. When authorized individuals or agencies did look at a student's file, they were to leave a written notice in the file so that a parent would later know that that file had been examined.
6. When a student reaches 18 years of age or enters postsecondary school, the rights and consent authority regarding access to school records transfers totally and solely from the parents to the student.
7. No federal funds were to go to any school unless students 18 years of age or older or enrolled in postsecondary education were informed of these rights.
8. All materials related to educational research projects (e.g., books, tapes, manuals, teacher instructions) were subject to inspection by the legal guardian of a child involved in such research.
9. Finally, there was a limit placed on the withholding of federal funds in that if an educational institution refused to turn over personally identifiable student information to a federal agency, federal funds would not be automatically withheld due to noncompliance with the agency until it could be determined which rules should govern (i.e., the need to respond to government or to protect students' privacy).

Some of these Buckley Amendment tenets were revised by later amendments whereas others were largely untouched. As usual, the rules appended at the end of this chapter do not simply contain FERPA (the Buckley Amendment) or any other changes separately. Instead, the appended rules represent the current compilation of rules that were affected by all the changes. The material within the chapter gives the reader a more complete picture of what has happened to educational rules in the past decade than could be learned simply by reading the current rules in the appendix.

After the Buckley Amendment

As noted previously, the Buckley Amendment (August 21, 1974) was quickly altered by Section 2 of a joint resolution (PL 93–568, December 31, 1974) as a result of an intense series of lobbying efforts by professional educational organizations. Essentially, these new rules sought to open up student records, whereas the earlier Buckley Amendment had sought, in general, to close such records. As we will see later, Senator Buckley himself agreed with these changes to the point of cosponsoring the new legislation.

In general, Section 2 of PL 93–568 changed the Buckley Amendment in the following ways:

1. The regulations were changed in wording to apply to any educational agency.

2. The rules applied to *past* as well as current students.
3. Both public and private educational agencies were specifically named in the rules.
4. Rather than listing specific types of educational records in the rules (e.g., test scores), the wording was changed to the all-inclusive phrase, "education records."
5. To balance this broader scope, some specific types of records were exempted from the rules requiring parental consent for access, such as instructional, supervisory, and personnel records and psychological treatment records.
6. A category referred to as directory information was exempted from the rules to make it easier for schools to release or publish such information as a student's name, address, date of birth, major field of study, and other such data as often appear on sports program brochures or student yearbooks.
7. Finally, provision was made for older students to waive their right to access to their own records, primarily so persons writing letters of recommendation to postsecondary schools need not worry about students reading the letters themselves.

An argument can be made that students might be less pleased with some of the changes to the Buckley Amendment than would educators, but there can be no doubt that education lobbying efforts definitely modified the rather sweeping scope of the original Buckley Amendment. As we shall now see, there are several sides to the issue of student privacy, and these issues dictate the ease of access any educational researcher can expect when using education records in research projects.

Opposing Sides on Student Privacy

There are at least four groups or "sides" involved in the area of education privacy or research with students: students, parents, researchers, and the government. We have already summarized the role of the government and the parent, and the details of these parental consent rules can be seen in the rules appended to this chapter. Educational researchers are, in general, considered by the rules to be a third party, and rules about third-party access can be found in the appendix, although the single brief section on educational research will be highlighted in this chapter. Finally, the parent–student relationship is one that, in contradistinction to previous parental consent issues, may produce more conflict in matters of individual rights than we have seen before.

Parents versus Students

The question of who has more control over the student's records, a parent or the student himself, is somewhat different from the typical researcher's dilemma when securing parental permission for a minor to participate in research. As we saw in previous chapters, without the child's assent, the researcher would be unwise to use solely parental permission to involve a child in research. Even if the child is incapable (e.g., by reason of age or mental impairment) of granting assent, an advocate for the child will probably be required by an Institutional Review Board to grant (or refuse) permission for a child to participate in research. In educational research, however, where the researcher may simply desire access to records (e.g., for statistical analysis of different education programs), the parent–student dichotomy appears sharper.

In essence, for these types of records, parents may simply have more rights than students do, at least so far as federal rules are concerned. Zdeb (1975) noted that the constitutional right to privacy is itself a recent judicial position,[8,9] and in the case of education records it involves sensitive information on children where there are conflicts in demands for and uses of the data by parents, pupils, schools, and the community.[10] Zdeb concluded that the Buckley Amendment clearly gave primary control of student records to parents. This dominance could be a problem, especially in those instances in which a student wishes to prevent parental inspection and/or dissemination of information about the student. As the rules currently stand, there is no advocate or other means for students to use in balancing this parental authority. Students must simply wait until they are 18 or enter postsecondary education to assume control over their own records.

Educational Research

As noted before, there are some specific references to conduct of research in the current rules. Although the complete rules should be examined by the researcher, two self-explanatory areas are highlighted here, the first on how certain studies are explicitly allowed without written parental permission and the second on rights of pupils in research:

> Sec. 438(b)(1)(F) [access to student records is permitted for] organizations conducting studies for, or on behalf of, educational agencies or institutions for the purpose of developing, validating, or administering predictive tests, administering student aid programs, and improving instruction, if such studies are conducted in such a manner as will not permit the personal identification of students and their parents by persons other than representatives of such organizations and such informa-

[8]*Griswold* v. *Connecticut* (381 U.S. 479, 1965).

[9]*Roe* v. *Wade* (410 U.S. 113, 1973).

[10]Zdeb, M. J. A student right to privacy: The developing school records controversy. *Loyola University Law Journal*, 1975 (Spring), 6 (2), 430–445.

> tion will be destroyed when no longer needed for the purpose for which it is conducted.

As the reader may have guessed, this exemption[11] from parental permission was one of the new clauses added to the Buckley Amendment in December, 1974, following intense educator lobbying efforts after passage of the original Buckley Amendment. The clause once again opened the door for a wide range of education research.

This next section was left unchanged from the original Buckley Amendment, and is titled "Protection of Pupil Rights":

> Sec. 439. All instructional material, including teacher's manuals, films, tapes, or other supplementary instructional material which will be used in connection with any research or experimentation program or project shall be available for inspection by the parents or guardians of the children engaged in such program or project. For the purpose of this section "research or experimentation program or project" means any program or project in any applicable program designed to explore or develop new or unproven teaching methods or techniques.

Clearly, this second clause refers more to applied research than to educational practices and ensures that both parent and child will be familiar with all materials used in the research if they so desire. This requirement goes beyond the requirements we saw in previous chapters, where the primary method used to protect a subject's privacy was to construct a consent form sufficiently complete so that the subject could reasonably understand what was to be involved in the research. For those situations, the federal rules did not also require that researchers make available the various materials that might be used to affect the subject in some way during the research.

A very useful resource guide on the area of student rights was prepared by the Children's Defense Fund.[12] Although the format is a question-and-answer style aimed primarily at parents, the professional will find its straightforward answers helpful. One indication of how determined the Fund was to inform the public adequately can be seen from the unusual notice that appeared on the back of the pamphlet: "This pamphlet may be revised and/or reproduced for use in your community."[13] Obviously, the Children's Defense Fund staff wanted to make sure that persons affected by the legislation would be adequately informed about it. They were concerned enough to forego their copyright protection and freely grant others the freedom to make use of their efforts. As we will now see, a similar interest in informing citizens and seeking public response *before* enacting

[11] Public Law 93–568; December 31, 1974.

[12] *Your School Records: Questions and Answers about a New Set of Rights for Parents and Students.* Report prepared by the Children's Defense Fund, 1975. 1520 New Hampshire Avenue, N.W., Washington, D.C. 20036.

[13] Ibid., p. 11.

legislation was noticeably absent when the education rules were first passed into law.

Politics and Education Rules

As noted before, the passage of the original Buckley Amendment created such a furor within education circles that barely four months later Senators Buckley and Pell cosponsored revisions in the Buckley Amendment. Various professional education groups, representing educational researchers among others, thus managed to lobby successfully enough to open up—under various conditions—records that the original amendment closed (e.g., for survey and research purposes) and close other records (e.g., supervisory or counseling records) that educators felt should not be available to students or parents.

Without delving too deeply into the details of the legislative process, it would be instructive for readers to learn how this education legislation that affected so many people came to pass. As outlined by Mattesich[14] and Burcky and Childers[15] the following were the main steps:

1. The chief protagonist for the rules was Diane Divocky who had described a number of abuses in school recordkeeping systems in an article in *Parade Magazine* in March, 1974.[16]
2. Senator Buckley was alerted to the article by John Kwapisy, an aide, who was then asked by the senator to draft appropriate legislation on student records and privacy.
3. Only four weeks later, in May, 1974, Senator Buckley introduced the amendment for a floor debate. Some modifications were made, but it passed by a voice vote and no legislative hearings were held.
4. From June through December, 1974, groups such as the American Council on Education, the National School Boards Association, and related groups lobbied, successfully, to get the bill modified.

The point to made here is that a senator of presumably good intentions bypassed input from others and quite rapidly got a bill enacted that radically changed how education records are handled. But whose fault is that? Long active in the education field, Senator Buckley most probably felt that the best interests of all involved were served by his initial Buckley Amendment. The fact of the matter is that educators, including educational researchers, *must* be involved in

[14]Mattesich, C. M. The Buckley Amendment: Opening school files for student and parental review. *Catholic University Law Review*, 1975, 24, 588–603.

[15]Burcky, W. D., and Childers, J. H. Buckley Amendment: Focus on a professional dilemma. *The School Counselor*, 1976 (January), 23, 162–164.

[16]Divocky, D. How secret school records can hurt your child. *Parade Magazine*, March 31, 1974.

the political process, at least to some degree. This involvement does not necessarily require active personal lobbying by the individual researcher, but one must at least stay in touch with Congress and federal agencies lest one be surprised with new impediments to professional activities. Hints on how to influence the legislative process were provided by Burcky and Childers,[17] and the reader will find the last chapter of this book useful as well.

SUMMARY

In general, the rules currently in existence governing student records were instituted by the Family Educational Rights and Privacy Act of 1974, popularly known as the Buckley Amendment. These rules specify those situations in which written parental permission is needed for anyone wishing to view student records, regardless of the purpose. Subsequent changes in the Buckley Amendment have made it easier to gain access to such information for research and government program auditing purposes.

These rules go somewhat beyond the informed consent provisions discussed in earlier chapters in that parents have the right to view any instructional or related materials used by teachers in educational research where their child is involved. In other words, the informed consent document alone, with its description of procedures, confidentiality, and so forth is insufficient when applied research involving experimental education techniques is concerned.

A controversial aspect of student privacy is that control over student records rests with the parent (not the student) until the student reaches 18 or enrolls in postsecondary education. Conflicting views on such matters, including disagreements on who could have access to student records without parental permission, came to light during the lobbying by educators following passage of the original Buckley Amendment. As has been noted in the literature, the very process used for passing the student privacy rules can be used to warn professionals to be better informed and more involved in the legislative process. Failure to do so can only bring about a continuation of regulations impeding, or at least complicating, both treatment and research in the United States.

[17]Burcky and Childers, op. cit., p. 164.

APPENDIX 7.1

Protection of Human Subjects in Research Involving Students

[Note: The following text is a compilation of amendments that added Sections 438–440 to the General Education Provisions Act, or GEPA. See text of chapter for a more complete legislative history.]

Protection of the Rights and Privacy of
Parents and Students

Sec. 438. (a) (1)

(A) No funds shall be made available under any applicable program to any educational agency or institution which has a policy of denying, or which effectively prevents, the parents of students who are or have been in attendance at a school of such agency or at such institution, as the case may be, the right to inspect and review the education records of their children. If any material or document in the education record of a student includes information on more than one student, the parents of one of such students shall have the right to inspect and review only such part of such material or document as relates to such student or to be informed of the specific information contained in such part of such material. Each educational agency or institution shall establish appropriate procedures for the granting of a request by parents for access to the education records of their children within a reasonable period of time, but in no case more than forty-five days after the request has been made.

(B) The first sentence of subparagraph (A) shall not operate to make available to students in institutions of postsecondary education the following materials:

(i) financial records of the parents of the student or any information contained therein;

(ii) confidential letters and statements of recommendation, which were placed in the education records prior to January 1, 1975, if such letters or statements are not used for purposes other than those for which they were specifically intended;

(iii) if the student has signed a waiver of the student's right of access under this subsection in accordance with subparagraph (C), confidential recommendations—

(I) respecting admission to any educational agency or institution,

(II) respecting an application for employment, and

(III) respecting the receipt of an honor or honorary recognition.

(C) A student or a person applying for admission may waive his right of access to confidential statements described in clause (iii) of subparagraph (B), except that such waiver shall apply to recommendations only if (i) the student is, upon request, notified of the names of all persons making confidential recommendations and (ii) such recommendations are used solely for the purpose for which they were specifically intended. Such waivers may not be required as a condition for admission to, receipt of financial aid from, or receipt of any other services or benefits from such agency or institution.

(2) No funds shall be made available under any applicable program to any educational agency or institution unless the parents of students who are or have been in attendance at a school of such agency or at such institution are provided an opportunity for a hearing by such agency or institution, in accordance with regulations of the Secretary, to challenge the content of such student's education records, in order to insure that the records are not inaccurate, misleading, or otherwise in violation of the privacy or other rights of students, and to provide an opportunity for the correction or deletion of any such inaccurate, misleading, or otherwise inappropriate data contained therein and to insert into such records a written explanation of the parents respecting the content of such records.

(3) For the purpose of this section the term "educational agency or institution" means any public or private agency or institution which is the recipient of funds under any applicable program.

(4) (A) For the purposes of this section, the term "education records" means except as may be provided otherwise in subparagraph (B), those records, files, documents, and other materials which—

(i) contain information directly related to a student; and

(ii) are maintained by an educational agency or institution, or by a person acting for such agency or institution.

(B) The term "education records" does not include—

(i) records of institutional supervisory, and administrative personnel and educational personnel ancillary thereto which are in the sole possession of the maker thereof and which are not accessible or revealed to any other person except a substitute;

(ii) if the personnel of a law enforcement unit do not have access to education records under subsection (b) (1), the records and documents of such law enforcement unit which (I) are kept apart from records described in subparagraph (A), (II) are maintained solely for law enforcement purposes, and (III) are not made available to persons other than law enforcement officials of the same jurisdiction;

(iii) in the case of persons who are employed by an educational agency or institution but who are not in attendance at such agency or institution, records made and maintained in the normal course of business which relate exclusively to such person in that person's capacity as an employee and are not available for use for any other purpose; or

(iv) records on a student who is 18 years of age or older, or is attending an institution of postsecondary education, which are created or maintained by a physician, psychiatrist, psychologist or other recognized professional or para-professional acting in his professional or para-professional capacity, or assisting in that capacity, and which are created, maintained, or used only in connection with the provision of treatment to the student, and are not available to anyone other than persons providing such treatment; provided, however, that such records can be personally reviewed by a physician or other appropriate professional of the student's choice.

(5) (A) For the purposes of this section the term "directory information" relating to a student includes the following: the student's name, address, telephone listing, date and place of birth, major field of study, participation in officially recognized activities and sports, weight and height of members of athletic teams, dates of attendance, degrees and awards received, and the most recent previous educational agency or institution attended by the student.

(B) Any educational agency or institution making public directory information shall give public notice of the categories of information which it has designated as such information with respect to each student attending the institution or agency and shall allow a reasonable period of time after such notice has been given for a parent to inform the institution or agency that any or all of the information designated should not be released without the parent's prior consent.

(6) For the purposes of this section, the term "student" includes a person with respect to whom an educational agency or institution maintains education records or personally identifiable information, but does not include a person who has not been in attendance at such agency or institution.

(b) (1) No funds shall be made available under any applicable program to any educational agency or institution which has a policy or practice of permitting the release of education records (or personally identifiable information contained therein other than directory information, as defined in paragraph (5) of subsection (a)) of students without the written consent of their parents to any individual, agency, or organization, other than to the following—

(A) other school officials, including teachers within the educational institution or local educational agency who have been determined by such agency or institution to have legitimate educational interests;

(B) officials of other schools or school systems in which the student seeks or intends to enroll, upon condition that the student's parents be notified of the transfer, receive a copy of the record if desired, and have an opportunity for a hearing to challenge the content of the record;

(C) authorized representatives of (i) the Comptroller General of the United States, (ii) the Secretary, (iii) an administrative head of an education agency (as defined in section 408(c) of this Act), or (iv) State educational authorities, under the conditions set forth in paragraph (3) of this subsection; and

(D) in connection with a student's applications for, or receipt of financial aid;

(E) State and local officials or authorities to which such information is specifically required to be reported or disclosed pursuant to State statute adopted prior to November 19, 1974;

(F) organizations conducting studies for, or on behalf of, educational agencies or institutions for the purpose of developing, validating, or administering predictive tests, administering student aid programs, and improving instruction, if such studies are conducted in such a manner as will not permit the personal identification of students and their parents by persons other than representatives of such organizations and such information will be destroyed when no longer needed for the purpose for which it is conducted;

(G) accrediting organizations in order to carry out their accrediting functions;

(H) parents of a dependent student of such parents, as defined in section 152 of the Internal Revenue Code of 1954; and

(I) subject to the regulations of the Secretary in connection with an emergency, appropriate persons if the knowledge of such information is necessary to protect the health or safety of the student or other persons.

(2) No funds shall be made available under any applicable program to any education agency or institution which has a policy or practice of releasing, or providing access to, any personally identifiable information in education records other than directory information, or as is permitted under paragraph (1) of this subsection unless

(A) there is written consent from the student's parents specifying records to be released, the reasons for such release, and to whom, and with a copy of the records to be released to the student's parents and the student if desired by the parents, or

(B) such information is furnished in compliance with judicial order, or pursuant to any lawfully issued subpoena, upon condition that parents and the students are notified of all such orders or subpoenas in advance of the compliance therewith by the educational institution or agency.

(3) Nothing contained in this section shall preclude authorized representatives of (A) the Comptroller General of the United States, (B) the Secretary, (C) an administrative head of an education agency or (D) State educational authorities from having access to student or other records which may be necessary in connection with the audit and evaluation of Federally supported education programs, or in connection with the enforcement of the Federal legal reguirements which relate to such programs: provided, that except when collection of personally identifiable information is specifically authorized by Federal law, any data collected by such officials shall be protected in a manner which will not permit the personal identification of students and their parents by other than those officials, and such personally identifiable data shall be destroyed when no longer needed for such audit, evaluations, and enforcement of Federal legal requirements.

(4) (A) Each educational agency or institution shall maintain a record, kept with the education records of each student, which will indicate all individuals (other than those specified in paragraph (1) (A) of this subsection), agencies, or organizations which have requested or obtained access to a student's education records maintained by such educational agency or institution, and which will indicate specifically the legitimate interest that each such person, agency, or organization has in obtaining this information. Such record of access shall be available only to parents, to the school official and his assistants who are responsible for the custody of such records, and to persons or organizations authorized in, and under the conditions of, clauses (A) and (C) of paragraph (1) as a means of auditing the operation of the system.

(B) With respect to this subsection, personal information shall only be transferred to a third party on the condition that such party will not permit any other party to have access to such information without the written consent of the parents of the student.

(c) The Secretary shall adopt appropriate regulations to protect the rights of privacy of students and their families in connection with any surveys or data-gathering activities conducted, assisted, or authorized by the Secretary or an administrative head of an education agency. Regulations established under this subsection shall include provisions controlling the use, dissemination, and protection of such data. No survey or data-gathering activities shall be conducted by the Secretary, or an administrative head of an education agency under an applicable program, unless such activities are authorized by law.

(d) For the purposes of this section, whenever a student has attained eighteen years of age, or is attending an institution of postsecondary education the permission or consent required of and the rights accorded to the parents of the student shall thereafter only be required of and accorded to the student.

(e) No funds shall be made available under any applicable program to any educational agency or institution unless such agency or institution informs the parents of students, or the students, if they are eighteen years of age or older, or are attending an institution of postsecondary education, of the rights accorded them by this section.

(f) The Secretary, or an administrative head of an education agency, shall take appropriate actions to enforce provisions of this section and to deal with violations of this section, according to the provisions of this Act, except that action to terminate assistance may be taken only if the Secretary finds there has been a failure to comply with the provisions of this section, and he has determined that compliance cannot be secured by voluntary means.

(g) The Secretary shall establish or designate an office and review board within the Department of Health, Education, and Welfare for the purpose of investigating, processing, reviewing, and adjudicating violations of the provisions of this section and complaints which may be filed concerning alleged violations of this section. Except for the conduct of hearings, none of the functions of the Secretary under this section shall be carried out in any of the regional offices of such Department.

(5) Nothing in this section shall be construed to prohibit State and local educational officials from having access to student or other records which may be necessary in connection with the audit and evaluation of any federally or State supported education

program or in connection with the enforcement of the Federal legal requirements which relate to any such program, subject to the conditions specified in the proviso in paragraph (3).

Protection of Pupil Rights

Sec. 439. All instructional material, including teacher's manuals, films, tapes, or other supplementary instructional material which will be used in connection with any research or experimentation program or project shall be available for inspection by the parents or guardians of the children engaged in such program or project. For the purpose of this section "research or experimentation program or project" means any program or project in any applicable program designed to explore or develop new or unproven teaching methods or techniques.

Limitation on Withholding of Federal Funds

Sec. 440. Except as provided in section 438(b)(D) of this Act, the refusal of a state or local educational agency or institution of higher education, community college, school, agency offering a pre-school program or other educational institution to provide personally identifiable data on students or their families, as a part of any applicable program to any Federal office, agency, department, or other third party, on the grounds that it constitutes a violation of the right to privacy and confidentiality of students or their parents, shall not constitute sufficient grounds for the suspension or termination of Federal assistance. Such a refusal shall also not constitute sufficient grounds for a denial of, a refusal to consider, or a delay in the consideration of, funding, for such a recipient in succeeding fiscal years. In the case of any dispute arising under this section, reasonable notice and opportunity for a hearing shall be afforded the applicant.

CHAPTER 8

Prisoners

The area of research with prisoners has been unusually controversial in comparison with most other types of subjects. This is perhaps best understood in a historical context, since physical abuse, torture, and even death have been reported for centuries in situations wherein helpless prisoners have been at the mercy of their jailors.

Although such atrocities may seem far removed from the daily routine of researchers in twentieth-century America, we do not have to turn back the pages of history too far to witness abuse of prisoners on a huge scale. In a basic premise on ethics for much of their work, the National Commission for the Protection of Human Subjects of Biomedical and Behavioral Research mentioned such instances in the Belmont Report,[1] as discussed earlier in this book. Specifically, the National Commission deplored the exploitation of unwilling prisoners as research subjects in Nazi concentration camps.[2] One outgrowth of these experiments was the development of the Nuremberg Code as a set of standards for judging physicians and scientists who had conducted biomedical experiments on prisoners in concentration camps.[3]

Of course, the types of research conducted on Nazi prisoners during World War II would appear to bear little resemblance to the kinds of research currently conducted in penal institutions in the United States. A variety of protections exist to insure that no such cruel or dehumanizing projects could be conducted in our prisons. Nevertheless, there are now a number of restrictions on research with prisoners. These restrictions have been implemented since 1978 and go considerably beyond previous restrictions. The rules for such research appear to have been passed partly because of a historical precedence for worrying about prisoner rights (a legacy of war crimes), but also because of the general debilitating

[1]*The Belmont Report:* Ethical Principles and Guidelines for the Protection of Human Subjects of Research, 1978. Report prepared by the National Commission for the Protection of Human Subjects of Biomedical and Behavioral Research. Washington, D.C.: U.S. Government Printing Office, Superintendent of Documents, DHEW Publication No. (OS)78-0012.

[2]Ibid., p. 9.

[3]Ibid., p. 1.

environment found in many prisons in America. Because of the overall negative conclusions drawn by various professionals about the ethical climate within prisons, and the concomitant likelihood of prisoners' being unable to give truly informed, voluntary consent, rather restrictive rules have been passed. We will briefly examine these causes for restrictive rules on research with prisoners in the sections on the Health and Human Services Department (HHS) and on the Food and Drug Administration (FDA).

Before we proceed to these sections, the reader should note that once again we have two separate sets of rules by HHS and by FDA governing research with prisoners. As discussed in earlier chapters, there were also differences between HHS and FDA for rules on operation of local institutional review boards and for informed consent rules. Again, as before, researchers wondering about which set of rules to follow should read the introductory portion of the rules themselves for explanations of applicability. The complete texts of the two sets of rules for research with prisoners are contained in the appendix to this chapter.

Health and Human Services

The current rules within HHS for research with prisoners were compiled when the agency was known as HEW, in 1978. To develop these rules, members of the National Commission for the Protection of Human Subjects in Biomedical and Behavioral Research visited prisons and associated research facilities, interviewed prisoners (including those who had previously served as research subjects and those who had not), conducted the usual array of public hearings, and obtained or wrote a variety of reports on the topic. Then, in January of 1978, HEW published proposed rules based on the National Commission's report.[4]

Negative Findings

In general, the National Commission was unfavorably impressed with the general conditions of prisons in the United States and therefore recommended very restrictive rules to HEW, citing a variety of reasons. The reaction of HEW at the time was to note that:

> The Department (HEW) has concluded that these requirements are so stringent that it is doubtful that any existing prison and few research projects could satisfy them. The Commission laid down these conditions because the Commission "did not find in prisons the conditions requisite for a sufficiently high degree of voluntariness and openness." In addition, the Commission stressed the "strong evidence of poor conditions generally prevailing in prisons and the paucity of evidence of any necessity to conduct research in prisons."[5]

[4]*Federal Register*, Vol. 43, No. 3, Thursday, January 5, 1978, pp. 1050–1053.
[5]Ibid., p. 1050.

In addition, the Commission cited four additional factors supporting a complete ban on research with prisoners[6]:

1. Research has already been prohibited in all federal prisons.
2. Such research has also been banned in eight states.
3. Research with prisoners is only conducted in about seven of the states that permit it or do not regulate it.
4. Such research is not conducted in countries other than the United States.

Essentially, then, the rules proposed by HEW in January of 1978 would have allowed biomedical or behavioral prisoner research (e.g., on the causes and effects of incarceration) *only* if no more than minimal risk was present in the research and if the research was intended to improve the health of the individual prisoner. These restrictions went beyond the scope of the limits suggested by the National Commission.[7]

Broad Scope of Restrictions

As can be seen up to this point, HEW was essentially forbidding most types of research in prisons, since even minimal risk research often does not actually improve the health of the subject. This was in contrast to the actual recommendations of the National Commission which would have also permitted nontherapeutic research if a variety of conditions were met. This type of research typically might not provide a tangible, direct benefit to each individual subject.

It should be noted that HEW itself went beyond the scope of restrictions recommended by the National Commission in at least one respect. Although the Commission defined *prisoner* according to the Omnibus Crime Control and Safe Streets Act of 1968 (42 U.S.C. 3781), HEW extended the definition to include even more types of individuals:[8]

> Sec. 46.303(c): "Prisoner" means any individual involuntarily confined or detained in a penal institution. The term is intended to encompass individuals sentenced to such an institution under a criminal or civil statute, individuals detained in other facilities by virtue of statutes or commitment procedures which provide alternatives to criminal prosecution or incarceration in a penal institution, and individuals detained pending arraignment, trial, or sentencing.

One problem with this broad definition is the phrase that includes persons who are sent to alternative placements instead of penal institutions, perhaps including persons not usually considered prisoners. In any event, this broader definition remained in the final rules on prisoner research published by HEW.

[6]Ibid.

[7]*Federal Register*, Vol. 42, January 14, 1977, p. 3076 (Report and Recommendations of the National Commission on Research Involving Prisoners).

[8]*Federal Register*, Vol. 43, January 5, 1978, p. 1052.

Current Rules

Later in 1978, on November 16, HEW published the rules that remain as the current rules directed by HHS for prisoner research.[9] As usual, the rules were based on the department's response to the comments received following publication of the proposed rules in January. Although these final rules essentially restrict most types of prisoner research, only 57 comments were received by HEW! As we have seen throughout this book, professionals and related associations could do a more thorough job of monitoring and influencing federal regulations that directly affect the performance of their professional responsibilities.

The full text of the current HHS rules is contained in the appendix to this chapter. In the interest of brevity, no more detail will be presented here on these rules other than to note that there was one major change from the rules that were proposed in January, 1978. Instead of permitting only research that was both of minimal risk (or less) *and* of direct benefit to a prisoner, the final rules did allow for two additional types of research:

1. Certain research on conditions particularly affecting prisoners as a class; and
2. Certain research intended to benefit the subjects but which involves the use of control subjects to whom such benefit may not be expected.[10]

These provisions clearly permit somewhat more research with prisoners, and the exact wording of the rules can be found in Section 46.306 ("Permitted research involving prisoners")[11] as contained in the HHS rules at the end of this chapter.

IRBs and Prisoner Research

As a final note on HHS rules for prisoner research, it should be pointed out that the rules also specify additional requirements for IRBs to heed when prisoner research is involved. There are several such additional duties for IRBs with this type of subject population, and the rules can be found in Sections 46.304 and 46.305 of the appended HHS rules. The reader can get an accurate view of the level of concern on the part of HHS for the lack of freedoms afforded prisoners by understanding that the membership of the IRB itself must be different from that of any we have previously discussed. That is, for IRBs that review prisoner research, a majority of the members of the IRB must have no association with the prison involved.

[9]*Federal Register*, Vol. 43, No. 222, Thursday, November 16, 1978, pp. 53652–53656.
[10]Ibid., p. 53652.
[11]Ibid., p. 53656.

Clearly, both the National Commission and HEW (HHS) felt strongly enough about sensitivity to coercion in prisons that they even removed the remotely possible source of conflict of board members' being associated with the prison. As we saw in previous chapters, this concern was so minimal that IRBs dealing with other populations need only make sure that they have at least *one* IRB member who is not associated with the institution sponsoring the research.

Food and Drug Administration

The extent and format of FDA rules for prisoner research are quite similar to those of HHS. However, an unusual aspect of these rules is that the filing of a lawsuit has blocked implementation of FDA's final rules and it may be some time before FDA's final rules take effect. Nevertheless, as an aid to the reader, the most recent FDA rules on prisoner research are appended to this chapter, but it should be noted that these rules are not yet in effect.

History

As was true for HEW (HHS), the Food and Drug Administration published proposed rules for prisoner research in 1978.[12] As for the separate rules on informed consent and IRBs, FDA felt that rules separate from HEW were necessary on prisoner research to govern research projects not regulated by HEW. For example, FDA rules would govern research where prisoners served as subjects for studies of substances covered by the Food, Drug, and Cosmetic Act and for related projects. These proposed rules followed the lead of HEW in that FDA proposed to permit only research that had a "reasonable probability"[13] of improving the health and well-being of the subjects.

Then, in May of 1980, FDA published its final rules,[14] based on comments received after publication of the proposed rules. These rules are presented at the end of this chapter. As did HEW, FDA also increased the scope of allowed research somewhat to permit research on general prison conditions and for control groups where not all subjects could expect to receive a direct benefit from the research. Similarly, FDA rules added that such research would require the review of FDA and outside experts in penology, medicine, and ethics. What is more, FDA would then need to publish in the *Federal Register* its intent of approving such research.

Such additional steps would, of course, significantly increase the amount of

[12]*Federal Register*, Vol. 32, No. 88, Friday, May 5, 1978, pp. 19417–19422.
[13]Ibid., p. 19421.
[14]*Federal Register*, Vol. 45, No. 106, Friday, May 30, 1980, pp. 36386–36392.

time needed to review and approve such research, since the local IRB would be more of a first stop in the review process rather than the final step. FDA also included in their rules, as did HEW, a number of extra precautions to be taken by an IRB when FDA-regulated substances were to be used with prisoners as research subjects.

In general, FDA was all but banning biomedical research in prisons. Comments received from the public on possible advantages to prisoners, such as earning money by participating in drug studies, did not convince FDA that such studies should ordinarily be permitted. FDA noted that the general social and economic deprivation of prisoners made them more susceptible to consenting to procedures that most ordinary citizens might refuse to experience.

Current Rules

An interesting development in this area was that the rules published in 1980, scheduled to go into effect in the summer of 1980, were stayed. That is, the rules appended at the end of this chapter for FDA have not yet taken effect, although they serve as a useful guide for this area.

In March of 1981, FDA issued a notice[15] that it was staying the effective date of the final rules. The reason given was that a lawsuit filed on July 29, 1980, had not yet been concluded and there existed the possibility that the rules could be declared invalid, depending on outcome of the suit. The suit (*Fante and the Upjohn Company v. Department of Health and Human Services et al.*, Civil Action No. 80–72778) was brought in the United States District Court for the Eastern District of Michigan to have the rules declared invalid. At that time, FDA decided simply to delay the effective date of the rules until five months following completion of the court case.

Then, in July of 1981, FDA again published another stay, but this one was indefinite.[16] Thus the rules appended at the end of the chapter are not actually enforceable, pending new action by FDA. In this latest notice, FDA also stated that it was likely that a whole new reproposal might be made by FDA for prisoner research, because the issue appears to be dwindling as prisoner research is decreasing in the United States. In December, 1981, FDA published a revision of part of the proposed rules.[17]

Thus, as was temporarily true for a different reason for rules for research with children as subjects, rules regarding prisoner research appear to be in a state of limbo. Researchers in this area should contact representatives of the FDA for the most up-to-date information.

[15]*Federal Register*, Vol. 46, No. 59, Friday, March 27, 1981, p. 18951.
[16]*Federal Register*, Vol. 46, No. 129, Tuesday, July 7, 1981, p. 35085.
[17]*Federal Register*, Vol. 46, No. 243, Friday, December 18, 1981, pp. 61669–61671.

Summary

As we have seen, the area of research with prisoners is unusual because of two interrelated factors: (1) prisoners experience such a deprived environment that their ability to consent freely can be questioned, and (2) the interest on the part of a private group (i.e., Upjohn) has been such that final regulations have been not yet gone into effect due to a lawsuit.

The history of mankind has clearly shown that prisoners, as a separate class of individuals, can be subject to inhumane treatment at worst and neglect at best in many settings. As a result of this legacy, HEW essentially forbade research with prisoners except when each individual prisoner could expect to receive direct benefit to his or her health and well-being. Subsequent federal regulations returned closer to the original National Commission recommendations and expanded this applicability somewhat by permitting research where prisoners served as control subjects (and therefore could not expect to receive any direct benefits from the research) or where the general conditions of life in prison were investigated.

Finally, we saw that HHS and FDA differ somewhat in their regulations on prisoner research. These discrepancies reflect the same differences we found earlier when HHS and FDA had separate rules for operation of IRBs and the construction of informed consent documents.

At the time of this printing, the HHS rules have gone into effect but the FDA rules have been blocked by a lawsuit. The following appendix contains the full text of prisoner research rules for both HHS and FDA, with the caution to the reader that the FDA rules are not yet enforceable.

APPENDIX 8.1

Regulations by the Department of Health and Human Services for Protection of Prisoners as Research Subjects

[Note: The following rules are taken from the *Federal Register*, Vol. 43, No. 222, Thursday, November 16, 1978, pp. 53652–53656.]

Title 45—Public Welfare

SUBTITLE A—DEPARTMENT OF HEALTH, EDUCATION AND WELFARE, GENERAL ADMINISTRATION

PART 46—PROTECTION OF HUMAN SUBJECTS

Additional Protections Pertaining to Biomedical and Behavioral Research Involving Prisoners as Subjects

AGENCY: Department of Health, Education, and Welfare
ACTION: Final rule
SUMMARY: These regulations stipulate additional requirements for Institutional Review Boards (Boards), provide for prisoners or representatives of prisoners on Boards when prisoners are involved, outline additional duties for Boards, and specify the conditions under which research involving prisoners is permitted.

[Note: Introductory material by HEW is omitted here.]

Subpart C—Additional Protections Pertaining to Biomedical and Behavioral Research Involving Prisoners as Subjects

Sec. 46.301 Applicability.

(a) The regulations in this subpart are applicable to all biomedical and behavioral research conducted or supported by the Department of Health, Education, and Welfare involving prisoners as subjects.

(b) Nothing in this subpart shall be construed as indicating that compliance with the procedures set forth herein will authorize research involving prisoners as subjects, to the extent such research is limited or barred by applicable State or local law.

(c) The requirements of this subpart are in addition to those imposed under the other subparts of this part.

Sec. 46.302 Purpose.

Inasmuch as prisoners may be under constraints because of their incarceration which could affect their ability to make a truly voluntary and uncoerced decision whether or not to participate as subjects in research, it is the purpose of this subpart to provide additional safeguards for the protection of prisoners involved in activities to which this subpart is applicable.

Sec. 46.303 Definitions.

As used in this subpart:

(a) "Secretary" means the Secretary of Health, Education, and Welfare and any other officer or employee of the Department of Health, Education, and Welfare to whom authority has been delegated.

(b) "DHEW" means the Department of Health, Education, and Welfare.

(c) "Prisoner" means any individual involuntarily confined or detained in a penal institution. The term is intended to encompass individuals sentenced to such an institution under a criminal or civil statute, individuals detained in other facilities by virtue of statutes or commitment procedures which provide alternatives to criminal prosecution or incarceration in a penal institution, and individuals detained pending arraignment, trial, or sentencing.

(d) "Minimal risk" is the probability and magnitude of physical or psychological harm that is normally encountered in the daily lives, or in the routine medical, dental, or psychological examination of healthy persons.

Sec. 46.304 Composition of Institutional Review Boards where prisoners are involved.

In addition to satisfying the requirements in Sec. 46.106 of this part, an Institutional Review Board, carrying out responsibilities under this part with respect to research covered by this subpart, shall also meet the following specific requirements:

(a) A majority of the Board (exclusive of prisoner members) shall have no association with the prison(s) involved, apart from their membership on the Board.

(b) At least one member of the Board shall be a prisoner, or a prisoner representative with appropriate background and experience to serve in that capacity, except that where a particular research project is reviewed by more than one Board only one Board need satisfy this requirement.

Sec. 46.305 Additional duties of the Institutional Review Boards where prisoners are involved.

(a) In addition to all other responsibilities prescribed for Institutional Review Boards under this part, the Board shall review research covered by this subpart and approve such research only if it finds that:

(1) The research under review represents one of the categories of research permissible under Sec. 46.306(a)(2);

(2) Any possible advantages accruing to the prisoner through his or her participation in the research, when compared to the general living conditions, medical care, quality of food, amenities and opportunity for earnings in the prison, are not of such a magnitude that his or her ability to weigh the risks of the research against the value of such advantages in the limited choice environment of the prison is impaired;

(3) The risks involved in the research are commensurate with risks that would be accepted by nonprisoner volunteers;

(4) Procedures for the selection of subjects within the prison are fair to all prisoners and immune from arbitrary intervention by prison authorities or prisoners. Unless the principal investigator provides to the Board justification in writing for following some other procedures, control subjects must be selected randomly from the group of available prisoners who meet the characteristics needed for that particular research project;

(5) The information is presented in language which is understandable to the subject population;

(6) Adequate assurance exists that parole boards will not take into account a prisoner's participation in the research in making decisions regarding parole, and each prisoner is clearly informed in advance that participation in the research will have no effect on his or her parole; and

(7) Where the Board finds there may be a need for follow-up examination or care of participants after the end of their participation, adequate provision has been made for such examination or care, taking into account the varying lengths of individual prisoners' sentences, and for informing participants of this fact.

(b) The Board shall carry out such other duties as may be assigned by the Secretary.

(c) The institution shall certify to the Secretary, in such form and manner as the Secretary may require, that the duties of the Board under this section have been fulfilled.

Sec. 46.306 Permitted research involving prisoners.

(a) Biomedical or behavioral research conducted or supported by DHEW may involve prisoners as subjects only if:

(1) The institution responsible for the conduct of the research has certified to the Secretary that the Institutional Review Board has approved the research under Sec. 46.305 of this subpart; and

(2) In the judgment of the Secretary the proposed research involves solely the following:

(A) Study of the possible causes, effects, and processes of incarceration, and of criminal behavior, provided that the study presents no more than minimal risk and no more than inconvenience to the subjects;

(B) Study of prisons as institutional structures or of prisoners as incarcerated persons, provided that the study presents no more than minimal risk and no more than inconvenience to the subjects;

(C) Research on conditions particularly affecting prisoners as a class (for example, vaccine trials and other research on hepatitis which is much more prevalent in prisons than elsewhere; and research on social and psychological problems such as alcoholism, drug addiction and sexual assaults) provided that the study may proceed only after the Secretary has consulted with appropriate experts including experts in penology, medicine and ethics, and published notice, in the *Federal Register*, of his intent to approve such research; or

(D) Research on practices, both innovative and accepted, which have the intent and reasonable probability of improving the health or well-being of the subject. In cases in which those studies require the assignment of prisoners in a manner consistent with protocols approved by the IRB to control groups which may not benefit from the research, the study may proceed only after the Secretary has consulted with appropriate experts, including experts in penology, medicine and ethics, and published notice, in the *Federal Register*, of his intent to approve such research.

(b) Except as provided in paragraph (a) of this section, biomedical or behavioral research conducted or supported by DHEW shall not involve prisoners as subjects.

APPENDIX 8.2

Proposed Regulations by the Food and Drug Administration for Protection of Prisoners as Research Subjects

[Note: The following proposed rules are taken from the *Federal Register*, Vol. 45, No. 106, Friday, May 30, 1980, pp. 36386–36392. These rules are not yet in effect due to a legal challenge, but they represent the most current policy views of FDA toward prisoner research. These proposed rules were amended, in part, in the *Federal Register*, Vol. 46, No. 243, Friday, December 18, 1981, pp. 61669–61671.]

DEPARTMENT OF HEALTH AND HUMAN SERVICES

Food and Drug Administration
21 CFR Part 50
(Docket No. 78N–0049)

Protection of Human Subjects;
Prisoners Used as Subjects in Research

AGENCY: Food and Drug Administration
ACTION: Final rule
SUMMARY: This document establishes regulations to provide protection for prisoners involved in research activities that fall within the jurisdication of the Food and Drug Administration (FDA). These regulations implement the recommendations of the National Commission for the Protection of Human Subjects in Biomedical and Behavioral Research (National Commission) on research involving prisoners. These regulations restrict the use of prisoners in research within the jurisdiction of FDA and establish requirements for the composition of and additional duties for institutional review boards when prisoners are involved in the research.

[Note: Deleted here are introductory explanations by FDA.]

PART 50—PROTECTION OF HUMAN SUBJECTS

Subpart A—General Provisions

Subpart A—General Provisions

Sec. 50.1 Scope.

(a) This part applies to all clinical investigations regulated by the Food and Drug Administration under sections 505(i), 507(d), and 520(g) of the Federal Food, Drug, and Cosmetic Act, as well as clinical investigations that support applications for research or marketing permits for products regulated by the Food and Drug Administration, including food and color additives, drugs for human use, medical devices for human use, biological products for human use, and electronic products. Additional specific obligations and commitments of, and standards of conduct for, persons who sponsor or monitor clinical investigations involving particular test articles may also be found in other parts (e.g., Parts 312 and 812). Compliance with these parts is intended to protect the rights and safety of prisoner subjects involved in investigations filed with the Food and Drug Administration pursuant to sections 406, 409, 502, 503, 505, 506, 507, 510, 513–516, 518–520, 706, and 801 of the Federal Food, Drug, and Cosmetic Act and sections 351 and 354–360F of the Public Health Service Act.

(b) References in this part to regulatory sections of the Code of Federal Regulations are to Chapter I of Title 21, unless otherwise noted.

Sec. 50.3 Definitions.

As used in this part:

(a) (Reserved)

(b) "Application for research or marketing permit" includes:

(1) A color additive petition, described in Part 71.

(2) A food additive petition, described in Parts 171 and 571.

(3) Data and information about a substance submitted as part of the procedures for establishing that the substance is generally recognized as safe for use that results or may reasonably be expected to result, directly or indirectly, in its becoming a component or

otherwise affecting the characteristics of any food, described in Sections 170.30 and 570.30.

(4) Data and information about a food additive submitted as part of the procedures for food additives premitted to be used on an interim basis pending additional study, described in Sec. 180.1.

(5) Data and information about a substance submitted as part of the procedures for establishing a tolerance for unavoidable contaminants in food and food-packaging materials, described in section 406 of the act.

(6) A "Notice of Claimed Investigational Exemption for a New Drug," described in Part 312.

(7) A new drug application, described in Part 314.

(8) Data and information about the bioavailability or bioequivalence of drugs for human use submitted as part of the procedures for issuing, amending, or repealing a bioequivalence requirement, described in Part 320.

(9) Data and information about an over-the-counter drug for human use submitted as part of the procedures for classifying these drugs as generally recognized as safe and effective and not misbranded, described in Part 330.

(10) Data and information about a prescription drug for human use submitted as part of the procedures for classifying drugs as generally recognized as safe and effective and not misbranded, described in this chapter.

(11) Data and information about an antibiotic drug submitted as part of the procedures for issuing, amending, or repealing regulations for these drugs, described in Part 430.

(12) An application for a biological product license, described in Part 601.

(13) Data and information about a biological product submitted as part of the procedures for determining that licensed biological products are safe and effective and not misbranded, described in Part 601.

(14) Data and information about an in vitro diagnostic product submitted as part of the procedures for establishing, amending, or repealing a standard for these products, described in Part 809.

(15) An "Application for an Investigational Device Exemption," described in Part 812.

(16) Data and information about a medical device submitted as part of the procedures for classifying these devices, described in section 513.

(17) Data and information about a medical device submitted as part of the procedures for establishing, amending, or repealing a standard for these devices, described in section 514.

(18) An application for premarket approval of a medical device, described in section 515.

(19) A product development protocol for a medical device, described in section 515.

(20) Data and information about an electronic product submitted as part of the procedures for establishing, amending, or repealing a standard for these products, described in section 358 of the Public Health Service Act.

(21) Data and information about an electronic product submitted as part of the procedures for obtaining a variance from any electronic product performance standard, as described in Sec. 1010.4.

(22) Data and information about an electronic product submitted as part of the procedures for granting, amending, or extending an exemption from a radiation safety performance standard, as described in Sec. 1010.5.

Subpart B—(Reserved)

Subpart C—Protections Pertaining to Clinical Investigations Involving Prisoners as Subjects

Sec. 50.40 Applicability.

(a) The regulations in this subpart apply to any clinical investigation involving prisoners as subjects that is regulated by the Food and Drug Administration under sections 505(i), 507(d), or 520(g) of the act, as well as any clinical investigation involving prisoners that supports any application for a research or marketing permit as defined by section 50.3(b).

(b) Nothing in this subpart shall be construed as indicating that compliance with the procedures set forth herein will authorize research involving prisoners as subjects to the extent such research is limited or barred by applicable State or local law.

Sec. 50.42 Purpose.

Because prisoners may be under constraints because of their incarceration which could affect their ability to make a truly voluntary and uncoerced decision whether or not to participate as subjects in research, it is the purpose of this subpart to provide additional safeguards for the protection of prisoners involved in research to which this part is applicable.

Sec. 50.44 Restrictions on clinical investigations involving prisoners.

(a) Any clinical investigation that is regulated by the Food and Drug Administration under section 505(i), 507(d), and 520(g) of the act, as well as any clinical investigation that supports an application for research or marketing permit as defined by Section 50.3(b), may involve prisoners as subjects only if the institution responsible for the conduct of the clinical investigation has certified to the Food and Drug Administration that the institutional review board has approved the clinical investigation under section 50.48; and

(1) The proposed clinical investigation involves solely research on practices that have the intent and reasonable probability of improving the health and well-being of the particular prisoners chosen. Subject to the approval of the institutional review board, prisoners may be assigned to control groups; or

(2) The institutional review board determines, after consultation with the Research Involving Human Subjects Committee, that the proposed clinical investigation involves research on conditions particularly affecting prisoners as a class (for example, vaccine trials and other research on hepatitis, which is much more prevalent in prisons than elsewhere). Subject to the approval of the institutional review board, prisoners may be assigned to control groups; or

(3) If the proposed clinical investigation involves research other than that described in paragraph (a) (1) or (2) of this section, the institutional review board determines, after consultation with the Research Involving Human Subjects Committee, that the following requirements are satisfied:

(i) The type of research fulfills an important social or scientific need, and the reasons for involving prisoners are compelling;
(ii) The involvement of prisoners in the type of research satisfies conditions of equity; and
(iii) A high degree of voluntariness on the part of the prospective participants and of accessibility on the part of the penal institution(s) to be involved characterizes the conduct of research.

(b) A sponsor that seeks approval of any clinical investigation that is regulated by the Food and Drug Administration under section 505(i), 507(d), or 520(g) of the act as well as any clinical investigation that supports a research or marketing permit as defined by section 50.3(b) shall present evidence to the institutional review board establishing that the proposed research meets the requirements of paragraph (a) (1), (2), or (3) of this section.

(c) Authorized representatives of the Food and Drug Administration may inspect at reasonable times and in a reasonable manner, any prison at which a research activity has been proposed or is being conducted, to assist the institutional review board in determining whether the requirements of this section are met.

(d) The institutional review board shall determine whether the requirements of this section have been met, and shall notify the sponsor of the decision to approve or disapprove the proposed research activity. If the institutional review board disapproves a proposed research activity, it shall include in its written notification to the research sponsor and the agency a statement of the reasons for the disapproval. The sponsor shall be given an opportunity to respond.

(e) Except as provided in paragraph (a) (1), (2), or (3) of this section, any clinical investigation regulated by the Food and Drug Administration under section 505(i), 507(d), or 520(g) of the act, as well as any clinical investigation that supports an application for a research marketing permit as defined by section 50.3(b), may not involve prisoners as subjects.

Sec. 50.46 Composition of institutional review boards where prisoners are involved.

In addition to satisfying any other requirements governing institutional review boards set forth in this chapter, an institutional review board, in carrying out responsibilities under this part with respect to research covered by this subpart, shall also meet the following specific requirements:

(a) A majority of the institutional review board (exclusive of prisoner members) may not be associated with the prison(s) involved, apart from their membership on the institutional review board.

(b) At least one member of the institutional review board shall be a prisoner, or a prisoner advocate with appropriate background and experience to serve in that capacity, except that if a particular research project is reviewed by more than one institutional review board only one institutional review board need satisfy this requirement.

Sec. 50.48 Additional duties of the institutional review boards where prisoners are involved.

(a) In addition to all other responsibilities prescribed for institutional review boards under this chapter, the institutional review board shall review each clinical investigation covered by this subpart and approve such clinical investigation only if it finds that:

(1) The research under review represents one of the categories of research permitted under Sec. 50.44(a) (1), (2), or (3);

(2) Any possible advantages accruing to the prisoner through his or her participation in the clinical investigation, when compared to the general living conditions, medical care, quality of food, amenities, and opportunity for earnings in prison, are not of such magnitude that his or her ability to weigh the risks of the clinical investigation against the value of such advantages in the limited-choice environment of the prison is impaired;

(3) The risks involved in the clinical investigation are commensurate with risks that would be accepted by nonprisoner volunteers;

(4) Procedures for the selection of subjects within the prison are fair to all prisoners and immune from arbitrary intervention by prison authorities or prisoners; unless the principal investigator provides to the institutional review board justification in writing for following some other procedures, control subjects must be selected randomly from the group of available prisoners who meet the characteristics needed for that research project;

(5) Any information given to subjects is presented in language which is appropriate for the subject population;

(6) Adequate assurance exists that parole boards will not take into account a prisoner's participation in the clinical investigation in making decisions regarding parole, and each prisoner is clearly informed in advance that participation in the clinical investigation will have no effect on his or her parole; and

(b) The institutional review board shall carry out such other duties as may be assigned by the Food and Drug Administration.

(c) The institution shall certify to the Food and Drug Administration, in such form and manner as the Food and Drug Administration may require, that the duties of the institutional review board under this section have been fulfilled.

CHAPTER 9

Fetuses, Pregnant Women, and in Vitro *Fertilization*

Perhaps one of the most emotionally charged topics involved in the protection of human subjects is the matter of pregnant women and fetuses and the related issue of *in vitro* fertilization, since it relates tangentially to numerous other behavioral, biomedical, and ethical questions. For example, although once existing only within the realm of science fiction, it now appears technically possible for scientists to conduct genetic engineering to the extent that selected traits of individuals could be encouraged. At a minimum, it is already possible, through amniocentesis, to identify the sex of an unborn child by removing a small amount of amniotic fluid from the pregnant women's uterus. The very possibility of such a technique worries those who fear that prospective parents may elect for abortion rather than give birth to a child of a nondesired sex.[1]

For our purposes, what must be remembered is that this is a clear instance in which research (i.e., on amniocentesis) has raised the possibility of moral and legal dilemmas, since it can lead to the question of abortion. Thus a scientific development has posed an ethical and judicial dilemma, depending on how the technique is used.

In keeping with the intent of Part II chapters of this book, we will only briefly highlight this controversial area, looking at research with fetuses, pregnant women, and *in vitro* fertilization.

Fetuses

Research involving therapeutic or nontherapeutic techniques on fetuses has posed a host of difficult questions. As a potential research subject, a fetus changes

[1]Baby research draws warning (UPI). In *Omaha World-Herald*, Monday, January 7, 1980, p. 8.

so substantially over time (e.g., from the first trimester of pregnancy to just before birth), that its very status as a human being for research purposes must be questioned. Depending on one's religious or philosophical convictions, a fetus may or may not be considered a human being in the fullest sense of the word at any given moment. Therefore, if an experimental technique might increase the likelihood of a later healthy life for the developing fetus, does a physician (or parent) have the right to consent to such treatment when the subject (i.e., the fetus) is incapable of assenting? On the other hand, do responsible persons actually have an obligation to consent to such procedures?

These questions are further complicated by the possibility that information derived from biomedical research procedures may lead to a decision to abort the fetus. This means that information from research permitted, or at least indirectly supported, a decision to (a) abort a creature not yet fully human or (b) kill a child not yet born, depending upon an individual's belief. Obviously, the outcome of such a decision is far more serious than worrying about someone learning about a subject's personality, as might happen in a research project in which an adult completes a psychological test.

Research involving fetuses is further complicated by other considerations. Since the fetus is not an independently functioning organism, does the mother's consent for research always prevail? What about the consent of the father, assuming the father is a responsible party? If the fetus has been aborted, either as a natural miscarriage or by any other means, can research be performed following expiration of the fetus? Should such research be permitted if it can lead to medical advances that might prevent such tragedies and permit treatment of fetuses to the point at which living a normal life later could be possible?

Research involving a fetus is perhaps the most complex topic within this area, both because of the moral and legal questions raised and because a fetus is a particular type of subject that is intrinsically and wholly dependent on others for life itself. A series of reports have been written for the federal government to help in construction of research rules. These rules are appended to this chapter, but a brief look at some portions of the reports should aid the reader for this chapter.

One of the earliest reports was by the National Commission for the Protection of Human Subjects of Biomedical and Behavioral Research in May of 1975.[2] One of the issues examined in this report was possible alternative means to achieve the same knowledge without conducting research on a fetus. The reader may remember that in the matter of research with children as discussed Chapter 6, the Commission recommended that alternative means be attempted where feasible before involving children as research subjects. Where children

[2]*Research on the Fetus: Report and Recommendations*, 1978. Prepared by the National Commission for the Protection of Human Subjects of Biomedical and Behavioral Research. In the *Federal Register*, Vol. 40, No. 154, Friday, August 8, 1975, pp. 33530–33551.

were to be involved in research, the rules require that, when possible, the research should be tried with animals first, adult humans second, older youths third, and finally with children. In the hope of exploring alternatives to fetal research as well, the National Commission contracted with Battelle Columbus Laboratories to analyze the cases in which significant medical advances were obtained through human fetal research.

The Battelle study traced the historical development of rubella vaccine; the use of amniocentesis for prenatal diagnosis of genetic defects; diagnosis, treatment, and prevention of Rh isoimmunization disease; and the management of respiratory distress syndrome.[3] In general, the study concluded that while not all studies used animals first, to have done so would have significantly delayed medical advances that demonstrably saved the lives of thousands of children. Tragically, it was found that although animal research indicated that rubella vaccine did not cross the placenta to harm animal fetuses (leading to its use on humans, where fetuses were harmed), human research should have been conducted. In one striking example of the financial cost of medical research, the study noted that the total funds used for research from 1930 to 1966 to develop a vaccine (Rho Gam—which has reduced the annual number of stillbirths from 10,000 to less than 5,000 for Rh disease) is equal to the present cost to society for the lifetime care of *six* children born with Rh disease-related brain damage.[4] And how can one even put a pricetag on the lives of children and families, now normal, because of the research done in the past?

Information such as this led to the lifting of the moratorium that had been placed on fetal research by HEW the year before, in 1974.[5] However, the area is still one of controversy. For example, when the rules went into effect in 1975, it was permissible for *ex utero* (outside the uterus) fetuses to be the subject of research even if they were deemed nonviable (would not live) if the research would provide knowledge to enable physicians to be better able to help such infants survive in the future. This was permissible even if artificial means were necessary to maintain the vital signs of such nonviable fetuses. This was permitted because "the Secretary (of HEW) is persuaded by the weight of scientific evidence that research performed on the nonviable fetus *ex utero* has contributed substantially to the ability of physicians to bring to viability increasingly small fetuses."[6]

In such instances, HEW presumed that "it is expected that no procedures will be undertaken which fail to treat the fetus with due care and dignity, or which affront community sensibilities."[7] Clearly, the intent was to conduct

[3]Ibid., pp. 33534–33536.
[4]Ibid., p. 33535.
[5]*Federal Register*, Vol. 39, August 27, 1974, p. 30962.
[6]*Federal Register*, August 8, 1975, p. 33528.
[7]Ibid.

research that could lead to life-saving techniques for the future of such fetuses. Nevertheless, this research was with dying fetuses and, as we will see, has been forbidden if artificial maintenance of the nonviable fetus's vital signs is conducted.

Pregnant Women

Judging both by the small number of pertinent rules and to the limited discussion in various reports on pregnant women as subjects, one must assume that pregnant women as a special class of subjects do not pose many new problems. The section (46.207) devoted to them in the rules is brief and essentially states that a pregnant woman can serve as a research subject if the need is for her health (and risk to the fetus is only as much as is needed for her health), or the risk to the fetus is minimal.

The general policy statement that precedes the rules (Sec. 46.102) also implies that institutional review boards should determine whether any proposed research even *might* pose a risk to a fetus *if* a female subject *might* become pregnant during the course of any type of research project.

Finally, although both mother's and father's consent are typically required to permit research on a fetus, the rules specify some conditions in which the pregnant women's consent will suffice alone:

> 46.207(b) . . . except that the father's informed consent need not be secured if: (1) the purpose of the activity is to meet the health needs of the mother; (2) his identity or whereabouts cannot reasonably be ascertained; (3) he is not reasonably available; or (4) the pregnancy resulted from rape.[8]

There are some other federal agencies that restrict or forbid research with pregnant or lactating women.[9] Obviously, women are intimately involved with fetal research and research on *in vitro* fertilization. Indeed, the interest shown by women in *in vitro* fertilization has been one of society's supports for continuing development of the technique. This can be seen from a report about the nation's first "test-tube" baby laboratory, approved in 1980 to be built at Norfolk General Hospital in conjunction with the Eastern Virginia Medical School.[10] The clinic had a waiting list of 2,500 women applying for the procedure even before the state had approved construction of the laboratory. With an estimated 560,000 infertile women in the United States, the demand for such a technique (and

[8]Ibid., p. 33529.

[9]*First Biennial Report on Protecting Human Subjects* by the President's Commission for the Study of Ethical Problems in Medicine and Biomedical and Behavioral Research. Washington, D.C.: U.S. Government Printing Office, 1981.

[10]Test-tube baby clinic swamped (AP). In *Omaha World-Herald*, Friday, January 11, 1980, p. 8.

associated research) is obviously great and will probably increase. But research on *in vitro* fertilization has remained controversial.

In Vitro Fertilization

In July, 1978, Louise Brown was born in England as the first publicly noted test tube baby in the world. This breakthrough came from the cumulative work of a number of researchers, but Steptoe and Edwards are generally credited with the significant achievement.[11]

One of the more thorough treatments of this issue was compiled by the Ethics Advisory Board (EAB) for HEW.[12] We have mentioned this board earlier in this book; the reader will remember that it was formed within HEW to advise the Secretary on various matters pertaining to protection of human research subjects. The Ethics Advisory Board and its functions are so closely allied to this area of protection of fetuses and pregnant women that a specific section of the regulations (46.204) is devoted to it.[13]

The report by the EAB dealt with a number of aspects of *in vitro* fertilization, including various types of fertilization and issues surrounding research in this area.

In Vitro *Fertilization and Embryo Transfer*

There are three primary techniques subsumed in this topic. First is *in vitro* fertilization without subsequent transfer of the embryo to the uterus of the female. Second, there is *in vitro* fertilization followed by actual embryo transfer to a uterus. Finally, there is embryo transfer from a uterus (*in utero*) following fertilization by mating or artificial insemination. Only the first two techniques have been used with human subjects.

For *in vitro* fertilization to take place, ova are obtained from the female through a technique called laparoscopy. This involves insertion of a needle through the woman's navel to the region of her ovaries where ova are removed. Placed in a laboratory medium and combined with male sperm, the resultant fertilized egg is grown in a second medium, then transferred to the uterus of a

[11]Biggers, J. D. *In Vitro Fertilization, Embryo Culture and Embryo Transfer in the Human*. Report prepared for the Ethics Advisory Committee, DHEW, September, 1978. Available from the Laboratory of Human Reproduction and Reproductive Biology, Harvard Medical School, 45 Shattuck Street, Boston, Masschusetts 02115.

[12]*Report and Conclusions: HEW Support of Research Involving Human In Vitro Fertilization and Embryo Transfer*. Prepared by the Ethics Advisory Board, May, 1979. Washington, D.C.: U.S. Government Printing Office.

[13]*Federal Register*, August 8, 1975, p. 33529.

female. Successful transfer leads to implantation in the uterus and normal development of the fetus until birth at term. Of course, this transfer can be to a uterus of someone other than the original donor, which further complicates the ethical issues involved.

There can be no question that this fertilization in a laboratory is considered repugnant by some, just as there can be no doubt that it holds promise for couples who otherwise would have no hope that the woman could bear a child that would truly be their genetic descendent.

It is also true that there are risks involved in the experimental procedure. It has been reported that four pregnancies have now been announced by Edwards and Steptoe following this procedure,[14] but this was out of 32 attempts. Two of the pregnancies later resulted in spontaneous abortions, or miscarriages. Of course, by the time this volume is published, these figures will have changed.

Obviously then, the fertilized embryo has been placed at risk in this endeavor by a procedure that, in its absence, would not even have witnessed the beginnings of an embryo. The mother too is placed at risk. But one conclusion reached in the EAB reports is that these processes are not at all dissimilar to the events (i.e., miscarriages and health risks for expectant mothers) that natural pregnancies also produce. In fact, the rate of fetal loss in natural insemination is much higher. That the artificial procedure can be done has been true since the late 1970s, but whether it *should* be done still poses a host of problems. Among these problems are legal complications of such activities.

Legal Complications

The 1979 EAB report summarized the conclusions of two papers regarding the legal issues surrounding human *in vitro* fertilizations[15]; one was by Dennis Flannery and colleagues of the Washington firm of Wilmer, Cutler and Pickering, and another report was by Barbara Katz, Office of Legal Affairs, University of Colorado Medical Center. Four topics were examined and are highlighted briefly here: (1) existing law (as of 1979) that might apply to *in vitro* fertilization and subsequent embryo transfer, (2) constitutional issues, (3) implications for tort liability for injuries, and (4) criminal law.

As far as law is concerned, there were no state laws or regulations then in effect that were applicable to *in vitro* fertilization, although 16 states had laws regarding research with live human fetuses. There were some examples from case law regarding the legitimacy of children conceived as a result of artificial insemination, and the trend was to declare such children as legitimate. Legit-

[14]*Report and Conclusions: HEW Support of Research Involving Human in Vitro Fertilization and Embryo Transfer*, 1979, p. 9.
[15]Ibid., pp. 60–80.

imacy has implications for inheritance rights and for claims for paternal support. The only federal rules in existence were the ones promulgated by HEW.

The constitutional issues involved are complicated, but essentially married couples might successfully contest government intervention to their efforts to have a child in such a manner by reference to a basic right to privacy and procreation within a marriage union. On the other hand, single persons using this approach might have a weaker case in objecting to government regulations or interference with using *in vitro* fertilization to become pregnant. With regard to research, it is likely that no constitutional *right* to conduct such research could be successfully defended, although such research could certainly be permitted.

The third legal area, that of tort liability, revealed that the United States (as an entity) would probably be liable for any negligence of HEW (now HHS) officers or employees who designed or conducted a research program involving *in vitro* fertilization. However, individual investigators, whose work was supported by HHS funds, would probably not be liable unless they violated federal regulations in the conduct of the research.

Finally, with regard to criminal law, it did not seem that allowing preimplantation human embryos (the result of *in vitro* fertilization but not yet implanted in a human uterus) to die would constitute the crime of "feticide," a form of homicide. However, state statutes governing fetal research may be worded so broadly that they *could* be applied in such cases. But "the law in this area is confused, at best."[16]

Regulatory History

The National Research Act of 1974 (PL 93–348) placed a moratorium on research with fetuses, which was then confirmed by HEW in August, 1974.[17] However, following the report on the subject by the National Commission, HEW lifted the moratorium in August of 1975 as it proposed special rules for this type of research population. These rules became the foundation for the rules used for the rest of the 1970s.

1975

In August of 1975, HEW published rules for research with fetuses, pregnant women, and *in vitro* fertilization.[18] These rules remain, changed only minimally, as the current rules. In November, 1975, a minor wording change to Sec. 46.206 extended the types of research allowable to include those wherein risk to

[16]Ibid., pp. 76–77.

[17]*Federal Register*, Vol. 39, August 27, 1974, p. 30962.

[18]*Federal Register*, Vol. 40, No. 154, Friday, August 8, 1975, pp. 33526–33552.

the fetus was greater than minimal if it was necessary to meet the health needs of the mother.[19]

1978

In early 1977, some amendments were published,[20] primarily on the definitions of *fetus* and *pregnancy*. More important, however, these amendments *deleted* the earlier (1975) provisions that would have permitted artificial maintenance of the vital functions of nonviable fetuses when the purpose of the research was to develop new methods for enabling fetuses to survive to viability. Despite this research aim, this type of research is now forbidden, since it raises the spectre of artificially maintaining a fetus that has no chance of surviving solely for the purposes of research. These amendments became final in early 1978.[21] These latter changes can be seen in the current wording for Sec. 46.209.

SUMMARY

We have seen that, above all, this area of protection of research subjects is a relatively new and confusing one. The topic of research with fetuses, whether they be *in utero* or *ex utero*, is emotionally charged and is at least indirectly linked with abortion in some instances.

Because of the unusual degree of helplessness of the subject population (i.e., the fetus), the types of research allowed must be directly linked to the health of the mother, the fetus, or both, or the risk to the fetus must be minimal. As we have also seen, research with a nonviable (i.e., dying) fetus is in general forbidden except in certain circumstances. Such research is not even allowed when the purpose is to develop biomedical knowledge that might help other such fetuses if, as part of the research, vital functions of the nonviable fetus are maintained through artificial means.

Pregnant women, as a separate class of research subjects, must be warned about any aspect of research that could harm them or their unborn child. Indeed, IRBs are encouraged to make certain that researchers warn female research subjects who *might* become pregnant if there are any features of the project which could pose risk to a fetus.

Finally, *in vitro* fertilization with humans is so new that the research facets of the technique have yet to be explored on a wide scale. There can be no doubt that thousands of women appear to be willing to risk the procedure as their sole

[19]*Federal Register*, Vol. 40, No. 215, Thursday, November 6, 1975, p. 51638.
[20]*Federal Register*, Vol. 42, January 13, 1977, p. 2792.
[21]*Federal Register*, Vol. 43, January 11, 1978, pp. 1758–1759.

hope of conceiving children who are genetically the offspring of them and their respective husbands. The current HHS rules regulate research in such situations in two respects. First, HHS's Ethical Advisory Board must approve any project involving such a technique if HHS funds are involved. Second, since an embryo so developed would be at least temporarily *ex utero*, the rules for such fetal research would also apply for any research in which the resultant embryo is the actual subject of research.

Fetuses, as a research population, pose a unique problem since authorities in various fields appear to disagree on when the fetus is fully human and thus due all the rights and privileges under applicable law, regulations, and custom.

APPENDIX 9.1

Regulations of the Department of Health and Human Services for Research with Fetuses, Pregnant Women, and In Vitro *Fertilization*

[Note: These rules are taken from the *Federal Register*, Vol. 40, No. 154, Friday, August 8, 1975, pp. 33528–33530; as amended in the *Federal Register* by Vols. 42—January 13, 1977—and 43—January 11, 1978.]

Subpart B—Additional Protections Pertaining to Research, Development, and Related Activities Involving Fetuses, Pregnant Women, and Human *In Vitro* Fertilization.

Sec. 46.201 Applicability.

(a) The regulations in this subpart are applicable to all Department of Health, Education, and Welfare grants and contracts supporting research, development, and related activities involving: (1) The fetus, (2) pregnant women, and (3) human *in vitro* fertilization.

(b) Nothing in this subpart shall be construed as indicating that compliance with the procedures set forth herein will in any way render inapplicable pertinent State or local laws bearing upon activities covered by this subpart.

(c) The requirements of this subpart are in addition to those imposed under the other subparts of this part.

Sec. 46.202 Purpose.

It is the purpose of this subpart to provide additional safeguards in reviewing activities to which this subpart is applicable to assure that they conform to appropriate ethical standards and relate to important societal needs.

Sec. 46.203 Definitions.

As used in this subpart:

(a) "Secretary" means the Secretary of Health, Education, and Welfare and any other officer or employee of the Department of Health, Education, and Welfare to whom authority has been delegated.

(b) "Pregnancy" encompasses the period of time from confirmation of implantation (through any of the presumptive signs of pregnancy, such as missed menses, or by a medically acceptable pregnancy test), until expulsion or extraction of the fetus.

(c) "Fetus" means the product of conception from the time of implantation (as evidenced by any of the presumptive signs of pregnancy, such as missed menses, or by a medically acceptable pregnancy test), until a determination is made, following expulsion or extraction of the fetus, that it is viable.

(d) "Viable" as it pertains to the fetus means being able, after either spontaneous or induced delivery, to survive (given the benefit of available medical therapy) to the point of independently maintaining heart beat and respiration. The Secretary may from time to time, taking into account medical advances, publish in the *Federal Register* guidelines to assist in determining whether a fetus is viable for purposes of this subpart. If a fetus is viable after delivery, it is a premature infant.

(e) "Nonviable fetus" means a fetus *ex utero* which, although living, is not viable.

(f) "Dead fetus" means a fetus *ex utero* which exhibits neither heartbeat, spontaneous respiratory activity, spontaneous movement of voluntary muscles, nor pulsation of the umbilical cord (if still attached).

(g) "*In vitro* fertilization" means any fertilization of human ova which occurs outside the body of a female, either through admixture of donor human spern and ova or by any other means.

Sec. 46.204 Ethical Advisory Boards.

(a) One or more Ethical Advisory Boards shall be established by the Secretary. Members of these board(s) shall be so selected that the board(s) will be competent to deal with medical, legal, social, ethical, and related issues and may include, for example, research scientists, physicians, psychologists, sociologists, educators, lawyers, and ethicists, as well as representatives of the general public. No board member may be a regular, full-time employee of the Department of Health and Human Services.

(b) At the request of the Secretary, the Ethical Advisory Board shall render advice consistent with the policies and requirements of this Part as to ethical issues, involving activities covered by this subpart, raised by individual applications or proposals. In addition, upon request by the Secretary, the Board shall render advice as to classes of applications or proposals and general policies, guidelines, and procedures.

(c) A Board may establish with the approval of the Secretary, classes of applications or proposals which: (1) Must be submitted to the Board, or (2) need not be submitted to the Board. Where the Board so establishes a class of applications or proposals which must be submitted, no application or proposal within the class may be funded by the Department or any component thereof until the application or proposal has been reviewed by the Board and the Board has rendered advice as to its acceptability from an ethical standpoint.

(d) No application or proposal involving human *in vitro* fertilization may be funded by the Department or any component thereof until the application or proposal has been reviewed by the Ethical Advisory Board and the Board has rendered advice as to its acceptability from an ethical standpoint.

Sec. 46.205 Additional duties of the Institutional Review Boards in connection with activities involving fetuses, pregnant women, or human *in vitro* fertilization.

(a) In addition to the responsibilities prescribed for Institutional Review Boards under Subpart A of this part, the applicant's or offeror's Board shall, with respect to activities covered by this subpart, carry out the following additional duties:

(1) Determine that all aspects of the activity meet the requirements of this subpart;

(2) Determine that adequate consideration has been given to the manner in which potential subjects will be selected, and adequate provision has been made by the applicant or offeror for monitoring the actual informed consent process (e.g., through such mechanisms, when appropriate, as participation by the Institutional Review Board or subject advocates in: (i) Overseeing the actual process by which individual consents required by this subpart are secured either by approving induction of each individual into the activity or verifying, perhaps through sampling, that approved procedures for induction of individuals into the activity are being followed, and (ii) monitoring the progress of the activity and intervening as necessary through such steps as visits to the activity site and continuing evaluation to determine if any unanticipated risks have arisen);

(3) Carry out such other responsibilities as may be assigned by the Secretary.

(b) No award may be issued until the applicant or offeror has certified to the Secretary that the Institutional Review Board has made the determinations required under paragraph (a) of this section and the Secretary has approved these determinations, as provided in Sec. 46.115 of Subpart A of this part.

(c) Applicants or offerors seeking support for activities covered by this subpart must provide for the designation of an Institutional Review Board, subject to approval by the Secretary, where no such board has been established under Subpart A of this part.

Sec. 46.206 General limitations.

(a) No activity to which this subpart is applicable may be undertaken unless:

(1) Appropriate studies on animals and nonpregnant individuals have been completed;

(2) Except where the purpose of the activity is to meet the health needs of the mother or the particular fetus, the risk to the fetus is minimal and, in all cases, is the least possible risk for achieving the objectives of the activity;

(3) Individuals engaged in the activity will have no part in: (i) Any decisions as to the timing, method, and procedures used to terminate the pregnancy, and (ii) determining the viability of the fetus at the termination of the pregnancy; and

(4) No procedural changes which may cause greater than minimal risk to the fetus or the pregnant woman will be introduced into the procedure for terminating the pregnancy solely in the interest of the activity.

(b) No inducements, monetary or otherwise, may be offered to terminate pregnancy for purposes of the activity.

Sec. 46.207 Activities directed toward pregnant women as subjects.

(a) No pregnant woman may be involved as a subject in an activity covered by this subpart unless: (1) The purpose of the activity is to meet the health needs of the mother and the fetus will be placed at risk only to the minimum extent necessary to meet such needs, or (2) the risk to the fetus is minimal.

(b) An activity permitted under paragraph (a) of this section may be conducted only if the mother and father are legally competent and have given their informed consent after having been fully informed regarding possible impact on the fetus, except that the father's informed consent need not be secured if: (1) The purpose of the activity is to meet the health needs of the mother; (2) his identity or whereabouts cannot reasonably be ascertained; (3) he is not reasonably available; or (4) the pregnancy resulted from rape.

Sec. 46.208 Activities directed toward fetuses *in utero* as subjects.

(a) No fetus *in utero* may be involved as a subject in any activity covered by this subpart unless: (1) The purpose of the activity is to meet the health needs of the particular fetus and the fetus will be placed at risk only to the minimum extent necessary to meet such needs, or (2) the risk to the fetus imposed by the research is minimal and the purpose of the activity is the development of important biomedical knowledge which cannot be obtained by other means.

(b) An activity permitted under paragraph (a) of this section may be conducted only if the mother and father are legally competent and have given their informed consent, except that the father's consent need not be secured if: (1) his identity or whereabouts cannot reasonably be ascertained, (2) he is not reasonably available, or (3) the pregnancy resulted from rape.

Sec. 46.209 Activities directed toward fetuses *ex utero*, including nonviable fetuses, as subjects.

(a) Until it has been ascertained whether or not a fetus *ex utero* is viable, a fetus *ex utero* may not be involved as a subject in an activity covered by this subpart unless:

(1) There will be no added risk to the fetus resulting from the activity, and the purpose of the activity is the development of important biomedical knowledge which cannot be obtained by other means, or

(2) The purpose of the activity is to enhance the possibility of survival of the particular fetus to the point of viability.

(b) No nonviable fetus may be involved as a subject in an activity covered by this subpart unless:

(1) Vital functions of the fetus will not be artificially maintained.

(2) Experimental activities which of themselves would terminate the heartbeat or respiration of the fetus will not be employed, and

(3) The purpose of the activity is the development of important biomedical knowledge which cannot be obtained by other means.

(c) In the event the fetus *ex utero* is found to be viable, it may be included as a subject in the activity only to the extent permitted by and in accordance with the requirements of other subparts of this part.

(d) An activity permitted under paragraph (a) or (b) of this section may be conducted only if the mother and father are legally competent and have given their informed consent, except that the father's informed consent need not be secured if: (1) his identity or whereabouts cannot reasonably be ascertained, (2) he is not reasonably available, or (3) the pregnancy resulted from rape.

Sec. 46.210 Activities involving the dead fetus, fetal material, or the placenta.

Activities involving the dead fetus, mascerated fetal material, or cells, tissue, or organs excised from a dead fetus shall be conducted only in accordance with any applicable State or local laws regarding such activities.

Sec. 46.211 Modification or waiver of specific requirements.

Upon the request of an applicant or offeror (with the approval of its Institutional Review Board), the Secretary may modify or waive specific requirements of this subpart, with the approval of the Ethical Advisory Board after such opportunity for public comment as the Ethical Advisory Board considers appropriate in the particular instance. In making such decisions, the Secretary will consider whether the risks to the subject are so outweighed by the sum of the benefits to the subject and the importance of the knowledge to be gained as to warrant such modification or waiver and that such benefits cannot be gained except through a modification or waiver. Any such modifications or waivers will be published as notices in the *Federal Register*.

CHAPTER 10

Persons Institutionalized as Mentally Disabled

Among the populations to be studied by the National Commission for the Protection of Human Subjects of Biomedical and Behavioral Research was that described as the institutionalized mentally infirm. Accordingly, as for other special subject populations described previously, the National Commission followed its usual series of contracted reports, public meetings and testimony, and commission debate to arrive at a set of proposed recommendations for HEW. In turn, HEW developed a set of proposed rules governing research with this population. Due at least in part to the freeze on government regulations, these proposed rules remain the ones for us to heed at present. It is not currently known whether these rules will be revised or made final by HHS in view of federal policy to limit regulatory activities and reduce government paperwork. In the author's opinion, the proposed rules that do exist, as they appear appended to this chapter, should still serve as a useful guide for researchers in this area. Unless unusually substantial changes are made in the proposed rules, final rules (when published) should draw heavily on the proposed rules as appended to this chapter. Once again, the cautious researcher should consult federal agencies (e.g., the NIH Office for Protection from Research Risks) for current rules for this area still in a state of limbo.

Regulatory History

Briefly, the development of these regulations can be traced as follows. In February of 1978, the National Commission released a 125-page report[1] on

[1] *Report and Recommendations: Research Involving Those Institutionalized as Mentally Infirm*, 1978. Prepared by the National Commission for the Protection of Human Subjects of Biomedical

research with this population. Acccompanying the report was a lengthy appendix[2] that contained copies of reports and other materials used by the National Commission in their deliberations. Among these appendix materials was a survey conducted by the Survey Research Center at the University of Michigan under the direction of Tannenbaum and Cooke.[3] This survey provided the commission with data from interviews with 151 researchers who worked with this population and from subjects or proxies. Thus the commission, as it did with other subject groups, drew upon actual experience to construct recommendations on future such research.

In March, 1978, HEW then published the full text of the report (minus the appendix) in the *Federal Register*,[4] seeking public comment before developing actual HEW rules based on the National Commission's recommendations.

Finally, following comments from about 100 organizations, institutions, legal and medical practitioners, and private citizens, HEW published the proposed rules[5] that remain our best guide for research in this area. Based on the number of responses HEW received, it can be concluded that human service researchers and practitioners were doing a better job of keeping abreast of federal rule-making activities by 1978 than had been true in the mid-1970s. This increased awareness has been aided by such publications as Reed Martin's *Law and Behavior*, which published a summary of these rules in 1978 to alert human service professionals to these developments with rules for the mentally infirm.[6]

Definition of Mentally Infirm

Before we proceed with the highlights of these rules, let us examine the nature of the subject population. Although the original mandate to the National Commission was to develop recommendations for the "institutionalized mentally infirm," the commission itself noted that the term *mentally infirm* was not

---

and Behavioral Research. Washington, D.C.: U.S. Government Printing Office, DHEW Publication No. (OS) 78–0006.

[2]*Appendix to Research Involving Those Institutionalized as Mentally Infirm*, 1978. Prepared by the National Commission for the Protection of Human Subjects of Biomedical and Behavioral Research. Washington, D.C.: U.S. Government Printing Office, DHEW Publication No. (OS) 78–007.

[3]Tannenbaum, A. S., and Cooke, R. A. *Report on the Mentally Infirm*. Survey Research Center, University of Michigan; Contract No. N01–HU–6–2110 from the National Commission as contained in the *Appendix* (see Note 2).

[4]*Federal Register*, Vol. 43, No. 53, Friday, March 17, 1978, pp. 11328–11358.

[5]*Federal Register*, Vol. 43, No. 223, Friday, November 17, 1978, pp. 53950–53956.

[6]Martin, R. Regulation: Protection of human subjects. *Law and Behavior*, 3(3), Summer, 1978, pp. 3–8. Champaign, Illinois: Research Press.

in clinical use.[7] Further, the label suggests adherence to a medical or disease model, although the behavioral school of thought places more emphasis on learning and environmental factors as causes for such dysfunction.

Thus, rather than supporting any one treatment theory, and to allow for the fact that some persons are institutionalized through misdiagnosis or other error, HEW originally changed the category for this type of population to persons who are "institutionalized *as* mentally infirm." This phrase, therefore, refers only to a *process*, not a definition of what kind of person is involved.

Finally, the proposed rules published in the *Federal Register* in November, 1978, had the phrasing further modified to define persons "institutionalized as mentally disabled."[8] The reader should note that since current proposed rules thus are categorized as for persons "institutionalized as mentally disabled," the author has designated this chapter under such a heading rather than for persons "mentally infirm," on the basis of historical precedent.

Mentally disabled is currently defined in the proposed rules as:

> (c) "Mentally disabled" individuals includes those who are mentally ill, mentally retarded, emotionally disturbed, psychotic or senile, regardless of their legal status or the reason for their being institutionalized.[9]

Community-based Settings

Although the reader can examine the proposed rules appended at the end of this chapter for definitions of other terms for this population (e.g., *minimal risk, assent, consent auditor*), one aspect should be highlighted here. Due to the increase in community-based alternatives to institutionalization, the proposed rules have been expanded to cover this population if they reside in a variety of settings, not just in large institutions. The following definition from the proposed rules makes this extension of the applicability of the regulations quite clear:

> (d) "Individuals institutionalized as mentally disabled" means individuals residing, whether by voluntary admission or involuntary confinement, in institutions for the care and treatment of the mentally disabled. Such individuals include, but are not limited to patients in public or private mental hospitals, psychiatric patients in general hospitals, inpatients of community mental health centers, and mentally disabled individuals who reside in halfway houses or nursing homes.[10]

---

[7] *Report and Recommendations: Research Involving Those Institutionalized as Mentally Infirm*, 1978, p. xvii.

[8] *Federal Register*, November 17, 1978, p. 53954.

[9] Ibid.

[10] Ibid.

Consent Conditions

As noted in a previous chapter, children are considered by final HHS rules often to be incapable of granting informed consent, although they can grant assent if a responsible adult concurs and gives adult permission. Presuming a similar lack of capacity to furnish fully informed consent on the part of many mentally disabled persons, proposed HHS rules for research participation require additional consent procedures prior to involvement in research.

Like the final rules for research with children, the proposed rules for research with the mentally disabled distinguish between assent and consent, allowing for a responsible *legally authorized person* to grant consent for the mentally disabled to participate. In addition, however, the proposed rules for the mentally disabled also require, under certain conditions, that an *advocate* and/or a *consent auditor* be appointed by the IRB further to protect this population. Actual definitions for these roles can be found in the appended proposed rules; but, in brief, the consent auditor ensures the adequacy of the entire consent process and the advocate represents the best interests of the subject in general. Neither is supposed to have financial interests in, or association with, the institution where the research is being conducted.

Obviously, HHS is concerned with this population's ability to provide fully informed consent. The reader should realize that this population represents a uniquely vulnerable class of subjects. Combined in this group are all the problems of diminished capacity to consent (e.g., like the class of children) and the restrictions on normal freedoms (e.g., like the class of prisoners). Judging by the detail in the proposed rules, researchers would be well advised to work especially closely with their IRB when involved with subjects who are mentally disabled.

Risk Factors

One final area, that of levels of risk, underscores the concern over involving this particular subject population in research. Without going into the detail which can be obtained from the appended proposed rules, it is noteworthy that there are actually four types of research (according to level of risk) that are regulated by the proposed HHS rules.

First, there is the type of research that does not involve greater than minimal risk. In general, this type of research requires the least number of extra precautions and is permitted even if the subject actually objects to participation and is not really capable of providing informed consent. Various conditions must be met in such cases (e.g., the research holds out the prospect of direct benefit to the subject or is needed to monitor the well-being of the subject), and someone other than the subject—such as his or her representative—can provide informed consent for the subject to participate. See Section 46.505 of the appended proposed rules for more details.

Second, there is research that presents greater-than-minimal risk, but also presents the prospect of direct benefit to the subject. The conditions to be met in these instances are similar to those necessary for less-than-minimal-risk research (e.g., consent of another person if the subject objects), with the added distinction of different proposed rules, depending on whether the subject is an adult or a child. See Section 46.506 of the appended proposed rules for more details.

Third, research is allowed if it involves greater-than-minimal risk yet provides no prospect of direct benefit to individual subjects if the research is likely to yield generalizable knowledge about the subjects' disorder or condition. However, there are a number of conditions placed on such research by the proposed federal rules, including the fact that "the risk represents a minor increase over minimal risk."[11] See Section 46.507 of the appended proposed rules for more details of this type of research allowed with persons institutionalized as mentally disabled.

Finally, there is perhaps the most difficult (in the sense of ease of obtaining HHS approval) research categoried as:

> Research not otherwise approvable which presents an opportunity to understand, prevent, or alleviate a serious problem affecting the health or welfare of individuals institutionalized as mentally disabled.[12]

This type of research requires, among other conditions, that it first be approved by the Secretary of HHS following his or her consultation with a panel of experts from various fields. Further, there must be an opportunity for *public review* and comment on the individual proposed research project. Obviously, unless there was little urgency to the research and the investigators could afford to go through such steps before applying for federal research funds, this type of research can be expected to form a minority of research projects involving persons institutionalized as mentally disabled.

Summary

We have seen that although originally referred to as the "institutionalized mentally infirm," this special subject population has been more appropriately renamed as "persons institutionalized as mentally disabled." This category covers several types of disorders or problems (such as mental retardation) and the term *institutionalized* refers to residential placement in halfway houses as well as large institutions.

[11]Ibid., p. 53956.
[12]Ibid.

The strictures placed on research are perhaps more numerous for this population, when contrasted to others. The author has speculated that this may arise from the fact that this population often combines the lack of consent capacity of children with the restricted freedom of prisoners.

Still, the proposed rules do allow for a variety of types of research involving persons who are institutionalized as mentally disabled. Allowable research falls into four separate categories, essentially distinct according to the level of risk posed for the subject. As was true for children, an IRB must see that a *legally authorized representative* consents for research with this population if the subject is incapable of providing informed consent. Further, a *consent auditor* is to be appointed by the IRB to ensure the overall adequacy of the consent process used by the researcher.

The appended rules were still not final at the time of printing but remain the most useful guide available for researchers in this area. The subsequent chapter provides the reader with a variety of ways to stay abreast of changes in rules such as these.

APPENDIX 10.1

Proposed Regulations by the Department of Health and Human Services for Research Involving Persons Institutionalized as Mentally Disabled

[Note: The following proposed rules were taken from the *Federal Register*, Vol. 43, No. 223, Friday, November 17, 1978, pp. 53954–53956.]

Subpart E—Additional Protections Pertaining to Biomedical and Behavioral Research Involving as Subjects Individuals Institutionalized as Mentally Disabled.

Sec.46.501 Applicability.

(a) The regulations in this subpart are applicable to all biomedical and behavioral research conducted or supported by the Department of Health, Education, and Welfare involving as subjects individuals institutionalized as mentally disabled.

(b) Nothing in this subpart shall be construed as indicating that commpliance with the procedures set forth herein will in any way render inapplicable pertinent State or local laws bearing upon activities covered by this subpart.

(c) The requirements of this subpart are in addition to those imposed under the other subparts of this part.

Sec. 46.502 Purpose.

Individuals institutionalized as mentally disabled are confined in institutional settings in which their freedom and rights are potentially subject to limitation. In addition, because of their impairment they may be unable to comprehend sufficient information to give a truly informed consent. Also, in some cases they may be legally incompetent to consent to their own participation in research.

At the samc time, so little is known about the factors that cause mental disability that efforts to prevent and treat such disabilities are in the primitive stages. There is widespread uncertainty regarding the nature of the disabilities, the proper identification of persons who are disabled, the appropriate treatment of such persons, and the best approaches to their daily care. The need for research is clearly manifest. It is the purpose of this subpart to permit the conduct of responsible investigations while providing additional safeguards for those institutionalized as mentally disabled.

Sec. 46.503 Definitions.

As used in this subpart:

(a) "Secretary" means the Secretary of Health, Education, and Welfare and any other officer or employee of the Department of Health, Education, and Welfare to whom authority has been delegated.

(b) "DHEW" means the Department of Health, Education, and Welfare.

(c) "Mentally disabled" individuals includes those who are mentally ill, mentally retarded, emotionally disturbed, psychotic or senile, regardless of their legal status or the reason for their being institutionalized.

(d) "Individuals institutionalized as mentally disabled" means individuals residing, whether by voluntary admission or involuntary confinement, in institutions for the care and treatment of the mentally disabled. Such individuals include, but are not limited to patients in public or private mental hospitals, psychiatric patients in general hospitals, inpatients of community mental health centers, and mentally disabled individuals who reside in halfway houses or nursing homes.

(e) "Children" are persons who have not attained the legal age of consent to general medical care as determined under the applicable law of the jurisdiction in which the research will be conducted.

(f) "Parent" means a child's natural or adoptive parent.

(g) "Legally authorized representative" means an individual or judicial or other body authorized under applicable law to consent on behalf of a prospective subject to such subject's participation in the particular activity or procedure. An official serving in an institutional capacity may not be considered a legally authorized representative for purposes of this subpart.

(h) "Minimal risk" is the probability and magnitude of physical or psychological harm or discomfort that is normally encountered in the daily lives, or in the routine medical or psychological examination, of normal individuals.

(i) "Assent" means a prospective subject's affirmative agreement to participate in research. Mere failure to object shall not, absent affirmative agreement, be construed as assent. Assent can only be given following an explanation, based on the types of information specified in Sec. 46.103(c), appropriate to the level of understanding of the subject, in accordance with procedures established by the Institutional Review Board.

(j) "Consent auditor" means a person appointed by the Institutional Review Board to ensure the adequacy of the consent process, particularly when there is a substantial question about the ability of a subject to consent or assent or when there is a significant degree of risk involved. Consent auditors are responsible only to the Board and should not be involved with the research, nor should they be employed by or otherwise associated with the institution conducting or sponsoring the research, or with the institution in which the subject resides. They should be persons familiar with the physical, psychological, and social needs of the class of prospective subjects as well as their legal status.

(k) "Advocate" means an individual appointed by the Institutional Review Board to act in the best interests of the subject. The advocate will, although he or she is not appointed by a court, be construed to carry the fiduciary responsibilities of a guardian ad litem toward the person whose interests the advocate represents. No individual may serve as an advocate if the individual has any financial interest in, or other association with, the institution conducting or sponsoring the research, nor with the institution in which this research is conducted; nor, where the subject is the ward of a State or other agency, institution, or entity, may the advocate have any financial interest in, or other association with, that State, agency, institution, or entity. An advocate must be familiar with the physical, psychological, and social needs and the legal status of the class of individuals institutionalized as mentally disabled in the institution in which the research is conducted. (This definition will be retained in the final regulations if duties are assigned to "advocates.")

Sec. 46.504 Additional duties of the institutional review boards where individuals institutionalized as mentally disabled are involved.

(a) In addition to all other responsibilities prescribed for Institutional Review Boards under this part, the Board shall review research covered by this subpart and approve such research only if it finds that:

(1) The research methods are appropriate to the objectives of the research;

(2) The competence of the investigator(s) and the quality of the research facility are sufficient for the conduct of the research;

(3) Appropriate studies in nonhuman systems have been conducted prior to the involvement of human subjects;

(4) There are good reasons to involve institutionalized individuals as subjects of the research. In reviewing proposals to involve institutionalized persons in research, the Board should evaluate the appropriateness of involving alternative, noninstitutionalized populations in the study instead of, or along with, the institutionalized individuals. Sometimes, the participation of alternative populations will not be possible or relevant, as when the research is designed to study problems or functions that have no parallel in free-living persons, (e.g., studies of the effects of institutionalization or studies related to persons, such as the profoundly retarded or severely handicapped, who are almost always found in residential facilities);

(5) Risk of harm or discomfort is minimized by using the safest procedures consistent with sound research design and by using procedures performed for the diagnosis or treatment of the particular subject whenever possible;

(6) Adequate provisions are made to protect the privacy of the subjects and to maintain confidentiality of data. For example, data may be disclosed to authorized personnel and used for authorized purposes only; data should be collected only if they are relevant and necessary for the purposes of the research and analysis; data should be maintained only as long as they are necessary to the research or to benefit the subjects; and all data should be maintained in accordance with fair information practices;

(7) Selection of subjects among those institutionalized as mentally disabled will be equitable. Subjects in an institution should be selected so that the burdens of research do not fall disproportionately on those who are least able to consent or assent, nor should one group of patients be offered opportunities to participate in research from which they may derive benefit to the unfair exclusion of other equally suitable groups of patients;

(8) Adequate provisions are made to assure that no prospective subject will be

approached to participate in the research unless the health care professional who is responsible for the health care of the subject has determined that the invitation to participate in the research and the participation itself will not interfere with the health care of the subject;

(9) The Board shall appoint a consent auditor to ensure the adequacy of the consent procedures when, in the opinion of the Board, such a person is considered necessary, e.g., when there is a substantial question about the ability of a subject to consent or to assent or when there is a significant degree of risk involved; and

(In the event the Department decides that there should be consent auditors for all projects, the above paragraph will be appropriately modified.)

(10) The conditions of all applicable subsequent sections of this subpart are met.

(b) The Board shall carry out such other duties as may be assigned by the Secretary.

(c) The institution shall certify to the Secretary, in such manner as the Secretary may require, that the duties of the Board under this subpart have been fulfilled.

Sec. 46.505 Research not involving greater than minimal risk.

Biomedical and behavioral research that does not involve greater than minimal risk to subjects who are institutionalized as mentally disabled may be conducted or supported by DHEW provided the Institutional Review Board has determined that:

(a) The conditions of Sec. 46.504 are met; and

(b) Adequate provisions are made to assure that no subject will participate in the research unless:

(1) The subject gives informed consent to participation;

(2) If the subject lacks the capacity to give informed consent, the research is relevant to the subject's condition, the subject assents or does not object to participation, and the subject's legally authorized representative consents to the subject's participation; or

(3) If a subject, who lacks the capacity to give informed consent, objects to participation:

- (i) The research includes an intervention that holds out the prospect of direct benefit to the subject, or includes a monitoring procedure required for the well-being of the subject,
- (ii) the subject's legally authorized representative consents to the subject's participation, and
- (iii) the subject's participation is authorized by a court of competent jurisdiction.

(Consideration is being given to mandating that, in addition to the above requirements: (1) A "consent auditor" be appointed by the Institutional Review Board to ensure the adequacy of the consent process and determine whether each subject consents, or is incapable of consent but assents, or objects to participation, and (2) whenever the consent auditor determines that a subject is incapable of consenting, the subject may not participate without the authorization of an "advocate.")

Sec. 46.506 Research involving greater than minimal risk but presenting the prospect of direct benefit to the individual subjects.

(a) Biomedical and behavioral research in which more than minimal risk to subjects who are institutionalized as mentally disabled is presented by an intervention that holds out the prospect of direct benefit for the individual subjects, or by a monitoring procedure likely to contribute to the well-being of the subjects, may be conducted or supported provided the Institutional Review Board has determined that:

(1) The conditions of section 46.504 are met;

(2) The risk is justified by the prospect of benefit to the subjects;

(3) The relation of the risk to anticipated benefit to subjects is at least as favorable as that presented by available alternative approaches;

(4) Adequate provisions are made to assure that no adult will participate in the research unless:

(i) The subject gives informed consent to participation;

(ii) If a subject who lacks the capacity to give informed consent, the subject assents to participation, and the subject's legally authorized representative consents to the subject's participation; or

(iii) If a subject who lacks the capacity to give informed consent, does not assent, or objects to participation: (A) The intervention or monitoring procedure is only available in the context of the research, (B) the subject's legally authorized representative consents to the subject's participation, and (C) the subject's participation is authorized by a court of competent jurisdiction.

(Consideration is being given to mandating that, in addition to the above requirements: (1) A "consent auditor" be appointed by the Institutional Review Board to ensure the adequacy of the consent process and determine whether each subject consents, or is incapable of consent but assents or objects to participation, and (2) whenever the consent auditor determines that a subject is incapable of consenting, the subject may not participate without the authorization of an "advocate.")

(5) Adequate provisions are made to assure that no child will participate in the research unless:

(i) The subject assents (if capable) and the subject's parent(s) or guardian(s) give permission, as provided in section 46.409 of this part; or

(ii) If the subject objects to participation, the intervention or monitoring procedure is available only in the context of the research, the subject's parent(s) or guardian(s) give permission, and the subject's participation is authorized by a court of competent jurisdiction.

(b) Where appropriate, the Institutional Review Board shall appoint a consent auditor to ensure the adequacy of the consent process and determine whether each subject consents, or is incapable of consent but assents, or objects to participation. (This paragraph will be deleted if a consent auditor is required in all cases.)

Sec. 46.507. Research involving greater than minimal risk and no prospect of direct benefit to individual subjects, but likely to yield generalizable knowledge about the subject's disorder or condition.

(a) Biomedical and behavioral research in which more than minimal risk to subjects who are institutionalized as mentally disabled is presented by an intervention that does not hold out the prospect of direct benefit for the individual subjects, or by a monitoring procedure that is not likely to contribute to the well-being of the subjects, may be conducted or supported provided an Institutional Review Board has determined that:

(1) The conditions of section 46.504 are met;

(2) The risk represents a minor increase over minimal risk;

(3) The anticipated knowledge (i) is of vital importance for the understanding or amelioration of the type of disorder or condition of the subjects, or (ii) may reasonably be expected to benefit the subjects in the future;

(4) Adequate provisions are made to assure that no adult will participate in the research unless the following conditions are met:

(i) The subject gives informed consent to participation;
(ii) If the subject lacks the capacity to give informed consent, the subject assents to participation, and the subject's legally authorized representative consents to the subject's participation; or
(iii) If the subject lacks the capacity to assent but does not object, the subject's legally authorized representative and a court of competent jurisdiction consent to the subject's participation.

(The Department is considering the following additions to the above provisions:

In Sec. 46.507(a)(4)(B), with respect to subjects capable of consenting: (i) Adding the requirement that inclusion of each subject be approved by the Secretary based upon the advice of a panel of experts, or (ii) requiring the approval of an "advocate."

In Sec. 46.507(a)(4)(C), with respect to subjects incapable of assenting: (i) Prohibiting use of such subjects on the theory that there is no research which can be performed only with these subjects, (ii) requiring approval by the Secretary based upon the advice of a panel of experts, or (iii) requiring the approval of an "advocate.")

(5) If the subject is a child, the requirements of Sections 46.407 and 409 of subpart D (relating to research involving children) are satisfied.

(b) No subject may be involved in the research over his or her objection.

(c) The Institutional Review Board shall appoint a consent auditor to ensure the adequacy of the consent process and determine whether each subject consents, or is incapable of consenting but assents, or is incapable of consenting but does not object, or objects to participation. (This paragraph will be deleted if a consent auditor is required for all research covered by this subpart.)

Sec. 46.508 Research not otherwise approvable which presents an opportunity to understand, prevent, or alleviate a serious problem affecting the health or welfare of individuals institutionalized as mentally disabled.

Biomedical and behavioral research that the Institutional Review Board does not believe meets the requirements of Sections 46.505, 46.506, or 46.507 may nevertheless be conducted or supported by DHEW provided:

(a) The Institutional Review Board has determined the following:

(1) The conditions of Sec. 46.504 are met; and

(2) The research presents a reasonable opportunity to further the understanding, prevention, or alleviation of a serious problem affecting the health or welfare of individuals institutionalized as mentally disabled; and

(b) The Secretary, after consultation with a panel of experts in pertinent disciplines (e.g., science, medicine, education, ethics, law) and following opportunity for public review and comment, has determined either (1) that the research in fact satisfies the conditions of Sections 46.505, 46.506, or Section 46.507, as applicable, or (2) the following:

(i) The research presents a reasonable opportunity to further the understanding, prevention, or alleviation of a serious problem affecting the health or welfare of individuals institutionalized as mentally disabled;

(ii) The conduct of the research will be in accord with basic ethical principles of beneficence, justice, and respect for persons, that should underlie the conduct of research involving human subjects; and

(iii) Adequate provisions are made for obtaining consent of those subjects capable of giving fully informed consent, the assent of other subjects and the consent of their legally authorized representatives, and, where appropriate, the authorization of a court of competent jurisdiction (and if Sections 46.505, 46.506, and 46.507 require an advocate, the authorization of that advocate).

PART III

Keeping up with Changing Rules

Part III consists of a single chapter by James Sweetland, M.L.S., Ph.D. A former colleague from Boys Town, Dr. Sweetland has since been responsible for directing libraries in Louisiana and Wisconsin. A specialty interest of his is reference work in various fields, especially human services. His efforts at tracing the often labyrinthine paths of government regulations have been responsible for locating many of the federal rules presented in Parts I and II of this book.

The next chapter provides numerous hints to the professional on how to keep up to date with changing federal regulations. These hints should be especially useful to readers who lack ready access to an attorney who specializes in government regulations. In addition, suggestions are provided for the more active reader who might wish to influence the political processes that eventually produce the federal rules which can so substantially affect our professional lives. While no one particular political stance is adopted, techniques are presented to enable the reader—of whatever political persuasion or professional association—to become *involved*.

Such involvement on the part of human service professionals could go a long ways toward ensuring that rules designed to protect the public reflect the views of researchers and practitioners as well as others.

CHAPTER 11

How to Be Informed and Involved

James Sweetland

This chapter explains some of the complexities of government and describes how an interested citizen can find out what the government is doing. It does not claim to give legal advice, nor to replace lawyers or law librarians, even in the rather narrow area of protection of human subjects. However, armed with the information contained, one should be able to make sense of government agencies' activities, take personal action to affect them, and know enough to ask reasonably informed questions about the processes.

First, a number of terms must be defined. In dealing with the government, one finds both *laws* and *regulations*. The former is that which results from an action of an elected legislature. This includes *statutes*, as well as treaties and constitutions. Lawyers often call this body of material *conventional legislation*.

However, laws may not be clear, or they may be too general, or they may need detailed application to a number of different situations. Thus there is also a body of *subordinate* or *delegated* legislation, in which the lawmakers delegate a specific agency to take the detailed action needed by the law. This sort of thing can include both delegated legislation proper and *inherent authority*. In the latter, to use a common example, the Constitution authorizes the government to coin money. Specific statutes set the value of the coinage and the denominations of the money. A specific agency, the Mint, actually produces the money, chooses the designers, and so forth. The principle of delegated authority legislation especially applies to the administrative and regulatory agencies set up since the late nineteenth century, especially during the New Deal. The rules issued by these agencies are what are called *regulations*.

How a Bill Becomes a Law

Let us first examine the process by which an idea eventually becomes a law, keeping in mind that this textbook example is not always followed in every detail.

The primary intent of this section is to give a brief, general idea of the process, noting especially those actions which produce a public document that can be used by a citizen to affect the process.

Introduction of a Bill

The first step is the introduction of the bill. This may result from a campaign promise, a suggestion of the president, expressed needs of a pressure group, or an investigation of a congressional committee. Proposed laws may also be in the form of a *joint resolution* (passed by both houses of Congress under slightly different rules than a law), but the procedures for the two are similar and both result in the same type of product. In any event, the bill is introduced formally in Congress, referred to a committee, and printed for public distribution.

Bills are numbered for easy reference, with a prefix giving the house of origin, and a serial number beginning with each session of Congress. Thus a bill introduced in the House begins with "H.R." while one in the Senate starts with "S." Joint resolutions begin with either the "H" or "S" followed by a "J.Res." This number serves to identify the bill forever. Normally, but not always, the introduction of a bill will be noted, at least by title, in the *Congressional Record.*[1]

The following example assumes a bill introduced in the House of Representatives, although the Senate process is quite similar. Once introduced, the bill is given to a specific committee. Usually, a subcommittee of this committee actually considers the bill, if it takes any action at all. In fact, most bills are never taken up and die in committee. Even a bill of extreme interest to a particular pressure group might thus disappear, except for a summary in the "Daily Digest" section of the *Congressional Record.* In the minority of cases, the subcommittee discusses the bill. Occasionally, hearings will be held, at which expert witnesses or other interested parties may be called to testify and support or attack the bill. All such hearings are announced, usually one week in advance, in the "Daily Digest."

Although a few hearings may be broadcast, most are actually not very public. The law requires that hearings be open to the public, but in practice this means only that people on the scene be permitted to attend. Normally, hearings are not even printed, and the only formal record is a typescript of the proceedings kept in the office of the clerk of the committee in Washington. The relatively few published hearings are listed in the *Monthly Catalog of Government Publications* (*Monthly Catalog*),[2] and sent to most depository libraries.

After the hearings, the committee meets again, normally in public, and

[1]United States Congress. *Congressional Record: Proceedings and Debates of the Congress.* V. 1—, 1873—. Washington, D.C.: United States Government Printing Office.

[2]*Monthly Catalog of United States Government Publications.* Washington, D.C.: United States Government Printing Office, 1951—.

marks up the bill. In this process the original proposal is modified, amended, and otherwise changed to meet the committee's desires. When this step is completed, the committee issues a report on the bill back to the full house, including both the new version of the bill and explanations of any changes made in the original. These reports are extremely valuable in determining what went on in the committee and what is actually meant by the wording of the bill. Should the law later come to court, these reports are often cited by lawyers as evidence of what the law is supposed to mean and as important evidence are one of the major parts of a legislative history.

Reports are given a number for easy reference. Since 1969, this has been a serial number prefixed by the number of the Congress and the initial of the House of Congress. For example, "H. Rep. 94–6" is the sixth committee report to have been issued by a House of Representatives committee in the 94th Congress. All such reports are printed and listed in the *Monthly Catalog*.

Debate in Congress

Once the bill has been reported, several actions can result, again with slight differences between the House and the Senate. Assuming no one objects, the bill may merely pass. Of reported bills, about nine out of ten pass in this manner. If there is not complete agreement, the bill is put on the calendar in order of its reporting, where it remains until its turn for full debate comes up. Since this may take months, a number of ways to beat the calendar have been evolved, all requiring action of the Rules Committee to bend the rules. This factor explains the extreme power of the chairperson of the Rules Committee.

One way or another, the bill may come before the whole house for debate. The formal record of debate is printed in the *Congressional Record*, sometimes with a full text of the bill. Theoretically, of course, the *Congressional Record* is supposed to be a complete verbatim transcript of the debate. As is well known, however, any member of Congress may extend or edit their remarks before the document is printed. This leads to situations in which stirring speeches which were never made appear in the debate. However, one can assume that the printed remarks at least indicate the views of the Congressperson. It should also be remembered that the "debate" may be nothing more than a reading of the bill, with no further discussion.

After this reading of the bill, with or without the debate, amendments may be added and in their turn debated. Finally, the bill as amended is voted upon, usually without a roll call. Even with the roll call (usually reserved for important or controversial bills) it is not always possible to tell how a person voted. In addition to a list of those voting for, against, and "present," there is often a list of "pairs." These result from an agreement between two people on opposite sides of the bill, who will not be present for the voting. In effect, they cancel out each other's vote. In the process, members can choose not to reveal on which side they would have voted.

Assuming the vote was favorable, the bill is now called an *act*. It is printed as such, but still with its original bill number, and sent to the other house. Here, the entire process is repeated, complete with debates, reports, and possible hearings.

Given this procedure, it is possible for each house to come up with rather different versions of the original. When this happens, both houses send managers to a conference committee. This *ad hoc* group works out the differences in the two versions and eventually makes a report with copies sent to both houses.

Once an agreeable version has been produced, both houses take a final vote—again with or without debate. If the act has passed, it is formally enrolled (copies with all changes and additions properly made, doublechecked and printed) and sent to the president for signature.

Presidential Action

The president has several basic choices. He can sign the act, at which point it becomes law (although it may not take effect until a specific date). In signing, he may or may not make any comments or may use the occasion for a full-fledged speech. Or he can refuse to sign the bill, in which case it becomes law in ten days without his signature. If Congress adjourns before the ten days are up, the bill is dead. This is the "pocket veto."

The fourth option is the formal veto. Here, the president refuses to sign the act, and returns it to Congress with his reasons for the action. Congress can vote again to override the veto, but most vetoed bills are dead.

The Bill Becomes Law

Having successfully passed all these hurdles, the act now becomes a law and is printed in pamphlet form, called a "slip law." At this point, it is given a public law number consisting of the initials "P.L." and a serial number composed of the number of Congress and the number of bills passed in the session.

With the law-making process over, the law-printing process begins in earnest. In addition to the slip laws, more permanent versions are also prepared. At the end of each session of Congress, all the laws are collected and published with editorial apparatus in the *Statutes at Large*.[3] This serial is the official permanent record of all laws passed in each Congress. The editorial apparatus includes

[3]United States. *United States Statutes at Large*. v. 1—, 1798–1845—. Washington: United States Government Printing Office.

references to all the various permutations of the bill, and notes in the margin referring to the *United States Code.*[4]

The *United States Code* is a compilation of all the laws of the United States. Unlike the *Statutes at Large,* which is arranged by Congress, the *United States Code* is arranged by subject matter in fifty titles. Each title represents the state of the law at the time it was published. The whole code is reprinted in full every six years and supplemented at the end of each Congress by a cumulative set of changes for each title. Although technically speaking the *Statutes* are the official record of the law, most lawyers usually refer to the *Code.*

It must be remembered that this outline can have many practical variations. Sometimes, identical bills are introduced in both houses at the same time, to speed up the process. Occasionally, several bills on the same general subject are introduced at about the same time and are collected by the relevant committee for hearings and consolidation into a single bill. Often, if there is no objection, a given step can be eliminated. However, every law passed will have been, in succession, a *bill* (with a bill number), a *public law* (with a public law number), a section in the *Statutes at Large,* and a part of a section in the *United States Code.*

How a Regulation Becomes a Regulation

Regulations, unfortunately, are much more difficult to discuss than laws. The actual details vary among agencies, but all are alike in that they are less regular and generate much less documentation than do laws.

Prior to the New Deal, there was no single place where anyone could find all regulations in force; not even the agencies who had made them. This situation was remedied as a result of the "hot oil" case, in which a company was charged with violation of a regulation of which it was unaware and which turned out not to have been in force when it was charged with the violation.

Currently, all regulations are governed by the Administrative Procedure Act (Title 5 of the *U.S. Code*) as modified by several executive orders, notably order number 12044. Executive orders are numbered sequentially from the first one issued and continue the numbering from one president to the next. Each major part of each government agency must issue a semiannual agenda of topics on which it might make regulations. The next step is to publish proposed rules and seek public comment. Then, as the agency actually holds hearings and meetings on proposed regulations, it must make them open to the public and issue timely notices of such meetings. Generally, few people attend these hearings, other

[4]United States. *United States Code.* Congress, House, Office of the Law Revision Counsel, 1940—.

than the officials involved and occasional lobbyists for pressure groups. After these steps, and modification of proposed rules as a result of public comments, an agency adopts the final formal rules.

The Federal Register

The basic course for following regulation making is the *Federal Register*,[5] which appears each business day. In addition to the text of regulations, it is the official alerting source for hearings and the agendas of agencies that develop regulations.

After listing the proposed regulation on an agenda and after due deliberation by a committee, with more or less input from the public, an agency will issue a proposed regulation. This normally appears in the *Federal Register* at least 60 days before it is to go into effect. This gives time for the public to comment on the rules; because these comments often lead to changes and corrections, this longer period is a great improvement over the former 30-day notice period. The announcement includes the text of the rule, usually some explanation of the text, and full information on whom to contact for more information or to make comments.

Having received comments, the agency then issues a final rule, again in the *Federal Register*. Final publication normally includes not only the revised text, but a rather long explanation of it, part by part, giving reasons for each part. If comments have been received, they are summarized, with the reasons why the agency has or has not followed the suggestions. Naturally, this final announcement also includes the date on which the regulation becomes effective and a reference to the relevant part of the *Code of Federal Regulations*.[6]

Code of Federal Regulations

The *Code of Federal Regulations* (abbreviated CFR) is published annually and is analogous to the *U.S. Code*, described previously for laws. It also is arranged in subject sections and represents the state of regulations on the subject as of the cover date. Different parts are updated at different times during the year, on a regular schedule.

Looseleaf Services

Unfortunately, regulations tend to change very rapidly. Since the *CFR* is issued only once a year, it is necessary to check the *Federal Register* for the

[5]*Federal Register*. v. 1—, 1936—. Washington, D.C.: Office of the Federal Register, National Archives and Records Service, General Services Administration.

[6]*Code of Federal Regulations*. Washington, D.C.: Office of the Federal Register, National Archives and Records Service, General Services Administration.

current state of a regulation. This process can be very time-consuming, even with the number of finding aids provided by the government. A number of legal publishers have thus produced their own updates to the *CFR*, complete with finding aids in particular subjects, especially business questions and taxes. Each of these gives basic regulations and updates the basic text by issuing new pages to replace those no longer current. These "looseleaf services" (so-called from their issuance in punched pages to be put in binders) are normally quite expensive, and their use can be somewhat complicated, even with the detailed instructions given by the publishers. Currently, there is no such service aimed directly at the subject of protection of clients or research subjects, but there is one on the general topic of freedom of information. This biweekly is called *Access*[7] and includes the basic laws and regulations with samples of more general relevant law. The service also includes a newsletter.

How Laws and Regulations Are Cited

In this section we discuss how to find a given law or regulation, assuming it is already in force. As we have seen, a law can have several different designators in its official guides, among which are citations to the *Statutes at Large*, to the *United States Code*, and to the public law number. Additionally, it will always have a formal name, and often a popular one. And, it may also have a *truly* popular name, as opposed to its official popular name. It is thus not unusual for an information seeker to approach the library with three or four varying citations, all of which turn out to refer to the same document. To confuse the issue further, it is also common for people in a given field, notably education, to refer to a part of a law by a special name in addition to all those listed above.

To give an example of the phenomenon, we can use the previously discussed Buckley Amendment. This is section 122g of Title 20 of the *U.S. Code* (cited 20 USC 1232g). It is also Public Law 93–380, as amended by PL 93–568, which is the same as saying that it is the section beginning in volume 88 of the *Statutes at Large* on page 571 (i.e., 88 Stat. 571) as amended by 88 Stat. 1858. Its official name is the *Education Amendments of 1974*, Title V, as amended by *Joint Resolution* (*S.J. Res. 40*) "To authorize and request the President to call a White House Conference on Library and Information Services not later than 1978, and for other purposes." Its official popular name is the *Family Educational Rights and Privacy Act of 1974*, often called *FERPA* (pronounced "ferpah"), and its true popular name is the Buckley Amendment.

[7]*Access Reports/FOI Newsletter* (Including Access Reference File). Washington, D.C.: Washington Monitor, Inc., v. 1—, 1975—.

In this example it is worth reminding the reader that the original amendment was quickly amended again by a "rider" on an almost wholly unrelated bill regarding libraries.

Fortunately, both the government and private publishers have produced a number of finding aids which permit people to find what they are actually looking for, even if the citations in the source they have are different from that for which they are looking. On the assumption that the reader is still willing to pursue the subject, a brief description of the principles of legal citation follow. For those interested in more detail, standard legal research texts go into much more detail.

Anyone familiar with higher education in any of its forms is familiar with one or more of the several style sheets for citing, including the *Publication Manual*[8] of the American Psychological Association, Kate Turabian's *Manual for Writers*,[9] the *Modern Language Association Handbook*,[10] and others. The law, on the other hand, has essentially only one basic form, making use of citations, even with their abbreviations, much simpler. Undoubtedly, the reader welcomes this news. Most legal texts have lists of the most common abbreviations, which need not be repeated here.

Legal citation differs from that used in the social sciences in a few ways. Normally, the volume comes first, then the title abbreviation, then the section or page number, as is evident from the tour through the Buckley Amendment.

Since there may be a number of compilations for laws, it is common for citations of some kinds to refer to section or title numbers rather than pages. Thus references to the *U.S. Code* (or *U.S.C.*, or *Code*), giving sections and titles, may be used with any of the several versions of the *Code*. The style of citation will be familiar to anyone who has ever used a standard edition of Shakespeare's plays or Plato's dialogues and is not much different from "*Hamlet*, Act IV, Scene I."

Finding Federal Laws

Armed with this information, you can now find the text of a law, given some information about it. Normally, you are likely to have some version of the title. With luck, this is all you need. The *U.S. Code* itself contains a list of popular names of laws. Should this fail, or be unavailable, you can use *Shepard's*

[8]American Psychological Association. *Publication Manual* (3rd ed.). Washington, D.C.: American Psychological Association, 1983.

[9]Turabian, K. A *Manual for Writers of Term Papers, Theses, and Dissertations* (4th ed.). Chicago: University of Chicago Press, 1973.

[10]M.L.A. *Handbook for Writers of Research Papers, Theses, and Dissertations*. New York: Modern Language Association, 1977.

Acts and Cases by Popular Name,[11] which includes both case and statutory law. Either of these will give the citation both to the *Code* and the *Statutes at Large*. This use of more than one access point to the same text is known as *parallel citation*.

Once you have the citation, you have several sources for the full text, depending on your need. The *Statutes*, remember, give the text of a particular law as it was passed. The *Code* gives the text at a particular time after it was passed. Typically, people seeking legal assistance want to know the current state of the law and will use the *U.S. Code*. Historians and political scientists, among others, often want the text of the original. The Buckley Amendment provides a useful example here: since it was amended soon after passage, reference to the *Code* of 1982 will not give the original version; reference to the original *Statutes* will not give the version being enforced in 1982.

However, since the *U.S. Code* is revised only once every six years, you must also check the cumulative supplement issued by each Congress. Given the citation by title and section, this is a simple matter of checking the basic volume and then the supplement under the same numbers. However, the law may have changed during the current session, and still further references may be necessary. In addition to changes in the law as such, it is also possible that it has been modified by court action, or even voided. Connections among the various court reporters with laws passed can be rather complicated, and is beyond the scope of this chapter.

Fortunately, there are some ways of doing much of this work without having to consult a lawyer. Two commercial publishers have issued annotated versions of the law, *The U.S. Code Annotated*[12] and the *United States Code Service*.[13] These republish the law with the addition of case notes and other interpretive information. Because of this added matter, one or the other of these compilations, rather than the official *U.S. Code*, is likely to be found in libraries and law offices.

One useful feature of these services, and many other legal publications, is the "pocket part." Each volume has a pocket in the back cover which is designed to hold a pamphlet of some size. From time to time, cumulative revisions for a given part of the code are issued in such a pamphlet, which is inserted in the pocket, the previous one being discarded. With one volume in hand, the user also has all updates except the most recent material. Remember, with the *U.S. Code* citation you can find the law in any of the three versions, including the

[11]*Shepard's Acts and Cases by Popular Names, Federal and State*. Colorado Springs, Colo.: Shepard's; McGraw-Hill, 1968—.

[12]United States. *United States Code Annotated*. St. Paul, Minn.: West Publishing, 1927—.

[13]United States. *United States Code Service*. Rochester, New York: Lawyers Cooperative Publishing Co., 1972—.

pocket parts. Technically, of course, only the official version is acceptable law, but the publishers take great pains for accuracy and can usually be relied upon. In any event, if you are actually involved in litigation, you should retain an attorney.

There still remains the difficulty of keeping up with a particular session of Congress while it is still in session. This is readily handled by the *United States Code Congressional and Administrative News*,[14] produced by the publisher of the *United States Code Annotated*. Among other things, it indexes the recently passed slip laws and comes out regularly during the session. It does not give the full text of all laws but does provide the PL number and several subject indexes and cross reference tables to the *U.S. Code*. Basically, this service can be used as a current version of the *Code Annotated*, and in fact much of the information is annually codified in those volumes.

Thus far, we have assumed that you have the name of the law. Sometimes, however, you have the citation. If it is either the *Code* or the *Statutes*, the process is simple. The various versions of the *Code* have finding tables for translating one citation into the other. The *Statutes at Large* give the *Code* citations. In the same way, the *Code* has tables translating PL numbers into both *Code* and *Statutes* citations. Keep in mind also that the slip law gives this information as side notes.

Essentially, the principle is simple: with any of the three citations to the law, you can translate into all other citations. With the codified version you can go forward from passage of the law using supplements or pocket parts and then the *U.S. Code Congressional and Administrative News* for the most recent material.

A slightly more difficult situation is the popular name which does not appear to be indexed. Fortunately, the sources named previously, as well as others, give *subject* access as well.

One warning: try to get as much general information as possible. Remember, for example, that the public law numbering system begins anew with each session of Congress. Approaching a librarian or information service with a request for law number 37 will lead only to frustration. Along the same line, remember that different professions may use different popular names, to the point of developing their own jargon. In the early days of the law on sex discrimination, for example, hundreds of law and document librarians went quietly mad trying to find out which "Title nine" a user wanted, especially difficult in those cases wherein the user declined to mention that the law had to do with higher education sex discrimination. Since a large number of laws related to education

[14] *United States Code Congressional and Administrative News*. St. Paul, Minn.: West Publishing.

in the past 20 years have more than nine titles, as do many other laws, requests for title IX of "that education law" were not particularly helpful.

Thus far we have assumed that the law was passed. Often, however, especially in a volatile area like privacy, one cannot be sure of this. The searcher may want to know the form in which a bill is passed, or whether it did pass, or where it is in the process. Once again, there are a number of ways to get this information. It is extremely helpful to have a rough idea of the date, keeping in mind that bills are numbered successively with each new Congress. In addition, the name of the Congressman or Senator who introduced the bill is very useful.

If the bill was before the recent Congress, the job is easy. The final bound set of the *Congressional Record* contains a list of all bills for that session of Congress and their legislative history from the time of introduction. Here you will find the date the bill was introduced, when it was referred to committee, when debated, and the like. For more recent bills, a number of sources noted later in this chapter will give similar information.

Finding Federal Regulations

In many ways, finding a particular regulation is much easier than finding a law, since there are only two sources to use. *The Code of Federal Regulations* (CFR) is issued annually and thus is current in the same sense as the *U.S. Code*, but without the need for supplements or pocket parts. Rarely will anyone want the older text of regulations. If one should, larger research libraries will retain older copies of the *CFR* in either the original or microfilm. The *CFR* is kept current by the daily *Federal Register*, which also may be retained in larger libraries, providing the original text for researchers.

Since regulations rarely have popular names, the process of finding them is simple. Normally, the searcher has a citation to the *CFR*. If so, merely check the appropriate title and section for the state as of printing date. To update, merely check the *Cumulative List of Sections Afffected*[15] issued by the Office of the *Federal Register*. This is a separate pamphlet, arranged by *CFR* citations, giving pages in the current year's *Federal Register*. A similar table, updated monthly and cumulating during the month, is printed each day in the *Federal Register* itself. Subject access is by the general index to the *CFR*, or by the indexes provided in each title of that code. Better subject coverage is possible through the *Prentice-Hall Regulatory Week*,[16] a new looseleaf service providing summaries of new rules, with subject, *CFR*, and other indexes.

[15]*Code of Federal Regulations: LSA. List of CFR Sections Affected*. Washington, D.C.: Office of the Federal Register, National Archives and Records Service, General Services Administration. 1977—.

[16]*Prentice-Hall Regulatory Week*. Englewood Cliffs, N.J.: Prentice-Hall, 1979—.

Keeping Current

In addition to finding the text of a particular law, the researcher is also interested in keeping current. Fortunately, a number of sources, governmental and commercial, provide such information. This section deals with a number of them, discussing by source, rather than by type of information.

Congressional Record

The official record of events on the floor of Congress is, of course, the *Record*, produced daily while Congress is in session and indexed biweekly. Business transacted in the House and Senate is listed separately in chronological order. Use of this tool is rather straightforward, with the warning that the annual bound version is a completely new edition. Thus, for example, citations in current newspaper articles to the daily *Record* will not usually correspond with the annual.

Calendars

Since much of the important business of Congress is in committee meetings, which are not covered by the *Record*, other sources must be used. The basic one is the calendar[17] issued by each house, although that of the House of Representatives is usually referred to when an unspecified calendar is mentioned. This lists all forthcoming business for Congress, gives a brief legislative history, and lists approaching committee meetings. It is issued daily while Congress is in session and usually includes a subject index each Monday. One final edition covers the whole session, with the last one for the Congress covering all sessions.

Many committees also issue calendars of their own, which often include lists of committee publications as well. These latter calendars are not likely to be held in the average library, whereas the House of Senate calendars may be.

Federal Register

This is a primary source for regulations and includes, besides the text of regulations, executive orders and similar documents, and a daily list of upcom-

[17]United States, Congress, House. *Calendars of the United States House of Representatives and History of Legislation*. Washington, D.C.: Superintendent of Documents, United States Government Printing Office. Also see United States, Congress, Senate. *Calendar of Business*. Washington, D.C.: Superintendent of Documents, United States Government Printing Office. Many individual committees also publish calendars available through the Superintendent of Documents.

ing meetings of various agencies, commissions, and other bodies. The table of contents is arranged by agency, and includes a daily alerting of selected subjects on the front page of the issue, giving a sort of subject approach. Unfortunately, the subject terms are rather loose. Usually, "privacy" refers to material regarding the Privacy Act rather than personal privacy. Regulations and hearings on human rights tend to be listed under the type of group, such as "prisoners," and "medical experimentation." With the adoption of a standardized vocabulary in the near future, this situation may improve.

Another way to use the *Federal Register* to keep current is by the "List of Parts Affected." A quick glance at the daily (as opposed to the cumulative) list of parts will provide notice of regulations. Unfortunately, hearings and other such meetings are not covered by this list.

Commercial Services

For keeping up with events in government, commercial services tend to give better coverage. Usually these are faster and give better access points than do government publications. The major disadvantage is that they cost much more as well.

U.S. Code Congressional and Administrative News.[18] Commercially prepared, this monthly record not only gives finding aids for bills but also provides the full text of all laws passed and a selection of committee reports connected with the more important bills. The cross-indexing between bill number, law number, and various forms of citation is excellent and quite easy to use. The publication's major drawback is that is is only issued monthly.

Congressional Index.[19] This includes bills, resolutions, laws, and their legislative history, as well as voting records of individual members of Congress and a considerable amount of information about the bills. Access is by bill's author and by date. Included are lists of "companion bills" (similar bills introduced in both houses) and much else of use. This source is especially useful when the searcher knows only the bill's sponsor or its subject matter. The service is issued weekly and includes a separate newsletter of general materials, *The Week in Congress.*

Washington Monitor.[20] Perhaps the most complete coverage of events in Washington is provided by the Washington Monitor organization which issues a set of newsletters listing current and forthcoming events for the next two to three months. The *Congressional Monitor*[21] includes all committee and subcommittee hearings, both in Washington and in the field. Listings begin up to three

[18]*United States Code Congressional and Administrative News.* St. Paul, Minn.: West Publishing.
[19]*Congressional Index.* Chicago: Commerce Clearing House, 1937/38—.
[20]The Washington Monitor, Inc., 499 National Press Building, Washington, D.C. 20045.
[21]*The Congressional Monitor.* v. 1—, 1978—. Washington, D.C.: Washington Monitor, Inc.

months before the meeting. This source also includes a summary of actions on the floor of Congress arranged by bill number. The same material is available in a lower-priced weekly edition and a more expensive daily. With either service one also has access to a phone service for immediate questions. A similar source for regulations is the *Weekly Regulatory Monitor,*[22] which notes new agency rules and lists coming hearings, proposed rules, and deadlines for comments. Each notice includes a brief abstract of the rule or notice and a citation to the *Federal Register.* The two services together give good coverage of the official actions of the federal government. The publisher also offers seminars on the congressional legislative and regulatory processes.

Congressional Quarterly.[23] Another set of materials providing somewhat slower coverage over a somewhat narrower field but in greater depth is provided by this firm. The main publication is called the *Congressional Quarterly Service Weekly Report.*[24] Information in this journal is cumulated and issued annually as the *Congressional Quarterly Almanac,*[25] known as the *CQ Almanac.* Generally, in addition to brief notes of business in Congress, this service covers a few items in depth. A convenient feature is the weekly status table of major bills in Congress.

Federal Information Centers.[26] The telephone is extremely useful for keeping up with events in government. For those within a reasonable distance of Washington, there are dozens of likely numbers to call and a number of directories. For information about regulations, telephone numbers are listed in the back pages of the *Federal Register,* and for bills, those of one's own congressional representatives are most useful. A developing network of Federal Information Centers can also be of use, especially for those far from the capitol. These centers are staffed neither by professional lawyers nor librarians, but personal experience has shown that they are extremely courteous and willing to help. They are at their best when given the most information, and are especially good in providing the current status of identified bills. However, a word of caution: a request for a government hearing vaguely mentioned in last night's newspaper will take much longer for response than a request for the current state of bill number such-and-such introduced by Senator X last week.

Political Research, Inc.[27] Once again, commercial firms provide more ser-

[22]*Weekly Regulatory Monitor.* v. 1—, 1977—. Washington, D.C.: Washington Monitor, Inc.

[23]Congressional Quarterly, Inc., 1414 22nd Street, N.W., Washington, D.C. 20037.

[24]*Congressional Quarterly Service Weekly Report.* v. 1—, 1945—. Washington, D.C.: Congressional Quarterly Service.

[25]*Congressional Quarterly Almanac.* v. 1—, 1948—. Washington, D.C.: Congressional Quarterly Service.

[26]Normally the local number may be found in the telephone book under United States Government listings, as well as in the regular business section under Federal Information Center. A commercial directory of Washington numbers is *Researcher's Guide to Washington.* Washington: Washington Researchers, in annual editions.

[27]Political Research, Inc., 16850 Dallas Parkway, Dallas, Texas 75248.

vice, for a fee. Political Research publishes *Taylor's Encyclopedia of Government Officials.*[28] The book, often found in public and school libraries, is a brief glossy reference that comes complete with colored pictures. A subscription includes access to a toll-free telephone line staffed by professional information specialists able to give brief information, do some research, or obtain copies of documents even if they are given very little information. Often, the information will be available after a short wait on the telephone.

There are a number of similar services, including the previously discussed Washington Monitor, which provide document delivery and telephone information. The major advantage of Political Research is that its product is likely to be held by smaller libraries which do not subscribe to the others. Unfortunately, many of the staff of these libraries are unaware of the potential of the toll-free number. There are some limits (such as length of document available without extra charge), but the information is usually good and timely. The questions must be about either national or state governments and are currently taken only between 10:00 A.M. and 2:00 P.M., Texas time. This service also includes a monthly newsletter aimed at the high school market, which does help in keeping up on major events in Washington. One useful feature is the "Bureaucratic Breakdown," which gives biographies of major agency chiefs and brief descriptions of what they and their agencies do.

It has been noted that a number of firms can supply documents. This fact is not to be undervalued. Although the *Monthly Catalog* does eventually list hearings, reports, and similar items, it is typically several months behind issue date. One often needs an immediate copy of a committee report within a few days of issue. Political Research and the Federal Information Centers can usually get a copy within a week. Typically, the Federal Information Center will get you an exact citation and refer you to one of your congresspeople; the commercial services will get you the exact document. A number of services have grown up in the last few years strictly for document delivery and for a proportionately larger fee can obtain and deliver material almost overnight. Most libraries with on-line searching services have lists of such firms, whose charges for immediate delivery of a 4-page slip law can reach $25.00 or more.

Congressional Information Service (CIS).[29] This is a commercial service which specializes in producing its own copies of government documents along with excellent indexing. Essentially, this service collects a copy of everything Congress produces and makes it available in microfiche (flat microfilm roughly 4″ by 6″ holding up to 96 pages of text) and indexes it within a few weeks of publication. The index comes out monthly and is cumulated annually.

This system is perhaps one of the best products to have been introduced into

[28]Clements, John (ed.). *Taylor's Encyclopedia of Government Officials, Federal and State.* Dallas: Political Research, Inc., 1967/68—.

[29]Congressional Information Service, Inc., 4520 East-West Highway, Bethesda, Maryland 20814.

the library and information retrieval field in the past generation. Each document issued by Congress is listed with full bibliographic information, complete with an abstract. For example, each witness in a committee hearing is named and identified (for example, by occupation) and his or her testimony summarized in one or two sentences. Additional information (e.g., reprints of magazine articles) which may be included in the testimony are also abstracted. All of these data, including the witness names, are indexed, as are bill numbers, public law numbers, and the names of government agencies. Access is thus not only more frequent and more timely than anything the government produces, but is also substantially more detailed. The annual index for all abstracts for the year also includes a legislative history. This *CIS Index and Abstracts*[30] is more likely to be found in larger libraries.

CIS also supplies the microfiche of the documents in complete sets or individual orders. Most documents cost a minimum of $2.25, but delivery is fairly rapid. All libraries with a subscription to *CIS Index and Abstracts* should have information and order blanks. Any organization may obtain the necessary arrangements for ordering documents, even if it does not possess the indexes.

The Monthly Catalog. This often-mentioned document is the official government list of all productions of the Government Printing Office and many other government agencies. Listings tend to come much later than production of the documents. For example, bicentennial material was still being listed for the first time in 1978. However, this source is relatively inexpensive and is likely to be found in many libraries which do not subscribe to any of the commercial services. By law, it will be found in any official depository library. It includes indexes by name and by subject, among others. Its primary value, in competition with other sources mentioned, is that it began in 1895, providing the greatest length of coverage. It is relatively comprehensive and the most likely tool to be found in the average library.

Capitol Services, Inc.[31] This is another useful commercial service. Among other things, it issues abstracts of both the *Congressional Record* and the *Federal Register*, in both complete editions, and in a number of subject-related sections.

Retrieving Documents

If an individual has been following the events in Congress and the agencies and has used one or more of the above tools, he or she may need the exact text of

[30]Congressional Information Service. *CIS Index*. Washington, D.C.: Congressional Information Service, 1970—.

[31]Capitol Services, Inc., 511 2nd Street, N.E., Washington, D.C. 20002. This organization is often abbreviated "CSI," not to be confused with "CIS."

a particular document. Often this will be available within a few weeks of issue at a local library. Unless there is tremendous hurry and the document has been issued only a few days ago, use of the library will often save quantities of aggravation, not to mention money. However, libraries are complex institutions.

To use a major library effectively, one must again begin with a few definitions. First, *document* as used in libraries usually has a specific meaning: an item produced by a government agency. *Document librarians* are those librarians in charge of these items; most larger libraries will have at least one such person. Government publications are often separated from the general library collection because of peculiarities in issuance and handling. Specifically for federal and often for state documents, libraries will be designated *depositories*. This means that the library receives certain classes of documents automatically from the government free of charge. Certain types of libraries, such as each state's state library, most larger public and university libraries, and most law school libraries, are depositories. By law, any library which is a federal depository *must* make its documents collection available to any citizen, regardless of its policies on access to the rest of its collections.

In addition to the ordering system, federal documents have many peculiarities in their handling. Since they are normally indexed in the *Monthly Catalog*, libraries tend not to enter them in the card catalogue but rather to arrange them by the system used by the Superintendent of Documents. In other words, a major portion of a library's holdings will *not* be in the card catalogue and among this portion are government documents.

The arrangement for federal documents is by issuing agency as indicated by the Superintendent of Documents. For example, all hearings of a given House committee will be together, regardless of subject. The Monthly Catalog and the CIS indexes give the "SuDocs" (Superintendent of Documents) numbers, as do many bibliographies. People unused to the system may confuse these numbers with Library of Congress call numbers because they contain both letters and numbers. The SuDocs number will contain colons, dashes, slashes, and other such punctuation; the Library of Congress (L.C.) number will have only decimal points. This number is not to be confused with the *order number* or the *stock number* often listed with document citations. These latter are used only for ordering publications and are useless in nearly all library applications for the user.

With the SuDocs number the search is quite easy. Since most libraries with document collections use the system, you do not need to look up the item again in each library (as you would if the item were cataloged). Also, in the off-hours, when a librarian is less likely to be on duty, the number will greatly shorten your search.

To simplify the issue (or, from another point of view, to complicate it further), libraries with large collections of Congressional documents will often have House and Senate reports and documents not in their original form, but in the *serial set*. This refers to a specially bound set of congressional documents, also sometimes called the *sheep set* from its older bindings. A House or Senate report or document number can be translated into a serial set number with different tables. Generally, but not always, the material will be bound in serial order, with each volume merely a convenient size.

To summarize, the simplest approach is to ask for the particular report or document by number; to have the Superintendent of Documents number available; and if neither of these numbers will help you, ask if the library has the serial set. If so, an attendant should be able to translate a report number into a volume number. If not, ask a librarian for help.

Libraries which are not depositories, especially the smaller ones, will probably not have a separate document collection. Perhaps such a library will have a vertical file of recent bills and may even have a legislative history file with all documents related to one bill in one folder. This latter is typical of many law libraries and other specialized collections.

Yet a third possibility is possession of the document in microfiche, typically from a commercial supplier, although the government is experimenting with this approach. If the library has documents in this format, they may be with government publications or they may be with the microforms, audiovisual media, or nonprint media. Here, again, they may be filed by a number of methods. Libraries with a major collection of CIS materials may use their accession number rather than the SuDocs number. Any competent library clerk can help find the material, once you have a regular citation.

Once the document is in hand, you may copy as much as you wish. Thus far, anything produced by the government is free of copyright. If the library has a collection of slip laws, it may be cheaper to copy them rather than to buy them, and faster to copy than to order from your Congressperson. Keep in mind, however, that the *U.S.C. Annotated* and the *U.S. Code Service* are commercial, except for the actual text of the laws, and extensive copying from them requires permission from the publisher.

The goal of this chapter has been to try to avoid confusion. Given the nature of legal documents generally and government publications specifically, it is likely that different people will approach them from different access points. The various approaches detailed in this chapter are the result of government's and others' attempts to make finding information easy for all users. Much of the apparent confusion is the result of the numerous choices available.

Computerized Systems

Since the early 1970s, a new way of keeping current as well as doing retrospective searches has developed: on-line bibliographic systems. Libraries and information centers have been able to offer the capability to search various indexing and abstracting services (and, more recently, complete documents) through computer. In brief, essentially one enters the subjects, authors, citations to the *CFR*, and other items of interest into the computer, which then searches a file of information, matches up the terms, combines them according to your instructions, and gives the citations to relevant documents, often with an abstract. Depending on the system, one can search indexes to the *Federal Register*, the *Congressional Record*, the *Washington Post*[32] (which reports much governmental news), and the full abstracts of the Congressional Information Service. In certain legal contexts, one can also search the text of lawyer-oriented legal services.

It is possible to store a search which has proved successful and merely repeat it as the data base is updated, obtaining only the new references. This routine updating of a search based on a specific profile of interests is usually called *selective dissemination of information* (SDI). Depending on the local situation, such an SDI system on the subject of rights of clients and research subjects could be provided for as little as $10.00 to $15.00 a month. This appears to be a substantial sum but is not so outlandish when compared to the cost of subscriptions to the printed indexes and the time needed to search them. This sort of service might be especially useful in the more isolated localities, where easy access to a substantial library is not available. Currently, other than the potential difficulties of cost, the main objection to such a system is that the indexing of the documents tends to be a week to a month after publication. It is quite likely that this situation will improve in the future.

Influencing the Process

Although discussion has centered on how to learn what the government is doing, this is but a prelude to trying to influence what it does. Since the first step is to keep informed, one should take seriously the duty to keep up with events in Congress and the regulatory agencies. Typically, the only response a rulemaker hears is from organized pressure groups. A few intelligent, informed comments from the right people at the right time can often make a large difference in the final action of government.

Ideally, one would regularly read the *Federal Register* and some source of congressional information, such as the *Congressional Record*, or keep up through some of the abstracting services noted above and the *CQ Weekly*. Many of the professional newsletters have begun carrying columns on privacy, rights of subjects, reporting requirements, and other government actions relevant to their readers. The *Chronicle of Higher Education*,[33] a weekly newspaper aimed at

[32]*Washington Post*. v. 1—, 1877—. Washington, D.C.: Washington Post, Inc.
[33]*The Chronicle of Higher Education*. v. 1—, 1966—. Baltimore: Editorial Projects for Education.

college administrators and faculty, includes a section on current bills and debates in Congress. The *APA Monitor*[34] and the ASA *Footnotes*[35] also carry regular items on rules and laws relating to research, and the *Washington Post* has begun a weekly column on the subject. Unfortunately, many of the professional newsletters give rather scanty information and tend to come out too infrequently.

Another useful way to keep informed is to get on the right mailing lists. For example, during the course of hearings of the National Commission for the Protection of Human Subjects of Biomedical and Behavioral Research, it was possible for one to receive regular notices of all actions and issuances of the commission, with brief abstracts and easy ordering information to get the full text.

Former President Carter's actions to open up the regulatory process should also help in the work of keeping informed. With the requirement to publish agendas in advance, it may become possible to dispense with daily perusal of the *Federal Register*. The point to remember, especially in rule making, is that most rules are made with *very* little public participation. Keeping current with the regulatory process may provide you with the chance to be one of the two dozen or so nonlobbyists who send in comments in time. There appears to be a tendency for many agencies to revise proposed regulations at least to a degree in response to these comments.

Just as there are formal ways of keeping up with the law, there are ways of establishing personal contact. The standard tool for this is the annual *Government Manual*[36] (formerly *Government Organization Manual*), which gives brief overviews of each part of each major agency and committee, with the bulk of the book necessarily concerned with the executive branch of government. It includes mailing addresses and telephone numbers of most major administrators in each agency and normally includes regional agency heads. The *Congressional Directory*,[37] provided for the use of those in Congress, gives much more detail, especially on Congress, and includes much useful information on the agencies as well. Thanks to the *Congressional Quarterly* (CQ) staff, there is a new similarly detailed directory for the agencies, the *Federal Regulatory Directory*.[38]

Often, it is useful to talk to a staff person rather than to the representative or agency head. Conveniently, there is the *Congressional Staff Directory*[39] for such

[34]*APA Monitor*. v. 1—, 1970—. Washington, D.C.: American Psychological Association.

[35]*ASA Footnotes*. Washington, D.C.: American Sociological Association.

[36]*United States Government Manual*. Washington, D.C.: Office of the Federal Register, National Archives and Record Services, General Services Administration, 1973/74—.

[37]United States. Congress. *Official Congressional Directory*. Washington, D.C.: publisher varies, 1809—

[38]*Federal Regulatory Directory*. Washington, D.C.: Congressional Quarterly, Inc., 1979/80—.

[39]*Congressional Staff Directory*. Indianapolis: Bobbs-Merrill, 1959—.

purposes. Many agencies also produce detailed directories themselves. Such publications are conveniently listed in the annual compilation of Privacy Act documents in the *Federal Register* and also issued separately. A good general directory is the *Washington Information Directory*,[40] again produced by the CQ organization.

Unfortunately, there are problems with nearly all these publications, the major one being timeliness. Staffs are likely to change, especially after elections, and the regulatory agencies appear to be in a constant state of flux. The information in any of the directories can be quite out of date, even by publication. One can get around this problem by contacting a Federal Information Center for the current person in charge, or by using a commercial service, such as Political Research. The Washington Monitor service provides a looseleaf directory, updated regularly.

A few words, even if they seem obvious, are appropriate here. It is well known that form letters to government officials rarely work. The letter, telegram, or telephone call should be to a specific person, such as the sponsor of a bill. It is best to refer to the issue at hand by bill number, regulation citation, or other specific identifier. Specific arguments, backed up with evidence and stated succinctly, appear to work best. Although this should be grossly evident, make sure the facts are straight. Each Congress seems to produce at least one mythical rule which causes a furor in some pressure groups, leading to hundreds of letters and calls. Aside from providing amusement, references to such phantoms will not impress your representative.

State Laws and Regulations

Thanks to our federal system of government, there tends to very great variation in state laws on most subjects. This is especially the case in privacy-related issues, which have no long legal tradition behind them. As a result, no general discussion of the subject can usefully deal with the states' positions. The following is meant to provide the beginnings of the researcher's own study of his or her own state.

Most state legislatures issue documents in a manner roughly parallel to that of Congress. Thus the state will have a *Code*, arranged in some type of subject format, updated by supplements and/or pocket parts. This code may not have been officially updated for decades and will still be cited by its basic title (e.g., "X State Code of 1944, revised"), even though the law will have been changed thoroughly. Current legislation will be issued in some type of slip laws, with

[40]*Washington Information Directory*. 1st ed.—, 1975/76—. Washington, D.C.: Congressional Quarterly, Inc.

standard citation to the bills and the passed laws. There is no annual statutes at large for the typical state. Larger libraries will have looseleaf binders of all laws passed and may retain all bills introduced.

State regulations are even more difficult to deal with. Any logical compilation is rare. The best approach is to identify the relevant agencies and keep in touch with them.

Fortunately, it is not that difficult to keep up with state agencies. Political Research, remember, issues the *Encyclopedia of Government Officials* and will answer questions on state officials and agencies. Many state libraries or other similar agencies maintain "hotlines" at least while the legislature is in session; and the state library or the state legislative library will usually answer questions (at least giving you the right person to call) all year round. The Council on State Governments publishes a set of useful tools, including the general *Book of the States*[41] and directories of state administrative officials and of elected officials, the former classified by function.

Fortunately, the commercial publishers have again filled in some of the gaps. A most useful general index mentioned before is the *State Publications Index* issued by Information Handling Services (formerly *Checklist of State Publications*). This includes references gleaned from the various state-issued publications lists and those that are produced by the Library of Congress (the *Monthly Checklist of State Publications*),[42] as well as references to many items not listed in these sources and from states which do not produce such lists. For periodicals, access is by subject, author, state agency, or title. This tool does not include laws or statutes and lacks some other material, but it is by far the most nearly complete access tool to state hearings, reports, and related materials.

A detailed compilation of state laws, with comments and some legislative history examples, is the Senate Judiciary Committee's *Freedom of Information: A Compilation of State Laws*,[43] issued in 1978. With luck, similar compilations of state laws on research on human subjects will be issued. A recent guide to searching for state law compilations by Boast and Foster was in *Law Library Journal*.[44]

[41] *The Book of the States*. v. 1—, 1935—. Chicago: Council of State Governments. Also see *State Administrative Officials, Classified by Functions*. Lexington, Kentucky: Council of State Governments, 1977—. Also see *State Elective Officials and the Legislatures*. Lexington, Council of State Governments, 1977—

[42] United States Library of Congress. *Monthly Checklist of State Publications*. v. 1—, 1910—. Washington: United States Library of Congress, Exchange and Gift Division. (For sale by the Superintendent of Documents.)

[43] United States. Congress. Senate. Committee on the Judiciary. Subcommittee on Administrative Practices and Procedures. *Freedom of Information: A Compilation of State Laws*. Washington, D.C.: United States Government Printing Office, 1978.

[44] Boast, C., and Foster, L. Current subject compilations of State laws: Research guide and annotated bibliography. *Law Library Journal*, Spring, 1979, 72 (2), pp. 209–221.

This chapter has been rather complex, trying to give all needed information for the researcher without attempting to repeat a course in library science or legal bibliography. With the information given, however, it will be possible for the reader to update the information contained in the rest of this book as needed or desired. The reader should now be better prepared also to influence the process in the future to ensure that important laws and rules are shaped by professionals who are in a position to provide cogent and useful contributions.

Index